Radar Electronic Countermeasures System Design

For a complete listing of the *Artech House Radar Library*,
turn to the back of this book . . .

Radar Electronic Countermeasures System Design

Richard J. Wiegand
Westinghouse Electric Corporation (WEC)
Advisory Engineer
EW Development Department

Artech House
Boston • London

Library of Congress Cataloging-in-Publication Data

Wiegand, Richard J.
Radar electronic countermeasures system design
p. cm.
Includes bibliographical references and index.
ISBN 0-89006-381-8
1. Electronic countermeasures. I. Title.
UG485.W44 1991 91-7575
623'.043--dc20 CIP

British Library Cataloguing in Publication Data

Wiegand, Richard J.
Radar electronic countermeasures system design
1. Military operations. Use of electronics
I. Title.
623.043

ISBN 0-89006-381-8

685 Canton Street
Norwood, MA 02062

International Standard Book Number: 0-89006-381-8
Library of Congress Catalog Card Number: 91-7575

10 9 8 7 6 5 4 3 2 1

This book is dedicated to my father,
John R. Wiegand
(1912 to 1986),
who earned the great respect of
those who knew him.

Contents

Preface

The subject presented in this book is the design and operation of active electronic countermeasure (ECM) jammer systems intended to negate the effectiveness of hostile radar systems. This is an introductory text for readers with a general technical background.

One goal for this text was to present aspects of ECM system design that, though available in the open literature, either are not adequately addressed, or, in the author's judgment, are characterized by lack of clarity within the electronic warfare (EW) community. Our intention is to address this shortcoming within the context of presenting the broad scope of the subject. This text differentiates the jamming technique repertoire, on the one hand, and the hardware and software needed to implement such techniques, on the other hand; the latter is the primary subject of this book. The text is intended to rectify an inadequacy in the open literature: the practical and theoretical aspects of designing active radar ECM jammer systems. The text is intended to give an understanding of the principles and concepts to readers with a general technical background, but who otherwise have varied interests, needs, and levels of experience. Therefore, where necessary, certain specialized knowledge is presented as background to the primary emphasis. The text and the illustrations are intended to be both a learning tool (with emphasis on facilitating the reader's conceptualization) and a reference tool. All of the graphical and other computed results are based on the author's own analysis.

The original source material for this book was a set of view graphs used for an intensive short course titled, "Active Electronic Countermeasures," sponsored by Technology Service Corporation from 1985 through 1987. The course instructors were the author, who emphasized ECM system design, and his colleague, Mr. Edward J. Chrzanowski, who emphasized ECM techniques. Mr. Chrzanowski has written the text, *Active Radar Electronic Countermeasures,* published by Artech House (1990). Mr. Chrzanowski's book and this one are intended to complement one another. The organization of this text is different from that of the course, and a considerable amount of new material has been added during the process of writing.

Consideration was given to the best starting point from which to present this subject. The author concluded that there was no ideal starting point, in the sense that descriptions of each aspect of the subject seemed to need background knowledge of the other aspects of the subject. This dilemma was resolved by starting the text with an overview. This is followed by a chapter describing the components and devices available to ECM system engineers. The process of system engineering—at least for ECM systems—is more dependent on knowledge of such elements, how they can be applied, and how they can be made to interact with one another efficiently, as opposed to such system design analytical tools as network mathematics. Certain specialized knowledge is presented in the next chapter as background, and then the principles of system engineering are presented, with emphasis appropriate for ECM systems. This includes a summary of the determinants of system quality. Much of this theory should be applicable to other system engineering applications as well. Some of the specialized knowledge is also appropriate for the next chapter, which concentrates on EW receivers. In addition to describing different EW receiver implementations, several key issues, such as bandwidth, are addressed. This is followed by a chapter on the general subject of microwave feedback, including the application of microwave memory loops. Microwave feedback is presented in some detail, because the available literature generally does not present certain concepts, mathematical formulations, and computed results in a fashion suitable for the problems encountered in ECM system design. The last chapter completes the system design description.

Consideration was given to the level of detail that would be appropriate. By intent, the level of detail presented varies throughout the text. The goal was (1) to meet the needs of the reader without unnecessary bulk, and (2) to provide more detail where in the open literature it is lacking. Single paragraphs included here describe certain subjects which could easily be expanded into a full book. We have striven to achieve a proper balance. This meant considering whether the reader with a general technical background would already be cognizant of such details or would have ready access to such information in general-purpose technical literature not concerned with EW considerations.

Technical books are generally based on each author's personal knowledge or a synthesis of the literature that the author has researched. By and large, this text is based on the former. Any errors are the author's responsibility.

The author wishes to express his appreciation to Mr. Eugene Fox for his considerable assistance in editing the text. Mr. Fox is the former manager of Westinghouse Electric Corporation's EW Engineering department.

Chapter 1
Overview

1.1 INTRODUCTION

This chapter introduces the subject of active *electronic countermeasures* (ECM) systems hardware and software by providing an overview of such systems. When considering the best method to present this material, one choice was to traverse the hierarchy by starting with a description of the components, then the key subsystems, and finally describe ECM on a system level. Such an approach has advantages, but it appears better to present to the reader an overview of the systems first. As such, the overall plan for this text is as follows:

- system and hardware overview;
- subsystem and component descriptions;
- system engineering principles;
- receiver subsystems;
- microwave feedback and RF memory loop transponders; and
- system design, operation, and digital processing.

To optimize the learning experience, we suggest that the chapters be read in sequence. This first chapter, however, should be reread, or at least reviewed, before reading the last chapter.

1.2 SUBJECT AND THEMES

One increasingly important aspect of modern warfare is electronic combat. ECM will be employed in ground battle areas to disrupt radar systems, but especially to inhibit communications among the hostile forces. For *electronic warfare* (EW) associated with aircraft, naval surface craft, ground vehicles, and even buildings, the negation of hostile radar systems and associated weapon systems will be of great importance: the design of the active ECM systems employed for that role is the primary subject of this book. The protection of such craft by active ECM can be accomplished by ECM systems in the vicinity, such as stand-off or stand-in jammers, by mutually cooperative ECM jamming, and by self-protection jammers. Self-protection active ECM jammers, as shown in Figure 1.1, will be addressed in the most detail.

EW includes ECM, *electronic support measures* (ESM), and *electronic counter-countermeasures* (ECCM). ECCM is intended to negate hostile ECM and ESM functions. ESM systems are used to provide an alert function by sensing *electromagnetic* (EM) waves, usually automatically interpreting them, and, when appropriate, initiating some type of alarm or ECM. ECM includes *active* and *passive* countermeasures. Passive countermeasures reflect the radar EM waves in such a manner as to compete with the true target reflection; this countermeasure is usually achieved with corner reflectors or chaff. Active ECM systems negate threats by transmitting appropriate radio frequency (RF) EM waves. Modern active ECM systems often include sophisticated receivers, or alternatively they exchange data with an ESM system. The EW receivers used by active ECM systems, along with the associated *digital signal processors* (DSP), will be described, and this description is applicable to ESM systems and to a lesser extent *electronic intelligence* (ELINT) systems. ELINT systems are employed to acquire data on (propagating) EM signals, often called the signal environment.

This text is written from a system engineering viewpoint, where the "system" is defined for the purposes of this text as the equipment and software used to perform the active ECM function. Chapter two describes the components and subsystems that constitute an ECM system, because a system engineer needs to understand the properties of the components in order to structure the system properly. A good deal of system engineering skill resides in applying the knowledge of components and other resources and then considering the optimum structure to interconnect the selected components and subsystems. A chapter will be devoted to the system engineering process and the principles underlying it. We hope that the structure of this text will help novices to the field, as well as component vendor engineers and experienced system engineers, who will find this book useful. We expect that some readers will only wish to understand how such ECM systems work, while others will want to apply this text to their own system engineering. The latter will find several graphs and other reference materials useful, because the available literature generally does not present certain concepts, mathematical formulations, and computed results in a fashion suitable for solving the problems encountered in ECM system design. The presentation of several concepts is, in fact, considerably different from that given in most textbooks to which the reader may refer. Indeed, one goal of this text is to give an understanding of ECM principles and concepts to readers with varied interests and needs.

As stated, this text will describe the key components and subsystems. Such descriptions can often be meaningless to the reader, however, unless the scale (or

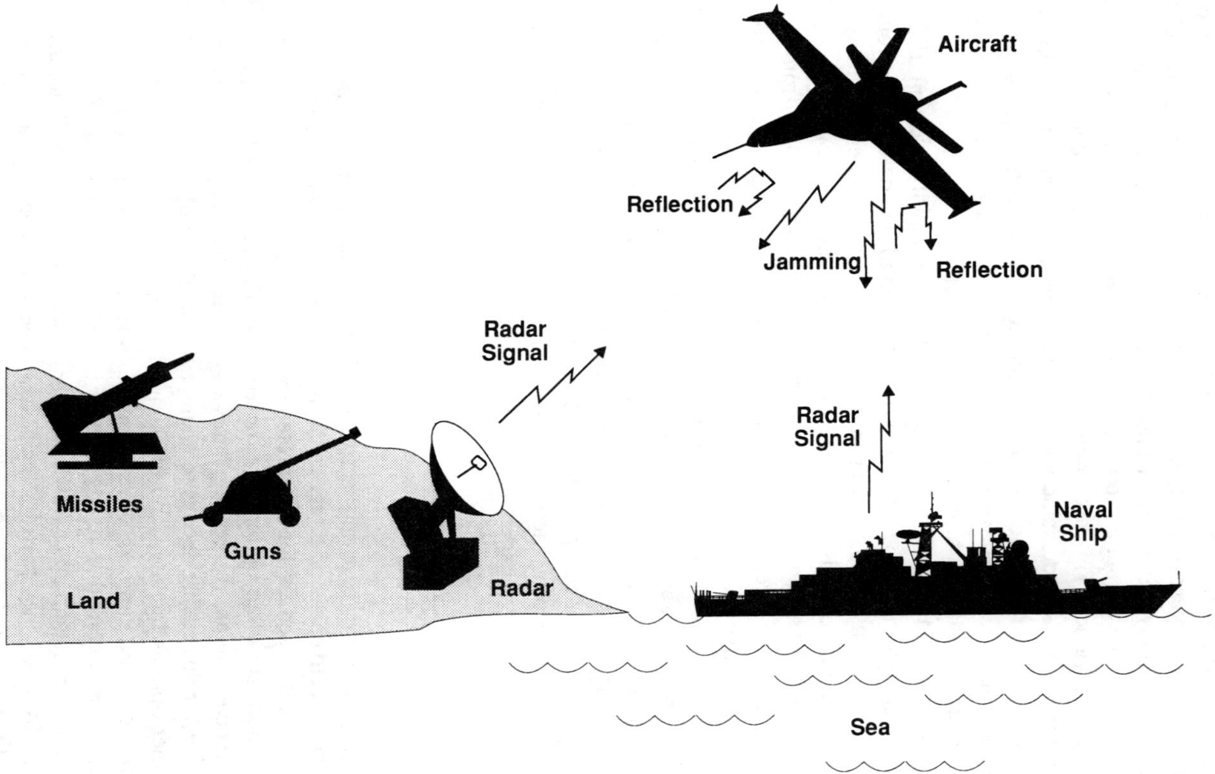

Figure 1.1 Self-protection active ECM.

at least the order of magnitude) of the values of the parameters are given. There are various problems for an author in doing so; for instance, (1) the technology may advance, making the value obsolete; (2) the author may not be cognizant of the capabilities of certain vendors; (3) other combinations of parameters are possible; and (4) the author may have certain unstated constraints in mind. Because giving "typical values" opens the author to criticism, he or she may simply avoid the problem by not giving values. This text, however, will give typical values in most cases to improve the reader's understanding. However, the reader is warned not to use them for system design purposes without checking with the actual component vendors. Specific components, subsystems, or systems will not be associated with the names of vendors or manufacturers that produce them.

The reader should understand that active ECM is a rapidly evolving field and, although the principles described herein will remain valid, the descriptions represent the art of ECM systems fabrication at just one point in time, albeit including some sense of what the future holds. Undoubtedly, a generation from now, some of these descriptions will appear quaint. Part of the impetus for change results from changes or improvements in the threat radar systems, which, in turn, are often a response to improvements in active ECM systems.

ECM systems design, both hardware and software, and ECM technique formulation are two distinct areas of expertise. The interaction between these two areas is considerable, and there needs to be substantial overlap of the knowledge base for experts of both types. The viability of an active ECM technique means not just its ability to negate the radar function, but also the practicality of producing the needed waveforms by a "real-world" active ECM system. The emphasis of this text is on active ECM system design expertise.

1.3 ECM TECHNIQUES

As generally used in this text, an "active ECM technique" refers to a method used to negate the effectiveness of threat radar systems by transmitting EM RF signals. The techniques discussed are for self-protection of a host craft unless explicitly stated otherwise. There are other options for the craft besides self-protection ECM, such as taking evasive actions, dispensing flares to counter *electro-optical* (EO) or chaff to obscure RF, or dispensing decoys to counter RF or EO. Within the realm of active RF EW, other options include stand-off jammers, escort jammers and multiple spatially diverse on-board jammers. A stand-off jammer is a large, powerful transmitter usually located on a craft that stands "out of harm's way" while transmitting ECM signals to protect engagement craft which either may not have, or will not use, self-protection ECM while they perform their engagement mission at a range closer to a threatening radar-directed weapon system. Similarly, an escort jammer is usually a craft outfitted with very capable and costly EW equipment that accompanies engagement craft (which, again, may not utilize ECM themselves) while they perform their engagement mission. Noise and false targets are often appropriate jamming techniques for stand-off, stand-in, or escort jammers. Spatially diverse jammers onboard multiple craft performing engagement missions in harm's way may be a cost-effective approach where self-protection ECM may not be effective in protecting lone craft, but where the combined ECM is highly effective. The cooperative ECM waveforms may be free-running, synchronized with accurate clocks, or with synchronizing communications. One ECCM technique against ECM noise jamming, one which denies range but allows angle tracking, is simply to fire a missile down the noise strobe with the expectation that, if the target is in range, the missile will eventually hit the target. However, if multiple craft with modest spatial diversity employ cooperative noise jamming by turning or modulating their noise jamming on and off with a half-period less than the time needed for the missile to settle on its "home-on-jam" flight path, this ECCM will be negated.

Active ECM systems generally have a set of jamming-deception techniques that are either fixed or set on a mission basis, or, for more modern active ECM systems, the system has a repertoire of techniques from which sophisticated DSP subsystems make selections and set parameters in real time—sometimes even including the phase of the waveforms. This text describes these techniques to the extent needed for an understanding of ECM system design. The unclassified techniques are well covered in the literature, and especially by the companion text by Chrzanowski (Artech House, 1989); selected references are listed in the bibliography. The parameter values given in this text for the waveforms associated with these techniques are based on the general physics of the situation, rather than on countering specific radars.

Suffice it to say that all active jamming techniques are based on the appropriate control of ECM EM waves in time, carrier frequency, carrier phase, amplitude, polarization, and direction. These parameters completely describe all propagating EM waves at a given location in space in the far field. This list of parameters is finite and leads to the potential for highly programmable ECM systems, with appropriate sensors and mission-related data, to support generically all potential ECM techniques. Historically, however, ECM hardware has not always been able to support new techniques and threat countermeasures; there has been a continual need for substantial ECM system evolutionary developments because of the

- creation of new techniques,
- unforeseen range of new technique parameters,
- constraints of cost and size,
- changing component technology base,
- need for more RF power or gain,
- need to transmit more efficiently,

- evolution of radar systems,
- diversity of RF bands,
- diversity of radar systems and modes,
- diversity of missiles and weapon systems, and
- increased threat density in space, time, and frequency.

Self-protection active ECM is considered highly desirable because it minimizes deployment logistics. Self-protection jamming techniques can generally be classified as appropriate for negating

- search radar functions,
- acquisition modes of tracking radars,
- radar tracking and guidance modes, and
- certain missile and munition functions.

Self-protection techniques to counter search radars include false targets and noise to inhibit the ability of the radar to ascertain the angle, range, range rate, and doppler. These techniques are also appropriate against radar acquisition modes; additionally switching between ECM modes or techniques is useful to inhibit the threat radar from acquiring or stabilizing track on the ECM signal in place of the true skin return target.

The ECM system's DSP must often apply limited resources against, first, the most critical threat missile functions, then missile guidance, and, finally, radar track functions rather than against the radar acquisition and search functions. Radar systems generally track their targets in angle and range; it is also increasingly prevalent to track in doppler, which is a measure of the target's relative velocity. The angle tracking may use a pair of servos for independent tracking in two dimensions, such as azimuth and elevation.

Angle jamming can be achieved by appropriate amplitude modulation of the transmitted RF signal. Figure 1.2 shows why the radar's measured angle becomes incorrect if the ECM can transmit enough power at the appropriate time. This can be an effective ECM (search) angle measurement or angle track negation technique, appropriate against the classic radar architecture, because it undercuts the assumption that the incoming signal is coming from the angle at which the radar's antenna is pointed. The radar design normally includes the means to normalize the received signals to the antenna pattern, such as including an *automatic gain control* (AGC), but a properly programmed ECM technique will take that into account. Figure 1.2 also illustrates the power contest between the ECM and the radar. The technique illustrated is inappropriate against more modern monopulse radars, which measure the angle independent of amplitude.

Radar range tracking can be defeated by transmitting RF signals before or after the true skin return echo pulse. Such ECM range jamming signals can be in the form

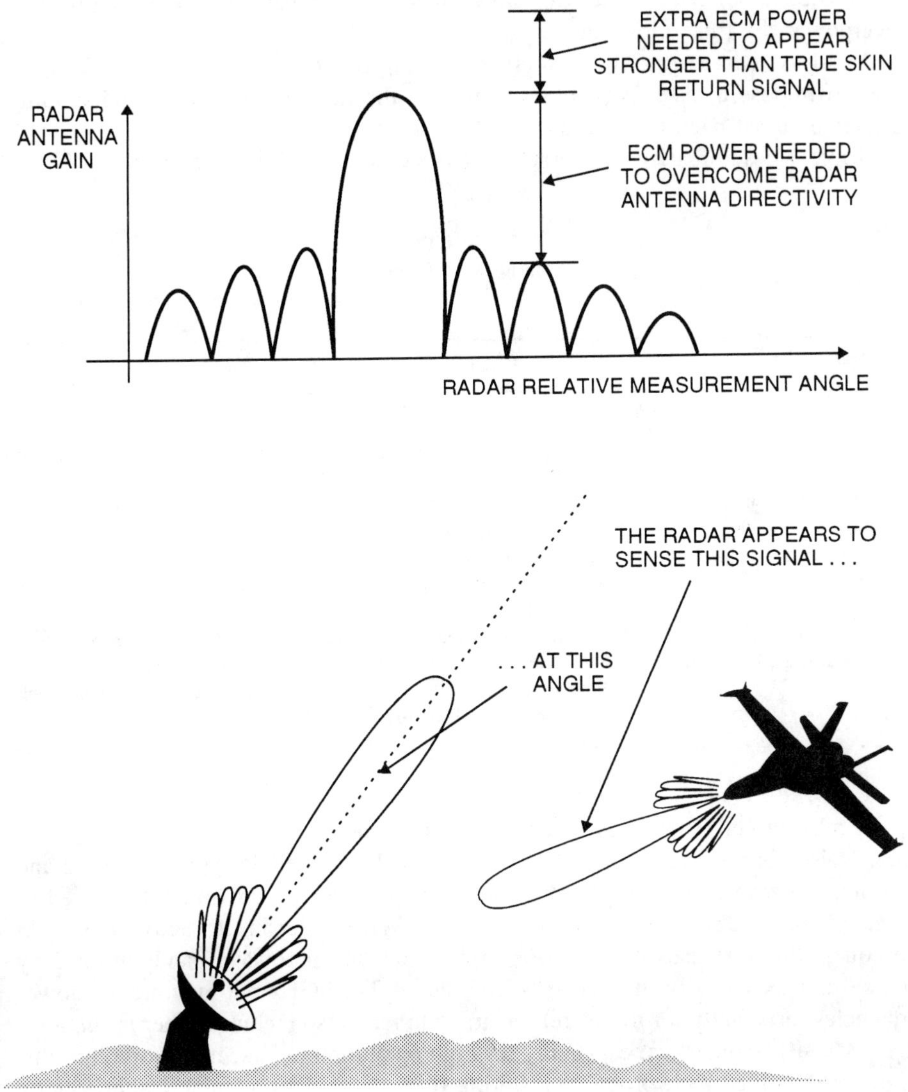

Figure 1.2 An angle deception technique.

of pulses that mimic the radar's own pulse, or it can be in the form of continuous or range (i.e., time) gated noise. Generally, these ECM signals must be strong enough to overpower the true skin return echo.

ECM systems negate radar doppler tracking by, for example, using RF modulators to impose a false frequency offset onto the radar's own signal and sending this altered signal back to the radar.

Table 1.1 summarizes some selected basic active ECM techniques.

Table 1.1
Some Basic ECM Techniques

To Counter Radar . . .	*. . . Use This Technique*
(1) Search	False targets or noise either from spatially diverse ECM sets or into sidelobes; see Figure 1.2
(2) Track acquisition	Blinking on and off or switching among jamming techniques.
(3) Conical scan angle tracking	Transmitting high power except near center of radar antenna main beam; see Figure 1.2.
(4) Range tracking	Transmitting noise or pulses around (just before, on, and just after) the true target range position.
(5) Doppler tracking	Phase-frequency modulation to cause a false frequency offset.
(6) Monopulse angle tracking	Range jamming (#4 above); once the range track has moved to a false range, use data-rate reduction to starve the radar angle track servo so that it drifts off the true angle.

A false target program is a deception technique and noise is a jamming technique. Noise obscures or distorts the radar signal. ECM includes both deception and jamming. However, the terms *ECM jammer* or just *jammer* are sometimes used as a comprehensive term that includes deception. When referring to active RF ECM techniques, this text uses the term *RF* to refer to the microwave bands utilized by radar systems. Generally, the range is from about 2–20 GHz but may include lower frequencies, especially for the search radars, and may also include higher *millimeter-wave* (*MMW*) frequencies, especially for high resolution radars. In our opinion, this use of "RF" is incorrect, probably resulting from the lack of a suitable abbreviation for "microwave," but because it is common practice, it is used in this text. Similarly, the terms *microwave* and *MMW* are misnomers in the sense that the "milli" wave has a shorter wavelength than the "micro" wave.

One question that should significantly influence ECM hardware and software design is: will the system be able to support all new techniques as well as revised and present parameters of existing techniques? A related question is: are there techniques that are truly universal? One goal of ECM technique experts should be to develop generic jamming deception techniques. Some jamming techniques are indeed appropriate against many types of radar, but the need to optimize the parameters and to handle the different radar types, modes, and functions (including various means to sense angle and range) lead to the conclusion that an ECM technique repertoire is indeed necessary.

The last chapter describes the technique formulation process. ECM technique formulation documentation should include the following:

- definition and quantification of the threat radar or weapon system intended functions;
- definition and quantification of the (sometimes unintended) threat vulnerabilities;
- definition, quantification and rationale for the ECM function; and
- calculations and analysis, incorporating the ECM hardware and software constraints, to show the range of viability.

The present text will describe the ECM system hardware and software and the nature and typical range of parameters of the common techniques, signals, or waveforms. The text will also describe the criteria for assessing the quality of an ECM system that can generate these various techniques. Another purpose of this text is to enable the reader to relate the specifics of any given ECM technique, and ECM techniques combined for a complex signal environment, to the system design needed to achieve the desired ends. The system constraints, including cost effectiveness issues, will also be discussed.

1.4 THE BASIC INTERNAL FUNCTIONAL STRUCTURE

A radar system transmits an RF signal and expects and utilizes the reflection of this signal off the target. The basic function of an active ECM system is to transmit appropriate RF signals in such a way as to thwart ultimate utilization by the radar of those return signals in order to harm the ECM's host craft. This determent can include directly negating weapons or munitions directed on the basis of information supplied by the radar, as shown in Figure 1.1.

A simple system structure achieves this basic active ECM function. A controlling modulator drives an RF generator that feeds a transmitter; this in turn feeds an antenna from which the EM wave propagates. In radar systems, the signal generator often feeds a high power signal directly to the transmit antenna, but ECM systems normally create the jamming signal at a low-power level and then feed it through a transmitter because of component bandwidth and other engineering constraints; hence, a low-level controllable RF generator and a high-power RF transmitter are both needed. In any event, only ECM systems dedicated for a rather specific role would have such a simple structure. For numerous practical reasons, the

ECM function usually requires a means to intercept the radar signal. Thus, the addition of a receive antenna allows the radar's own signal to be sent back with superimposed modulation intended to deceive the radar. The question of whether one antenna can perform the task of both intercepting and emitting EM waves is rather complex; most ECM systems use separate receive and transmit antennas. Having the means to intercept the threat radar signals allows digital signal processors to determine automatically if a response is required, select the appropriate response, and perhaps even evaluate the result.

The basic ECM internal functions are illustrated in the functional block diagram in Figure 1.3. The four basic functional blocks are

(1) electromagnetic propagation transducer (antennas)
(2) sensor (receivers)

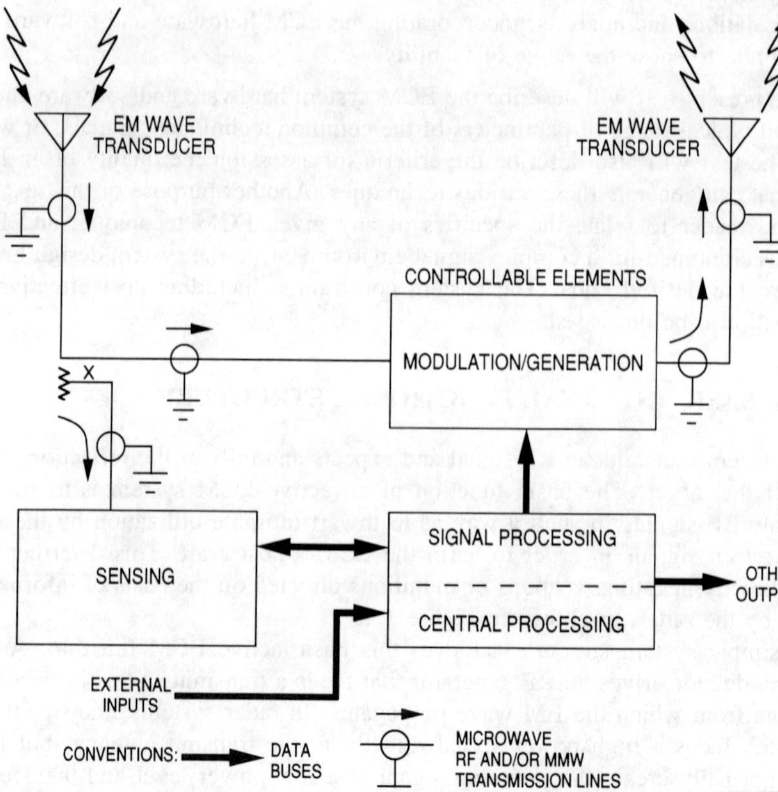

Figure 1.3 The four basic internal functions.

(3) digital signal processor and central processing unit (DSP, CPU)
(4) controllable elements (modulators or generators)

In a modern active ECM system, the signal processing is almost always accomplished with digital circuitry formed into a major subsystem known as the digital signal processor (DSP). The distinction between it and the *central processing unit* (CPU), which is a general-purpose software programmable unit, is tending to blur; indeed, the engineering community is working hard to eliminate this distinction, at least in hardware assets. Nevertheless, this text will continue to distinguish between the DSP and the CPU; the DSP is a dedicated unit that is structured with hard-wiring or firmware, and often has parametric programming provisions. The DSP typically must handle large amounts of data at high speed, and is designed to perform certain algorithms very efficiently. It is often convenient to have the CPU set the algorithm parameters of the DSP.

The antennas act as transducers between electromagnetic energy propagating through space and the signals on the RF transmission lines. Traditionally, the ECM antennas have been simply passive broadband broadbeam transducers, the function of which is indicated in Figure 1.3. The trend, however, is to incorporate the receive antenna more closely into the sense function (e.g., measure the incident *angle of arrival,* AOA), and likewise to utilize the transmit antenna as a device to modulate, or, more properly, to focus and re-focus, and otherwise to use it as an element for more actively creating the ECM technique.

Figure 1.3 shows a sensor and separate receiving and transmitting antennas, the typical implementation as stated. It may be appropriate to utilize a single antenna for both reception and transmission if the dual antenna option lacks sufficient isolation. An active ECM system with no sense capability may be viable if there is an on-board ESM system to turn the ECM on and off at appropriate times. However, modern ECM systems are required to respond "intelligently," as the situation demands, and to adapt to time-varying conditions as the host craft encounters various threats in turn. It is difficult for a separate ESM system to support such complex jamming functions adequately; thus, there is a need for a receive ECM antenna and dedicated sensor and signal processor that are tightly coupled to the technique waveforms so as to meet overall ECM needs.

The final function illustrated by Figure 1.3 is that of controlling RF signals, including signal generation and modulation. The hardware associated with this function includes all the controllable RF components and subsystems. The RF signal to be sent to the transmit antenna, or transducer can have its origin at the receiving antenna or at an internal RF signal source generator. Most such RF signals have substantial and distinctive modulations applied. Although there is no pure distinction between modulation and RF signal generation, from a practical point of view they are distinct because of considerable differences in mechanization.

1.5 BASIC JAMMING MODES

Figure 1.3 illustrates the four basic ECM system internal functions. Table 1.2 summarizes the definition of the three basic ECM jamming modes that such an ECM system operates in and shows an example for each. Figure 1.4 illustrates examples of the input and output RF signal envelopes for these three modes, where the cross-hatch represents an imposed modulation. For the transponder and noise modes, the RF signal is created, or at least replicated, within the system, while in repeater mode, the ECM system simply modulates the threat radar's own signal. The difference between transponder mode and noise mode is that the former mimics the threat radar's pulse signal, albeit with a range, or time, delay. The repeater mode is usually implemented with RF amplifiers and RF modulators. The transponder mode is generally implemented with an RF memory subsystem or a tuned set-on RF oscillator subsystem that is used to create one or more replicas of the RF pulse signal. The noise mode is usually also implemented with a tuned set-on RF oscillator subsystem, but with relatively narrow deviation noise FM for spot-noise mode and relatively wide preset FM deviation for barrage-noise mode. The repeater mode does not directly deny range to the threat radar, whereas the transponder mode causes the radar to measure a false range, and the noise mode obscures the true target return, thereby inhibiting the radar's range measurement altogether. The transponder mode of operation normally operates at a medium to low duty cycle. The noise mode tends to have a high duty cycle, although the need to jam several threats simultaneously may provide an incentive to hold down the noise duty cycle— the span in range—applied to each threat. The repeater mode of operation is generally useful regardless of the

Table 1.2
The Three Basic ECM Jamming Modes

Mode	*Definition*	*Example*
Repeater	Transmitting the amplified signal with amplitude/phase/frequency modulation (usually with constant gain)	Velocity gate pull-off (VGPO)
Transponder	Transmitting a (range) delayed signal replica with amplitude/phase/frequency modulation (usually at peak power)	Range gate pull-off (RGPO)
Noise	Transmitting a signal with no attempt at mimicry, to cover the true signal (usually extended in range and near maximum average power)	Barrage noise

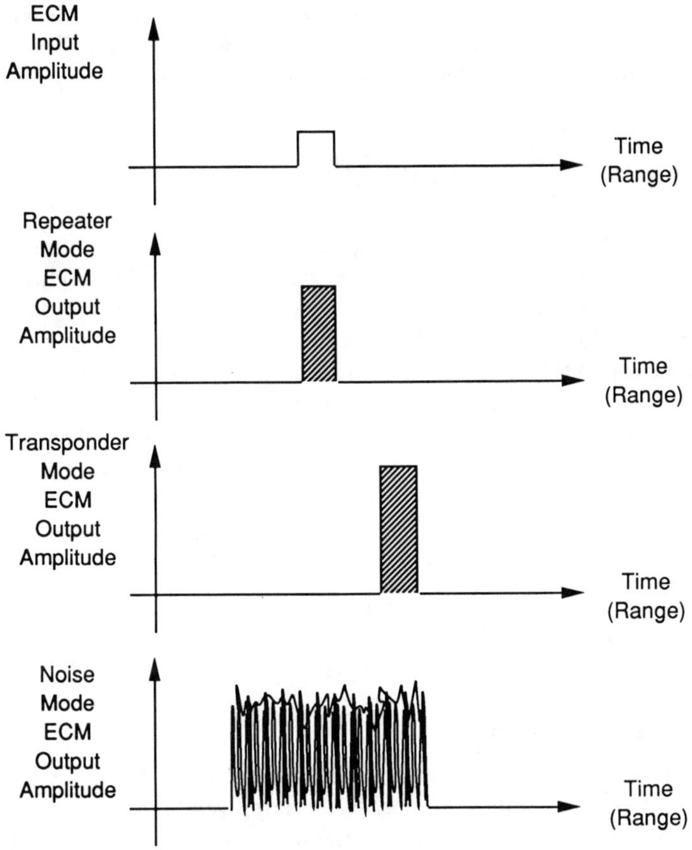

Figure 1.4 Input signal *versus* output *versus* mode.

duty cycle, and indeed can be used against both low duty cycle pulse trains and *continuous wave* (CW) signals. The reader should be sure to understand the distinction between these three basic modes; these definitions are quite useful even though a purist can undoubtedly find some examples that do not conveniently fall into one of these three categories.

The above definitions for repeater and transponder modes are similar, but not identical, to those used by the satellite communications community. A communications repeater receives a signal, then amplifies, modulates, and transmits it: such modulation is usually limited to frequency translation. The ECM repeater intercepts signals, then amplifies, modulates, and transmits them: in this case, the ECM modulation is much more varied. A communications transponder automatically transmits when the proper interrogation is received, usually at a different carrier frequency. The ECM transponder automatically transmits when it intercepts an RF pulse, which is nominally at the same frequency, albeit with varied modulations.

1.6 ECM HARDWARE ARCHITECTURE

The host vessels protected by active RF self-protection ECM systems include both fixed wing and rotary wing aircraft, naval vessels, and surface vehicles. The on-board ECM mounting is either of two classes: (1) in board or (2) out board. Traditionally, in board mounting has been preferred, but in the past, craft have often been defended with out board systems. The reason for this is (1) the rapidly evolving nature of ECM technology alluded to previously, and (2) the fact that ECM is added as an afterthought. The advent of modern programmable systems has made in board mounting a more viable approach. Stand-off and stand-in jammers use in board mounting since they are specifically designed for their EW mission.

A convenient partitioning of the physical hardware, appropriate from a mechanical engineering point of view, is

- the chasis, with shell for out board,
- the power supplies,
- the low level RF and video,
- the transmitter, and
- the digital circuit boards.

Figure 1.3 shows the four basic system functions, while Figure 1.5 shows the typical RF electronic structure of the hardware employed to perform these functions. The antennas of Figure 1.5 correspond to the transducers of Figure 1.3; the receiver corresponds to the sensor, and the amplitude-frequency-phase modulator, signal memory, signal source, and sometimes the power amplifier all belong to the controllable elements of Figure 1.3. The CPU and DSP hardware is not shown in Figure 1.5; it is conventional to draw such block diagrams by showing only the RF paths. The digital bus lines, however, are indicated with arrow tips showing the data flow direction. Note that the data flow on the digital bus is only into the modulator, while the data flow is both into and out of the receiver subsystem. In the event that both RF (microwave) and video (including digital) lines are shown on a block diagram, it is recommended that the RF lines be made clearly distinctive; this clear distinction is made in Figure 1.5 by using the coaxial symbols. The data busses are shown wide to indicate that they actually contain many lines. In general, the video lines have arrows to indicate the signal flow direction. As will be described in subsequent chapters, the video line signals are characterized by their voltages, whereas the RF line signals are best characterized by their power flow, making them qualitatively different. The small arrows near the coaxial symbols indicate the true RF power flow

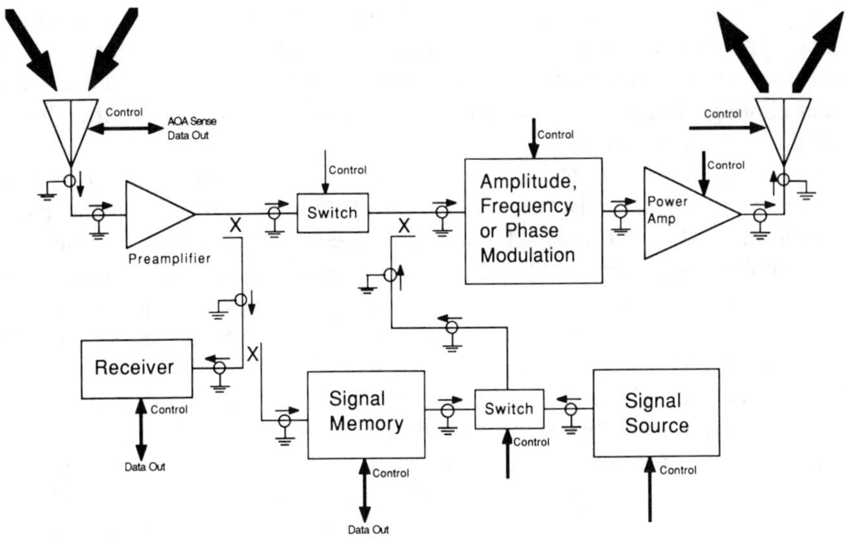

Figure 1.5 The typical ECM system architecture.

direction. For video signals in other words, it is more appropriate to think of a given signal as being present, whereas for RF signals it is more appropriate to think of signals flowing; this explains the difference in the symbols shown in Figure 1.5.

In Figure 1.5, the symbolic representation of RF transmission lines interconnecting subsystems has areas with close parallel lines having an *X* between them; the *X* represents a coupler. In pure video systems, wires can simply be connected to one another; this is shown symbolically on schematic wiring diagrams with dots at the intersections of lines. Interconnecting two RF transmission lines, however, generally requires a component called a coupler. For RF couplers with coaxial connectors, two pairs of ports are connected on a direct current (DC) basis, and these connections can be verified with a simple ohmmeter. In general, however, the EM wave RF power must transit a gap, or a distance with no dc connection, within the coupler in order to carry the RF power from one port to another. Couplers will be described more fully in the chapter on components, but the reader should already appreciate the connectivity information regarding couplers shown in the block diagram of Figure 1.5. Generally the straight-through path has less loss than the coupled path through the X, but, aside from that, the loss value of the coupling is not indicated on such block diagrams by just the symbol itself.

The preamplifier in Figure 1.5 is a key component in what is commonly referred to as the system's *front end*. The purpose of the preamplifier is (1) to increase the signal amplitude to a more conveniently useable value, and (2) to minimize the degradation of the *signal-to-noise ratio* (SNR) that the subsequent component losses will cause. In other words, the preamplifier makes the system sufficiently sensitive. Control lines to this preamplifier are not indicated in the figure, and indeed it is a common although not universal characteristic of such amplifiers that they have no provision for electronic control.

Figure 1.6 illustrates the linkage between components and subsystems via the digital data busses. Data only flows into controllable components, whereas the data flows both into and out of most subsystems. These busses pass both communication data type information (pure data) and command data type information (commands). Although the distinction between these types of data is not always important, this text will often distinguish between data and commands. For example, the digital data flow into the receiver may include command words to tune to a given RF, and the digital data flow out of the receiver may include a data word representing the amplitude measured at that RF. The digital data flow into the CPU is generally only pure data, whereas the flow into the DSP will include both commands and data; however, as stated above, the trend is to blur the distinction between the CPU and DSP. An instruction to the DSP to use a particular, perhaps hard-wired, algorithm is clearly a command, and RF signal measurement information is clearly pure data; parametric values that the algorithm will employ could be considered command type data.

Many of the more complex subsystems internally distribute the data to components within that subsystem in a manner transparent to the control of the main system data busses.

The following will describe the RF signal path through the typical ECM system structure of Figure 1.5 for the three basic ECM modes of Table 1.2; these paths are illustrated in Figure 1.7.

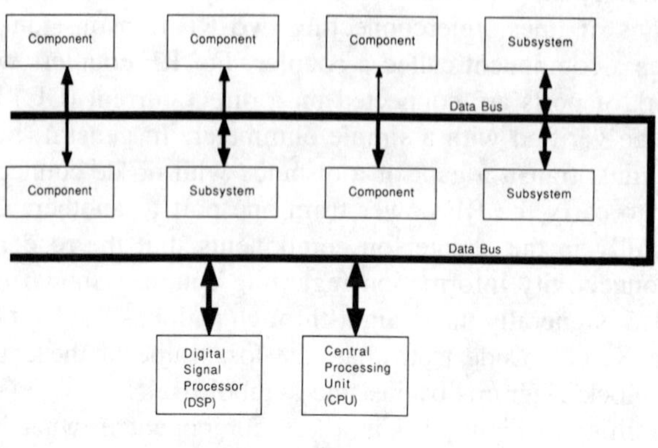

Figure 1.6 Communication and control via digital busses.

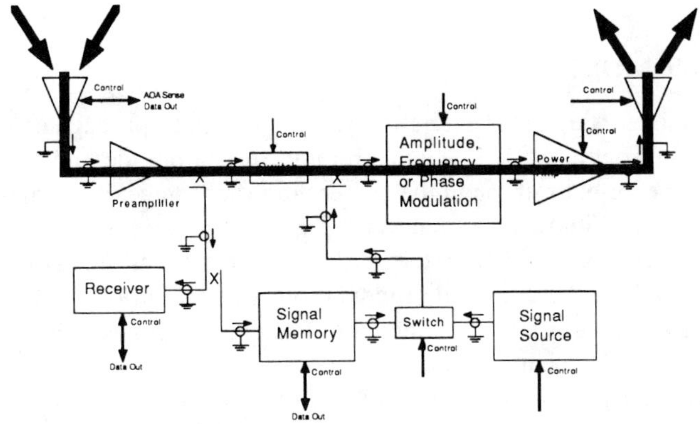

Figure A: Repeater Mode RF Signal Flow

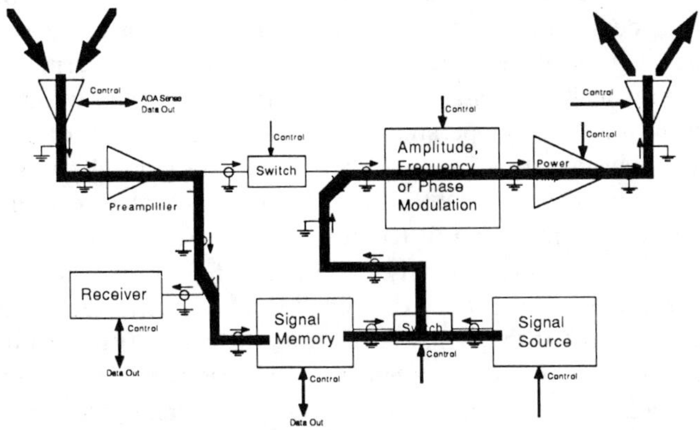

Figure B: Transponder Mode RF Signal Flow

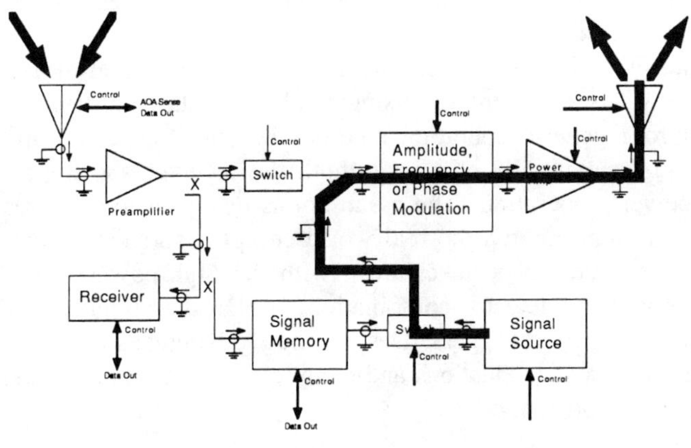

Figure C: Noise Mode RF Signal Flow

Figure 1.7 RF signal paths.

1.7 REPEATER MODE

In the repeater mode, the signal enters the receive antenna, propagates through the RF preamplifier, through the RF switch, through the amplitude and the frequency or phase modulator, or both, through the power amplifier, and exits from the transmit antenna, as shown in Figure 1.7a. That is the general description of the RF signal flow for the repeater mode, although the reader will understand from the subsequent discussion that much of the amplitude modulation is of the on-off type imposed via the RF switch. The signal can thus be thought of as being repeatedly blocked. Figure 1.4 illustrates a typical repeater mode input-to-output relationship.

For medium duty cycle, high duty cycle, or CW signals, the only CPU-DSP control in this repeater mode is via the RF modulator or RF switch, and perhaps the steering of one or both of the antennas. For a low duty cycle signal from a threat the DSP-CPU considers critical, however, it is quite common to pulse the output amplifier to a higher peak power, since the absolute power is the critical parameter for effective jamming at shorter ranges. At longer ranges, the gain, which determines relative power from the linearly operating power amplifier, is the critical parameter for repeater mode. Because such RF power amplifiers are generally fabricated with broadband *traveling wave tubes* (TWT) that cannot sustain the operating voltages continually needed to generate the full RF power, the power amplifier is pulsed on just long enough to transmit the low-duty-cycle pulse; Figure 1.5 shows the control bus feeding into the power amplifier. In order to pulse such a tube without a sophisticated DSP, a portion of the signal needs to be coupled off the main path prior to the RF switch but after the RF preamplifier, and directed into the receiver. For this role, the receiver needs to be no more than a broadband detector sensitive to the band the low-duty-cycle threat operates in. The receiver then triggers some logic which gates on the TWT power amplifier within a small percentage of the pulse width, usually for a preset nominal gate time.

Operating an ECM jammer in repeater mode with CW ouput tubes, assuming the tube's power level is sufficient and using a robust modulation technique, is one of the most foolproof generic operating methods because it uses the threat's own signal and because it potentially eliminates both the receiver and the assumptions regarding the receiver's operation. The assumptions designed into the receiver and DSP combination may not match the real world complex signal environment. Furthermore, great care needs to be taken even with the high-power, low-duty-cycle pulsed tube and broadband detector combination, so that the output tube duty cycle is not wasted because of false triggers. Despite these advantages of the simple low-cost CW repeater mode, many situations and many effective techniques require pulsed operation or one of the other modes.

1.8 TRANSPONDER MODE

The transponder mode RF signal paths are illustrated in Figure 1.7b. In the transponder mode, the signal enters the receive antenna, propagates through the RF preamplifier, is coupled off the main path, and enters the signal memory subsystem. The main repeater path should be programmed to have the RF switch in the off state, that is, at high attenuation. The signal memory subsystem holds the signal for a duration determined by the CPU-DSP control, and then outputs a pulse replica. The signal memory subsystem may be implemented with an RF tapped delay line, an RF recirculating memory loop, a digital RF memory, a *charge coupled device* (CCD), an *acoustic charge transport* (ACT) device or other implementation. This signal memory subsystem output transponder pulse replica is directed onto the main repeater path via the RF switch, propagates through the modulator path, propagates through the power amplifier, and exits from the transmitting antenna. Figure 1.4 and Figure 1.8, with more detail, illustrate the typical transponder mode input-to-output

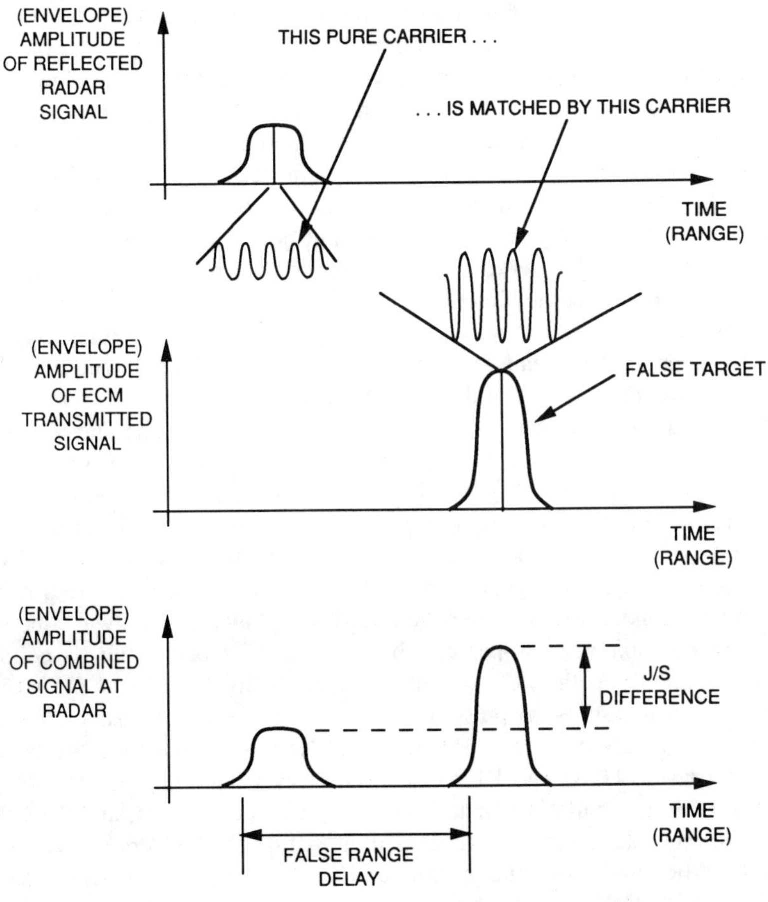

Figure 1.8 Transponder mode.

relationship. The transponder output pulse mimics the victim radar pulse by having nominally the same width and RF carrier. The output carrier tolerance must be significantly less than the threat radar's reception bandwidth; to deceive coherent radars, there are additional requirements on the output carrier described in Chapter 6. For deceiving threat tracking radars the programmed, perhaps time-varying range delay in the signal-memory unit of Figure 1.5 may be considerably more than a pulse width, but usually is much less than a *pulse repetition interval* (PRI), and therefore can operate against numerous tracking threats simultaneously. For techniques that deceive search or acquisition with false targets, the range delay is usually a significant portion of the PRI; unless the signal-memory unit of Figure 1.5 can hold more than one pulse at a time, long range delays will reduce the number of threats that can be simultaneously deceived.

The transponder mode modulation imposed by the RF switch and RF modulator of Figure 1.5 can be quite similar to that used in the repeater mode. The transponder mode, however, requires high rate amplitude modulation, often of the on-off type, and this modulation component is usually imposed directly by the signal-memory unit itself. The reason is that the generation of range delay and the maintenance of coherency are closely intertwined, and hence coherent range delay is most appropriately imposed directly in the signal-memory unit. However, the RF switch or RF modulator can impose range delay for noncoherent output pulses. In either case, during the delayed pulse transmission, the output power amplifier may be pulsed on to increase the power, just as discussed earlier for the repeater mode; the delayed pulse-up command will come directly from the signal memory unit instead of coming from the receiver. Such pulse-up operation is more likely in the transponder mode than the repeater mode, because the transponder mode is best suited and most effective for operating against low-duty-cycle pulsed threats.

There is an alternative transponder mode of operation. The alternative results from constraints on the RF signal-memory technology, including (1) constraints on storing multiple signals for long delays with regard to the PRI, (2) constraints on making several replicas of the input pulse, and (3) constraints related to the output transmission corrupting the input signal because of insufficient antenna isolation. This alternative utilizes the signal source unit shown in Figure 1.5. The flow for this alternative transponder operation may be traced as follows: the EM wave input signal enters the receive antenna, propagates through the RF preamplifier, is coupled off the main path and the path to the signal memory, and impinges on the EW receiver. The EW receiver measures the pulse parameters, and passes this data on to the DSP. Usually the most important two parameters, at least for transponder operation, are the *time of arrival* (TOA) and RF carrier frequency value. The pulse width is also useful. The DSP accumulates a history of these pulse descriptors, and the CPU uses this information to determine if a threat is illuminating the host craft to be defended. If so, it identifies the threat and assigns an operating mode to be used against the threat. If the CPU-DSP decides the transponder mode is appropriate, the CPU may then select the signal-memory unit as an appropriate resource. With the exception of needing certain free running modulations, such as range delay modulations, the signal-memory unit does not need to be directly controlled by the CPU; there are many conditions under which the signal-memory unit may run freely. However, if the CPU decides for any of the reasons given above not to select the signal-memory unit, it may choose to use the signal-source unit as the appropriate resource. The use of the signal-source unit for transponder mode of operation implies direct control, unless only a single threat is anticipated. The important parameter to be so controlled is the carrier frequency; the transmission time must also be precisely controlled if hardwired triggering is not used.

For present day technology, the key component in the signal-source unit is the RF *voltage controlled oscillator* (VCO). This has a transfer function that, although good linearity is difficult to achieve, provides an output frequency proportional to the input tuning voltage and a nominally constant amplitude *versus* output frequency. The CPU must tune the VCO to the threat's frequency. Normally, the CPU gives a control word which a *digital-to-analog converter* (DAC) uses to generate the correct voltage to drive the VCO to the threat's RF carrier frequency. The VCO is continually creating a CW RF carrier. The RF switch is pulsed on (i.e., to low attenuation) to create the transponder pulse from this CW VCO signal, just as for a noncoherent signal memory, and this pulse propagates through the modulator and the power amplifier and propagates out the transmit antenna, as shown in Figure 1.7b. The power amplifier may be pulsed as discussed above.

1.9 NOISE MODE

Whereas the repeater mode does not deny range to the threat radar, and whereas the transponder mode causes the radar to measure a false range, the noise mode is intended to inhibit the range measurement altogether, as shown in Figure 1.4. In the noise mode, the RF signal originates within the system, in the signal-source unit of Figure 1.5. As such, the signal-source unit is employed as an RF CW generator. This CW signal is characterized by complex frequency modulation (FM) intended to create amplitude or noise in the radar's receiver; the signal is often called barrage noise, especially when it is extended in range and covers the expected tuning band of the threat radar. This CW-noise signal passes through the RF switch and RF modulator of Figure 1.5, where it may become gated CW. The RF modulator will have similar angle deception modulations as repeater and transponder modes impose (e.g., maximum power into the threat's sidelobes, as in Figure 1.2), but it will not have the relatively high speed pulsewidth shaping modulation typical of the transponder mode. If the ECM transmission is being managed by the DSP-CPU for maximum efficiency, a function known as *power management*, then the CW-noise gating has a width typically ten to twenty times the threat radar's pulsewidth and the range measurement denial will also have this span.

The CW-noise signal propagates from the signal source through the gating switch, through the modulator, through the power amplifier, and out the transmit antenna, as shown in Figure 1.7c—the same path used by the alternate-transponder mode. Information from the receiver is used to tune the set-on noise center frequency in a manner similar to the alternate-transponder mode tuning. Because the range measurement denial is achieved by spanning the noise across range, it is generally difficult to pulse the power amplifier. However, if the power-managed gated noise is employed with not too long a gated width, it may be possible to pulse up for a very few threats; this is in contrast to repeater and transponder modes for which amplifier pulse-up gating can handle many threats virtually simultaneously, since the power amplifier will be pulsed-up for only the pulse width of each pulse.

In theory, the noise modulation can be imposed with either amplitude modulation (AM), FM, or phase modulation (PM), and if done properly the choice is of no consequence. However, because there usually is uncertainty regarding the threat's precise carrier frequency, it is generally deemed prudent to impose the noise-like characteristics with frequency modulation. Frequency modulation conveniently lends itself to generating a uniform band of power: a VCO can readily be employed to generate a flat spectrum across the intended band. Amplitude modulation and phase modulation, however, are not conducive to generating such a uniform spectrum. In other words, if the threat's carrier frequency is precisely known, the noise modulation can be created equally well with AM, FM or PM, but if there is a threat carrier frequency uncertainty that exceeds or is even a significant percentage of the threat radar's *instantaneous bandwidth* (IBW), then practical hardware constraints lead to the conclusion that FM is superior. The FM deviation performs the dual role of creating amplitude noise within the radar's receiver and allowing a tolerance for the nominal radar carrier frequency tuning, as shown in Figure 1.9.

The signal source of Figure 1.5 is generally an RF VCO, as stated. To create the set-on noise shown in Figures 1.4 and 1.9, as opposed to wideband barrage noise, the DC tuning voltage is adjusted so that the VCO carrier is at the center of the range of uncertainty of the threat's carrier. An *alternating current* (AC) noise waveform is then superimposed on this DC tuning voltage. The relationship between the FM rate and the FM deviation as well as their peak ratios and time functions, is a complex subject. The EW community has spent considerable effort researching the qualities of "good noise." Certain principles can be stated, however, based on the generic physics of receiver responses. To create large noise amplitude depth of modulation, as viewed by the threat receiver, the FM deviation parameter must exceed the IBW of the threat receiver. Since most threat receiver front ends include multiple filtering characterized by steep filter skirts, a deviation of two to three times the nominal threat receiver's IBW is adequate. Also, FM rate, as opposed to FM deviation, components that are extremely different from the threat receivers IBW are

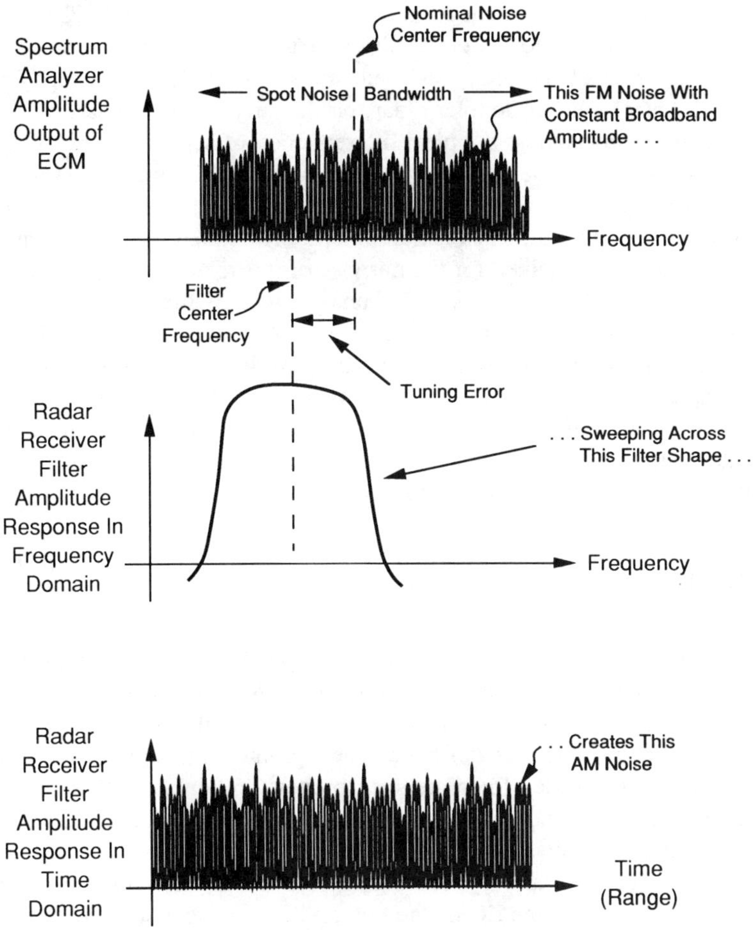

Figure 1.9 FM noise creating AM noise.

not conducive to generating good noise if they predominate. Too high an FM rate component will cause the threat receiver to average or null; too low an FM rate can be blocked by appropriate filtering, and also allows the threat a free look at the real pulse as the ECM's VCO is tuned to a different portion of the noise band. The use of FM deviation to cover the range of uncertainty above and beyond that needed to create good noise if the carrier were known perfectly has two serious consequences: (1) the noise power spectral density is reduced, and (2) difficult hardware requirements are imposed on the combination of the DAC, the tuning circuit, and the VCO. Ideally, FM noise should have a peak-to-peak deviation wide enough to create AM noise with sufficient depth of modulation and also to allow for VCO tuning tolerances, but not so wide as to lower the ECM efficiency.

The description of the three basic modes is now complete, except for a comment about doppler noise. A doppler radar, particularly a search radar, often displays the signal in two dimensions: doppler, or frequency offset, *versus* range or time in microseconds; the amplitude of the signal is represented by the intensity of the display. The noise described above will completely fill such a display screen, that is, it will show noise in both dimensions; the internal RF-signal flow is shown in Figure 1.7c. Doppler noise, as defined for the purpose of this text, passes the repeater mode signal through a modulator such as an RF phase shifter, which is controlled or modulated, by a noise waveform; the internal RF-signal flow is shown in Figure 1.7a. This will create noise on the radar display only in the doppler dimension. Doppler noise has characteristics that seem suitable for both repeater mode and noise mode; according to the definitions in Table 1.2, the doppler noise jamming technique is part of the repertoire of the repeater mode. This repeater mode noise denies measurement in the doppler frequency domain.

1.10 MODULATION PRINCIPLES

The preceding section described the input-to-output RF signal relationship, the RF signal flow, the points and sources of imposed modulation, and some of the characteristics of such modulation. This section describes with more detail (1) the modulation, (2) the principles that determine the optimum modulation characteristics, and (3) the hardware needed to create such properly modulated RF signals.

The first issue to be addressed concerns low-rate amplitude modulation. Figure 1.2 shows one purpose for such amplitude modulation: by transmitting a strong signal into the radar's sidelobes and a weak signal or none at all when the radar's main beam is pointed directly at the ECM, the radar will perceive its ECM-defended target to be at a false angle. One question is, what are the trade-offs between switching modulation and the imposition of a linear amplitude modulation function? Following are the key trade-off considerations:

- an RF switch is simpler than an RF or linear modulator;
- an on-off waveform is easier to generate;
- dynamic range considerations tend to negate smoothing;
- the on-off harmonic frequency components can be used to reduce the sweep range; and
- an RF switch lends itself to multiplexing.

The first two considerations simply state that an RF switch is simpler than a linear RF modulator, and the control waveform for the former is likewise simpler than that for the latter. The third consideration has to do with amplitude or dynamic range.

If the ECM fabrication cost is increased in order to produce a high dynamic range linear response on a log scale, many linear threat radar receivers will negate such efforts, as illustrated in Figure 1.10. The linear receiver's output voltage will only be large for the top few dB of the modulation. The fourth consideration has to do with exploiting the harmonic content of switching modulation and is illustrated with an example in Figure 1.11. This is an angle deception modulation example. Threat radars generally have angle scan rates on the order of 10s to 100s of hertz, as in the case of a conical scanning radar. Such values are by and large determined more by the laws of physics rather than by technology constraints; in this case the radar's PRI is determined by the maximum unambiguous range and the scan rate by the need to get an adequate number of hits, or pulses, within the main antenna beam. The ECM angle deception modulation needs to be on the order of that scan rate. Figure 1.11 illustrates how switching modulation will produce spectral components that are at multiple frequencies simultaneously, to mix time and frequency domain terms. For example, if the range of uncertainty that the angle modulation had to cover were 100–900 Hz, then by using a swept square wave and sweeping it from 100–300 Hz, the third harmonic will cover the remaining range all the way up to 900 Hz. The third harmonic will have 9.5 dB less power than the fundamental, but in many cases that power loss is less important than the need to cover the range of uncertainty. The third harmonic will also sweep three times as fast. Sweep rate is critical, because to be effective, the modulation rate must approximately equal or be equal to a multiple of the threat radar's lobing rate for a time interval at approximately the threat's angle-tracking servo time constant. This fact leads to the rule that, in order to cause a significant perturbation, the swept square wave sweep rate must be less than the square of the reciprocal of the threat's angle tracking servo time constant. Since the ECM sweep rate is therefore limited, it could be an advantage to have spectral components simultaneously present, provided such harmonics fall in a range of effectiveness. A linear modulation would not provide such harmonics.

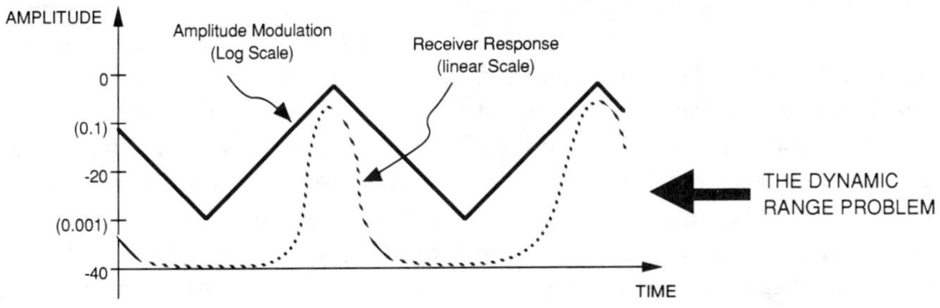

Figure 1.10 Linearity and dynamic range.

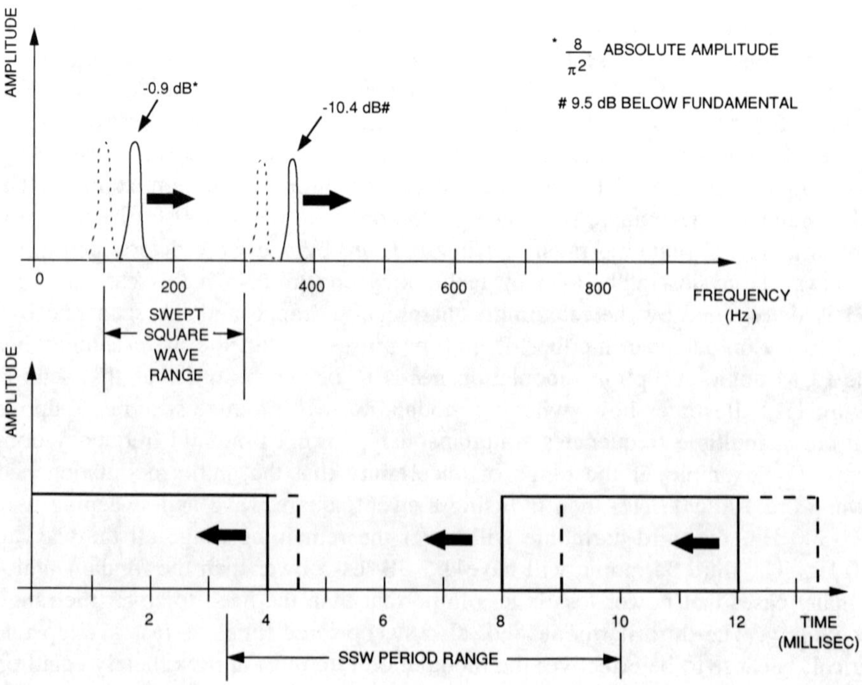

Figure 1.11 Exploiting harmonics.

The fifth item in the above list of switching AM *versus* linear AM considerations is the exploitation of the RF switching for a dual roll: simultaneously modulating and multiplexing. This is illustrated in Figure 1.12, where the RF modulator is shown before the switch; the switch is being used to impose the repeater-mode modulation, a square wave which can be seen to delete half the pulses. The down modulation time is not wasted, however, because when the repeater path is blocked, the signal source input is switched on and allowed to pass a barrage-noise signal. As long as the spectral components from the signal source are not in the same band that the repeater is operating in, the threat radar experiencing the repeater mode ECM modulated signal will not recognize the difference if the ECM system transmits barrage noise in another band when the down modulation occurs in the repeater path. The reader should be sure to understand this example, because it is intended to clarify a key principle, the means that ECM systems employ to operate efficiently. It also illustrates another advantage of on-off modulation with respect to linear analog modulation. The simultaneous jamming, in separate bands, of Figure 1.12 is only viable if on-off type modulation is employed.

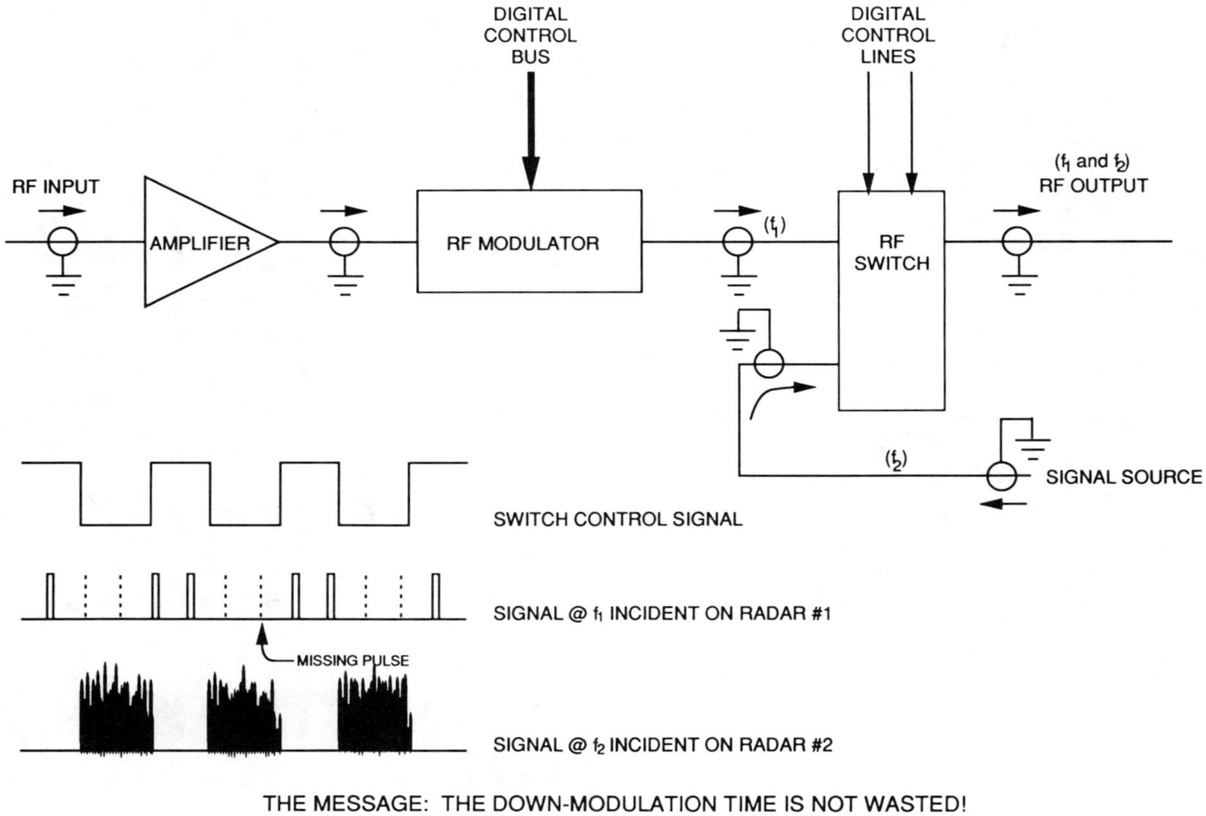

Figure 1.12 Modulating while multiplexing.

Finally, Figure 1.12 illustrates that the noise-mode barrage-noise jamming efficacy in the other band will not be influenced by the repeater mode or angle deception, amplitude modulation except to the extent that such influence lowers the average noise power within the radar's bandpass. The reduction of the ECM noise power received by the radar because of equal (50%) time multiplexing sharing will be between the extremes of 3 and 6 dB, depending on the relationship between the noise FM and multiplexing rates; the reduction is always 6 dB for a coherent signal. This assumes that the repeater AM is not near a harmonic of the barrage-noise jammed threat radar's lobing-scanning rate—but if it were, the result would be even better. The noise mode average power may thus drop as little as 3 dB for a square wave type of jammer multiplexing, where that square wave multiplexing modulation is in fact the modulation needed for repeater mode.

Some selected common ECM modulations and their characteristics are summarized in Table 1.3, excluding multiplexing considerations. The first one listed, AM, is appropriate for all three jamming modes against a conically scanning tracking radar. For swept AM, the percentage deviation is moderate to small, usually well under 100%; the time it takes to go through such a sweep is usually several seconds. The sweep period is determined by three considerations: (1) the range of uncertainty for the threat radar's lobing scan rate, (2) the maximum sweep rate, set by the above stated rule for significant perturbation, and (3) the need to make the sweep period as short as possible, because that is the time the threat has an accurate aim, less the possibly substantial time to recover from the previous perturbation.

AM intended for other goals, such as false angle targets against scanning search radars, or fouling the AGC of tracking radars, or blinking on and off for cooperative jamming purposes, would have as low or even lower modulation rates.

As shown in Table 1.3, the frequency or phase modulation is appropriate for repeater or transponder modes, and especially for coherent transponder mode. The frequency offsets should be compatible with relative target velocities and the RF band. Such modulations are not appropriate for noise modes, and in any case would be dwarfed by the relatively wide bandwidths of such noise. That is, the noise mode FM has deviation orders of magnitude larger than the values indicated in Table 1.3 for the frequency-phase modulation, making such modulation superfluous for noise mode. One partial exception to Table 1.3 frequency-phase modulation characteristics is repeater mode doppler-type noise, which would not have a sweep. The frequency-phase modulation sweep case can span both positive and negative frequencies relative to the original RF carrier. Because the sweep starts near zero frequency offset so as to steal the threat radar's doppler velocity gate, the percentage deviation is generally extremely large; however, the percentage deviation is quite small with respect to the absolute carrier value which is being modulated. This is often called *velocity gate pull-off* (VGPO) modulation.

The reader should understand that the distinction between phase modulation and frequency modulation is one of degree, not kind. Over a particular range of

Table 1.3
Non-Multiplexing Modulation Characteristics

Modulation	*Goals*	*Modes*	*Modulation Characteristics*
Amplitude	False angle	Repeater	On-off type
		Transponder	Hertz to hundreds of hertz
		Noise	On-off duty cycle ≤50%
			Max sweep rate equals the square of threat servo bandwidth
			FM sweep period of several seconds
			Small to moderate sweep deviation <100%
Frequency	False doppler	Repeater	Hertz to kilohertz offset (from RF carrier)
Phase		Transponder	Continuous (100% duty cycle)
			FM offset sweep period of seconds
			Sweep starts very near zero (large percentage deviation)
			Sweep both offset frequency polarities
			Sweep rate acceleration proportional to maximum-*G* track
Range	False range	Transponder	Repetitive burst patterns
	False range rate		Pulse widths of 0.1–10 microseconds
			Nil to hundreds of microseconds of delay
			Range delays ≪ AM periods
			AM periods ≪ range sweep periods
			Sweep periods of several seconds
			Sweep rate acceleration proportional to maximum-*G* track
			Sweep starts near zero delay (large percentage deviation)
			Negative sweeps are difficult
Noise	Obscure	Noise	Pseudo-random carrier FM pattern
			Analog or stepped carrier FM
			Pseudo-random FM rate
			FM rate < FM deviation
			Carrier FM deviations of 5–30 MHz
			Duration of tens of microseconds to CW

values, or for certain conceptual viewpoints, it may be more convenient to refer to FM rather than PM, and *vice versa,* but this judgement should be recognized as somewhat subjective. For example, a 1 kHz FM offset is equivalent to a 360° per millisecond phase saw-toothed modulation.

The next modulation listed in Table 1.3 is the range, or time delay modulation, which is generally only useful for transponder mode. (Detailed example descriptions will be given in Chapter 5.) However, there are several characteristics that should be noted, including the large magnitude of difference between the range delays, the AM periods, and the range sweep periods. The range sweep is known as the *range gate pull-off* (RGPO) modulation. The RGPO modulation has a very large percentage deviation; the generation of negative sweeps is quite difficult for range modulation, but not so for frequency modulation.

For AM, the sweep is intended to compensate for uncertainty rather than cause deception, and the maximum sweep rate is limited by the threat radar's angle track servo bandwidth. For FM-PM VGPO or RGPO, the sweep is a key part of the deception, and it is the maximum sweep rate acceleration that is limited. For the case of either the VGPO or RGPO, it is assumed that there is no initial-condition uncertainty to deal with, since the "walkoff" program starts at zero relative doppler or range. If the assumption is false, then the threat is not properly tracking the target, so lack of viability of the assumption is of no concern. For VGPO and RGPO sweep, the maximum acceleration rate is then simply the maximum *G* acceleration tracking capability of the threat. As for all the other sweeps, the recycle time for this walkoff sweep is an important measure of the effectiveness of the system; the sweep period is the worst-case time interval in which the threat can maintain an undisturbed aim at the target.

The noise modulation characteristics are also shown in Table 1.3. Again the reader should appreciate, as one of the themes of this text, that the noise characteristics are ultimately related to some rather basic considerations, especially the physical size of the target. That is, there could be orders of magnitude of difference in modulation parameters between ECM noise to protect a naval battleship and a jet fighter, but probably not between two jet fighters. The ECM parameters, although they often have a significant span, are not arbitrary, but are related to the threat radar parameters, many of which are in turn related to the propagating medium and the target's character, background, distance, and motion dynamics.

For these previously stated reasons, the noise is generated with FM that characteristically has a pseudorandom analog or stepped pattern intended to make the result less dependent on the precise value of the threat's reception IBW. The combination of the FM rate and the FM deviation determine the texture of the noise and its peak-to-peak amplitude. The rule for the deviation, as stated above, is that it should significantly but not grossly exceed the threat receiver's IBW. If the deviation is less than the receiver's IBW, then the average power within the bandpass will be higher, but the signal will lose its FM-PM based noise quality, and can even be filtered out by some types of radar receivers. The FM deviation characteristic listed in Table 1.3 is for narrow-band set-on noise, to be applied to a particular threat; wide-band barrage noise, to be applied to a class of threats, would have a much larger FM deviation. The rule for the FM rate (inverse step time) is that it should have a significantly lower percentage span than the deviation; higher rates are not compatible with either a flat, uniform spectrum or the rise time of the threat receiver. Both the FM deviation pattern and the FM rate pattern should, as the name implies, have the appearance of a random distribution. The values in the table are based on a radar IBW on the order of 10 MHz, which is suitable for a radar tracking a target with dimensions on the order of tens of feet. The repeater mode noise, as stated, is not shown in Table 1.3.

1.11 TIME SCALES

One key concept to the understanding of both the functioning and structure of ECM systems is the concept of time scales. One of the purposes of this book is to present to the reader many of the principles for system engineering that are especially appropriate for ECM systems. One such principle is that ECM system engineering should be based on time scale analysis studies.

One of the reasons that time scale analysis is so important to ECM systems is that the range of time scales employed is so huge. Table 1.4 illustrates this point. The shortest time scale, appropriate for the RF carrier periods, is a sub-nanosecond range, whereas the longest time scale appropriate for the range or velocity walkoff periods is on the order of seconds. Each example shown has a time range at least one order of magnitude different from the adjacent examples. It is interesting to note that most of these time scales are separated by at least two orders of magnitude.

Table 1.4
ECM Signal Modulation Time Scales

Time Scale	*Example*
Sub-nanosecond	Microwave carrier period
Hundreds of nanoseconds	Noise carrier frequency stepping
Microseconds	Range rate deception
Hundreds of microseconds	False targets
Tens of milliseconds	Angle deception modulation
Seconds	Range or velocity walk-off period

Because of these orders of magnitude or greater time scale differences, these signals and modulations do not interact to distort waveforms. That is, each layer of amplitude-frequency-phase time dependence modulation can be superimposed to serve its purpose without undue interaction from the modulation occurring on the other time scales. Furthermore, both of the statements, "modulations are simultaneously present" and "only one signal at an instant is present," can be true for a given ECM system; each statement is made in the context of a distinct time scale.

The frequency domain representation of this time scale analysis is shown in Figure 1.13. The operating frequencies are divided into two classes for this representation: the threat radar modulations and the ECM modulations. Note that there is a corresponding ECM frequency for many but not all radar frequencies. The ECM modulations are themselves divided into two sub-classes, based on whether they are truly free running. For example, the typical false target delay is the reciprocal of the false target operating frequency shown, no matter such a timing circuit needs to be triggered. That is, the range delay is only meaningful with respect to the radar pulse. In contrast is the angle deception modulation, which generally uses no trigger. Figure 1.13 should not, however be construed as implying that any of the ECM waveforms shown would not be more effective if synchronized with the appropriate radar waveform; they may be viable without such synchronization, whereas those marked with an asterisk are absolutely dependent on such synchronization.

1.12 INTRAMODAL MULTIPLEXING

The above text has described how modulations on vastly different time scales do not interact; it is appropriate here to apply time scale analysis to the issue of intramodal multiplexing. An example for noise mode is illustrated in Figure 1.14, showing frequency step dwells. In this example, the FM is intended to generate noise against two radar threats, operating at frequencies F1 and F2, simultaneously. What do we mean by generating power at F1 and F2 simultaneously? Can such simultaneous operation be achieved by a single VCO hardware resource as previously described? The answers to these questions hinge on the realization that if the VCO jumps from *F1* to *F2* and then back to *F1* in a time interval much less than the time constant of the radar receiver, this absence is not a meaningful time for such a receiver. For example, if the receiver has an IBW of 1 MHz, the time response is approximately 1 μs, or the reciprocal of the IBW. If a VCO generating a signal within the IBW would jump away for, say, less than 0.1 μs, and then return, the output of the receiver will not have changed significantly during such absence. Hence a VCO multiplexing between such 1 MHz IBW receivers, with dwells less than 0.1 μs, can indeed be at both frequencies simultaneously. Here simultaneity must be measured over a 1 μs interval.

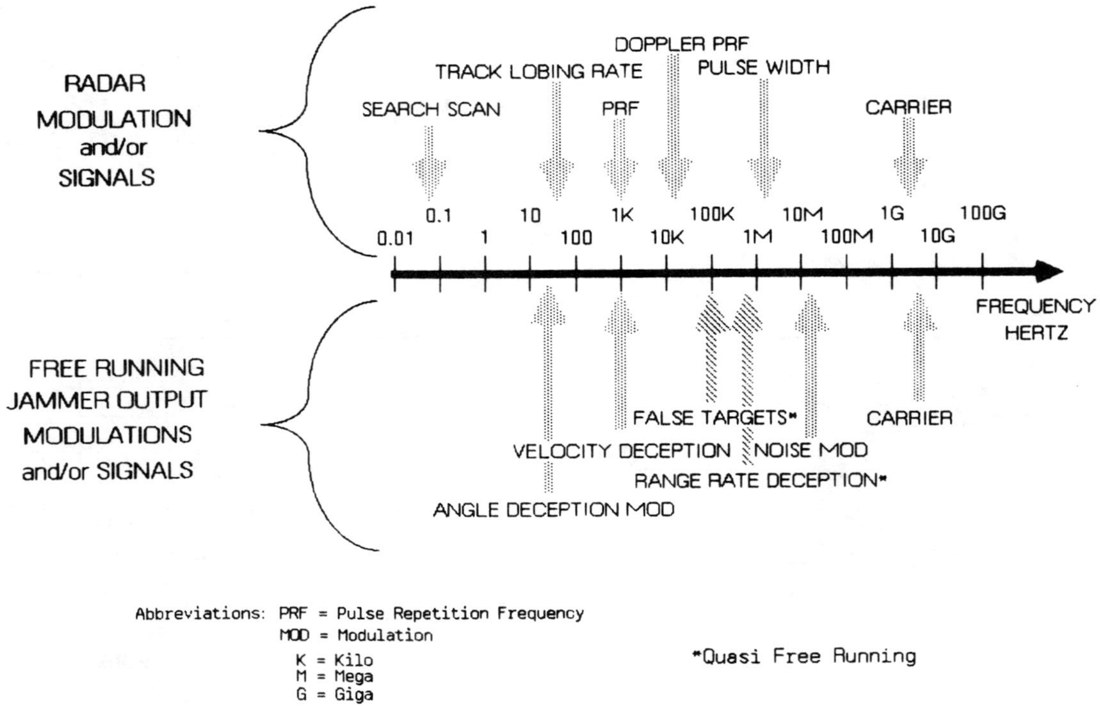

Figure 1.13 Operating frequencies.

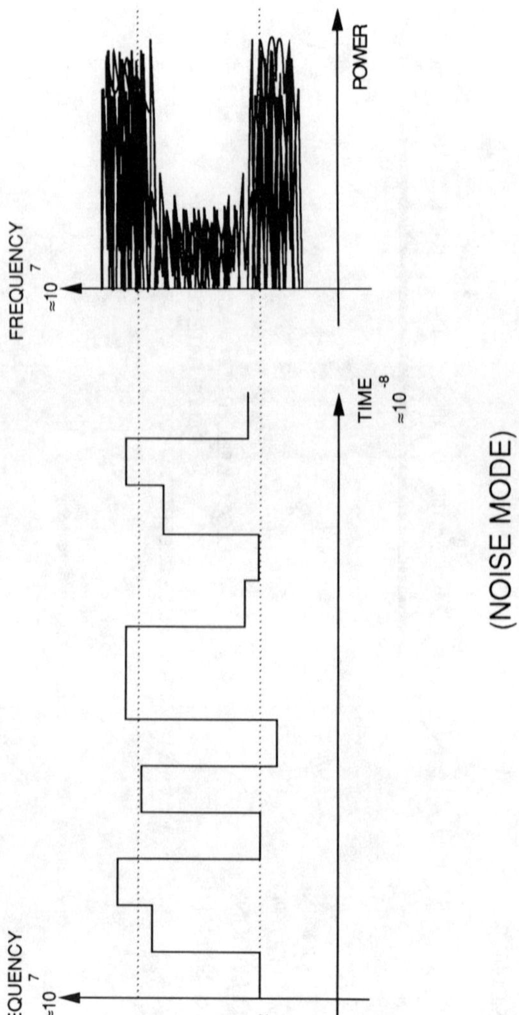

Figure 1.14 Intramodal multiplexing.

Two notes of caution are needed about the preceding discussion of simultaneous transmissions. First, no matter how fast the simultaneous multiplexing is done, the signal will experience duty cycle and perhaps spectral spreading loss. This loss, for sharing a coherent signal against two simultaneous receivers, is 6 dB, and for sharing two noncoherent noise patterns is usually 3 dB. Secondly, when a VCO is jumped away and then back to the original frequency, the signal output is generally not coherent. This is usually true for analog RF VCOs; however, the technology trend is toward digitally controlled RF sources, and for them this rule-of-thumb may or may not be valid. That is, a digital RF synthesizer may be quite capable of being jumped to a second frequency, and when commanded back to the first frequency will resume its oscillation at the same RF phase angle that would have existed if the VCO had not been tuned away, making the signal at the first nominal frequency coherent. It is unlikely that analog VCOs will ever be capable of accurately maintaining the phase, because there are a large number of cycles that occur in the typical dwell interval; hence, even a small percentage error means a large phase angle error.

In Figure 1.14, the reciprocal of the multiplexing dwell times can be thought of as the FM rate. The dwells should not all be equal in length, so that the receiver can respond to the longer dwells at its IBW related rise-time rate with a large amplitude change; for the shorter dwells, it just starts to respond. Such a range of dwells therefore influences the noise texture, which is also influenced by the FM deviation pattern. Note, in the figure, that the VCO is not simply jumped to the nominal frequency center at each dwell. Being slightly mistuned, as opposed to jumping well outside the receiver's IBW, influences the response in two ways: (1) the step transient response of a filter is different for mistuned signals, and (2) the steady state amplitude this transient response is aiming toward is generally lower than at the center. The receiver phase will also be different for somewhat mistuned carriers, but, as just stated above, the VCO will in any case have what effectively is a random phase angle for each dwell, so that this difference is of little consequence.

The type of noise mode multiplexing illustrated in Figure 1.14, which generates simultaneous noise in the form of spectral strobes against two receivers by multiplexing with dwells on the order of the receivers' time constants, can be extended to generate spectral strobes noise simultaneously against many radar receivers. Of course the more such strobes are generated, the more the *effective ECM power* is reduced against each radar. (See Chapter 3 for the definition of EEP.) There is another concern associated with sharing noise power between numerous threats simultaneously, which is not entirely related to the loss of average EEP. If, because of multiplexing against many threats, the VCO often stays away from each receiver's nominal frequency too long (several time constants) then the radar operator, or even the radar servo, may be able to see through these holes in the noise to observe the proper pulse that should be hidden by the noise.

1.13 INTRAMODAL MODULATION

Figure 1.15 shows another waveform quite suitable for time scale analysis: the RGPO waveform, quite commonly used in transponder mode. Figure 1.15 plots two different time scales. Angle deception intramodal modulation is quite effective when superimposed onto the RGPO modulation, but such modulation occurs on a different time scale than either the abscissa or ordinate scales shown in Figure 1.15, and hence is not visible except as a symbolic crosshatch. The crosshatch on the hook pulse is intended to symbolize the imposition of a low duty cycle gating modulation, which is an angle deception modulation.

The unusual characteristic of Figure 1.15 is that both the abscissa and ordinate scales are delineated in units of time. That is, the figure is literally a graph of time *versus* time! How can you have a graph of time *versus* time? It is possible when the time scales are so radically different, as illustrated in Figure 1.15. The abscissa is delineated in seconds while the ordinate scale is delineated in microseconds. If this still seems confusing to the reader, the ordinate scale can be thought of as range delay delineated in feet, yards or meters—500 feet per microseconds round trip.

The purpose of the waveform illustrated in Figure 1.15 is to "steal" the threat radar's range tracking gate. The hardware needed to perform that function is described in other chapters, but it is important that the reader understand the illustrated waveform, since it is a classic case of related modulations operating on different time scales. A constant acceleration walkoff, which has a parabolic shape on such graphs is shown. Over a time interval of several seconds, the range delay is continuously increased, to past 16 μs. The left-hand graph, or broadband detector response, illustrates the output waveform at one point in time on a time scale of seconds; this is what an oscilloscope synchronized with the threat pulse reception would display.

As stated above, because the layers of superimposed modulations have time scales that are so different, they do not cause waveshape distortion as normally envisioned. However, the AM and the RGPO both support each other's deception function. If the radar maintains good angle track, then the range track can soon be reestablished after being pulled off. If the radar maintains good range track, then the angle track is often difficult to break, because of the competing target return. The synergistic result of both can be quite effective: the RGPO allows the AM to break the threat's angle track without a competing signal before range reacquisition.

1.14 POWER MANAGEMENT

The application of the principles of power management provides a means for an ECM active jammer system to protect effectively and efficiently against several radar threats simultaneously. Power management is a specialized term that means the proper control of the parameters delineated in Table 1.5 to improve system efficiency and ef-

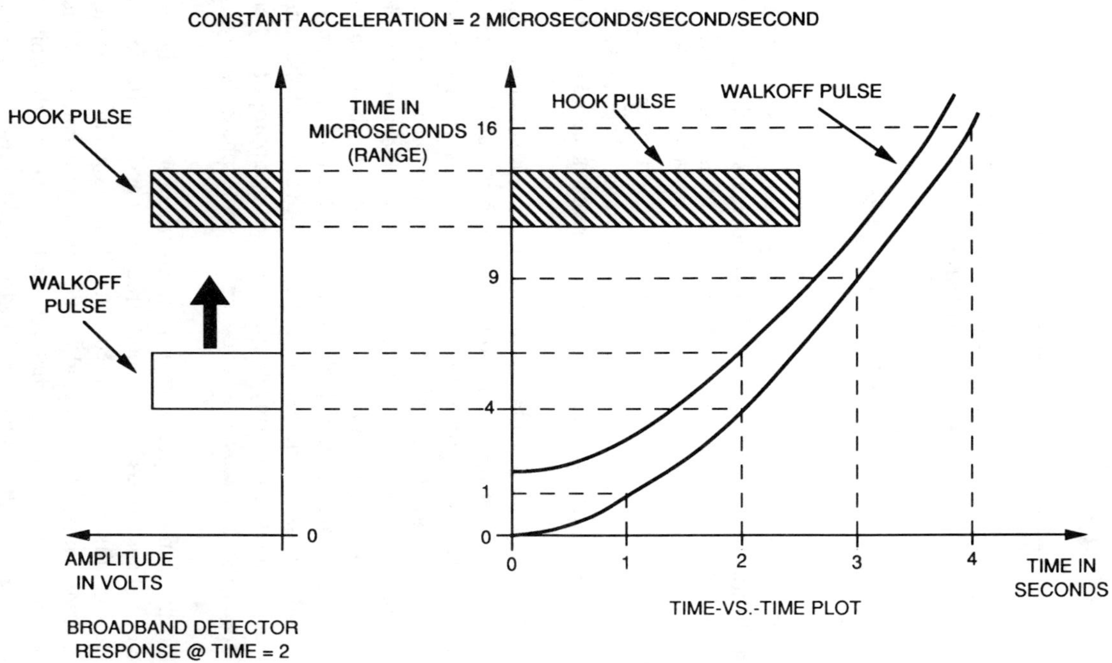

Figure 1.15 Range gate pull-off (RGPO).

Table 1.5
Power Management

Improving System Efficiency by Properly Controlling. . . *Parameter*	*Example*
Spectral	Frequency set-on
Temporal	Gated TOA windows
Spatial	Steered antenna beam
Polarization	Polarization control
Power	Duty cycle control

ficacy in a multi-threat environment. The three key parameters, in order, are frequency, time, and direction, with the two other parameters in the table sometimes having importance. Power management can be thought of as multiplexing the ECM resources against multiple threats, requiring either intramodal or intermodal multiplexing, or both. Subject to limited resources and conflicting requirements, the optimum mode and parameters are used against each threat. The above description of efficient noise power sharing against multiple threats is an example of power management.

A purist could argue that modulation in time and modulation in frequency are really the same thing. This is indeed theoretically correct, but because of the vast range of time scales referred to previously, there really is a valid distinction. The frequency control referred to usually means the microwave carrier frequency.

Figure 1.16 shows an example of *spatial power management*. In this illustration, the ECM system has a high gain antenna. This is a narrow beam antenna which is (1) either alternately aimed at threat Number 1 and Number 2 in an intramodal multiplexing fashion, similar to the examples given above, or (2) the antenna generates two precisely pointed beams truly simultaneously, by appropriately controlling the antenna to focus the EM waves at two angles; some phased arrays have this capability.

Figure 1.17 gives an example of power management in time. The figure plots time, on a scale of microseconds, *versus* both frequency and amplitude. At the left, a CW noise mode signal is being transmitted. The illustration shows the signal being interrupted by a transponder mode signal, which is a pure carrier. Note that the transponder output amplitude is larger than all but the peaks of the noise when the detector is band-limited; a full bandwidth detector would show the same amplitude for both noise and transponder modes, if, as is typical, all the noise is generated by frequency modulating a microwave VCO.

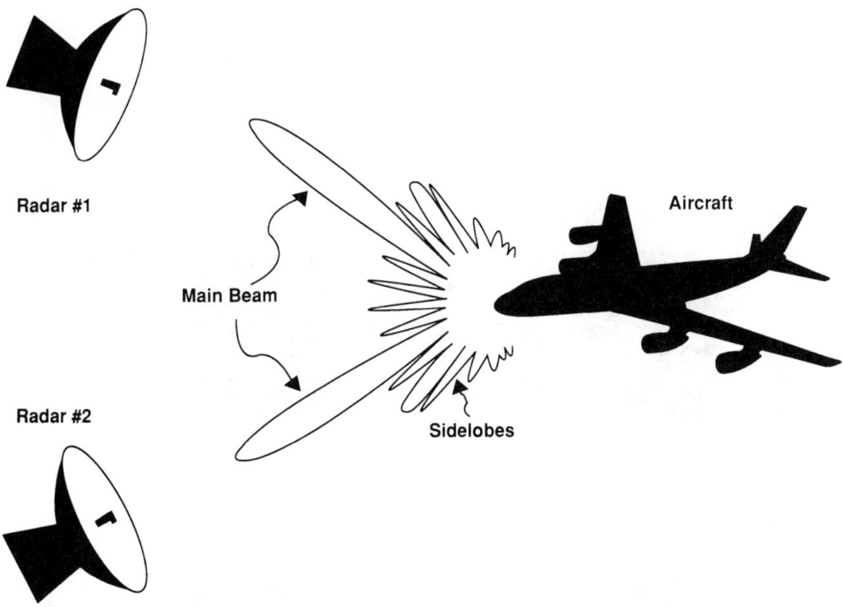

Figure 1.16 Spatial or angular power management

Thc figure shows the noise transmission resume when the transponder transmission ends. In other words, the transponder mode punched a hole in the noise mode transmission. Typically, for best efficiency, the ECM output tube generates maximum power in both modes, and only the character of the frequency control has changed. This type of multiplexing is efficient because it is reasonably consistent with the critical parameters of both modes: the noise mode (1) average power level per megahertz, (2) noise texture, and (3) gap look through; and the transponder mode (1) tuning stability, (2) tuning accuracy, and (3) timing.

As can be seen in the figure, the noise mode transmission is interrupted again, but this time by a repeater mode "window." This window is generated by a *pulse repetition frequency* (PRF) tracker that predicts the approximate position of the pulse. As is often the case for repeater mode, the repeated pulse is not at maximum amplitude.

Figure 1.17 illustrates typical intermodal time multiplexing employed for multisignal jamming by a modern power-managed ECM jammer that has the signal processing and CPU capability to perform this sophisticated, accurate, high-speed multiplexing as part of its power management capability.

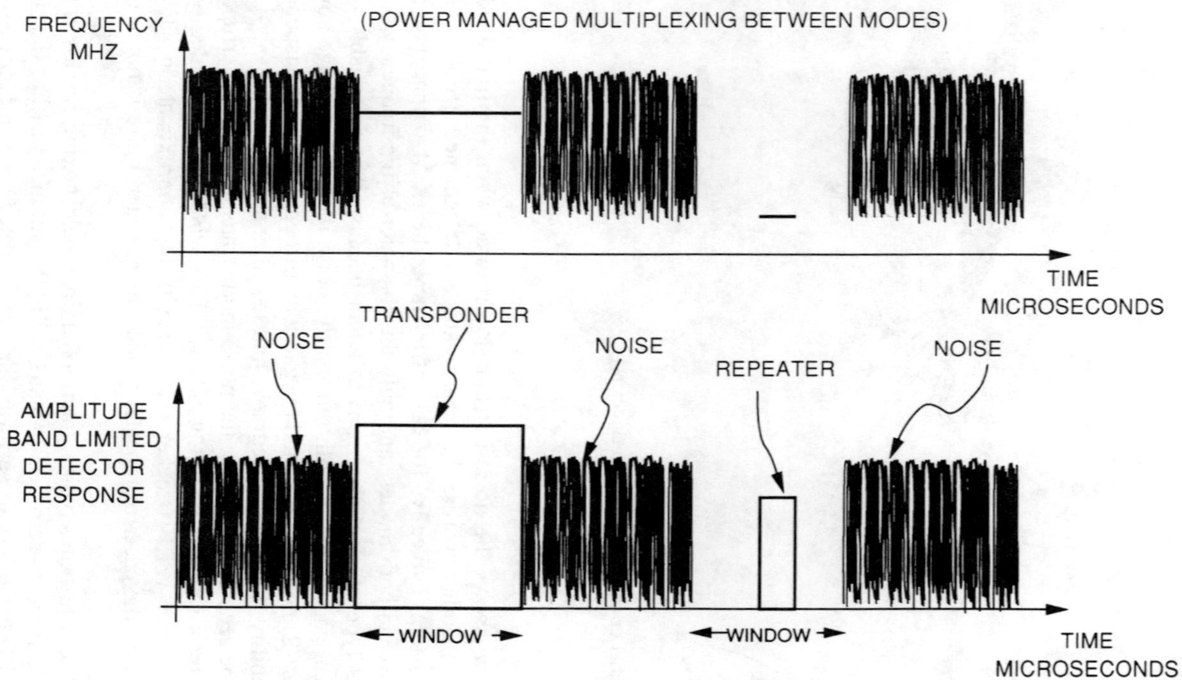

Figure 1.17 Power management in time.

1.15 ECM ARCHITECTURE VARIANTS

Figure 1.5 showed the basic ECM system architecture that was used as the basis for the above descriptions. This section will discuss some variants to that architecture.

Because resources such as the receiver, the signal memory, and the signal source shown in Figure 1.5 are so very expensive, and because, although they generally have wide bandwidths, they cover the full microwave band, neither instantaneously nor completely, the baseband structure shown in Figure 1.18 is employed. This structure folds all the micowave bands, such as the two shown in the figure, into one common baseband, where it is easiest to fabricate the critical subsystems: receiver, signal memory and signal source (VCO). This is accomplished by beating or mixing the RF input signal with a carefully chosen *local oscillator* (LO) to convert to baseband, and then beating any generated output baseband signal by the same LO to convert back to the original RF band. The output result is the desired microwave signal, since a double conversion using the same LO will return any signal to its original frequency. Such frequency conversions also create spurs, which are filtered out as shown. Quite often, there are several RF bands, each with their own antennas, preamplifiers, and transmitter power amplifiers, which are folded into or out of the baseband in an overlapping fashion.

Sharing the key subsystems among the different bands, by employing a common baseband, is cost effective to a significant degree.

Another variant to the basic ECM system microwave architecture is shown in Figure 1.19. This example is appropriate for repeater mode, and is typical of the jammers of the 1960s.

The problem the variant of Figure 1.19 addresses is the need to put distinctive modulations on the carriers of the different threats, based on threat type. The implementation of this variant is based on the assumption that the carrier value itself is a good discriminant to separate the radar threats. The figure shows the band multiplexed into several channels by using filter separators and filter combiners. The idea is that a common antenna and an RF preamplifier efficiently pass the signal into this RF carrier demultiplexer, or separator, following which the distinctive modulation is imposed, and the signals are then brought together again with a multiplexer or combiner, to transmit efficiently through one common power amplifier and transmitting antenna.

One problem with the variant shown is the fact that the channel frequencies are built into the RF circuitry; that is, the separators and combiners are fabricated to match a given set of filter parameters, and it is therefore difficult and expensive to alter this channelization subsequently in response to threat changes. Besides that problem, modern ECM systems have tended to abandon this type of structure because the threat signal environment has become more complex, with threat carrier ranges overlapping, so that the carrier is no longer considered a good discriminant.

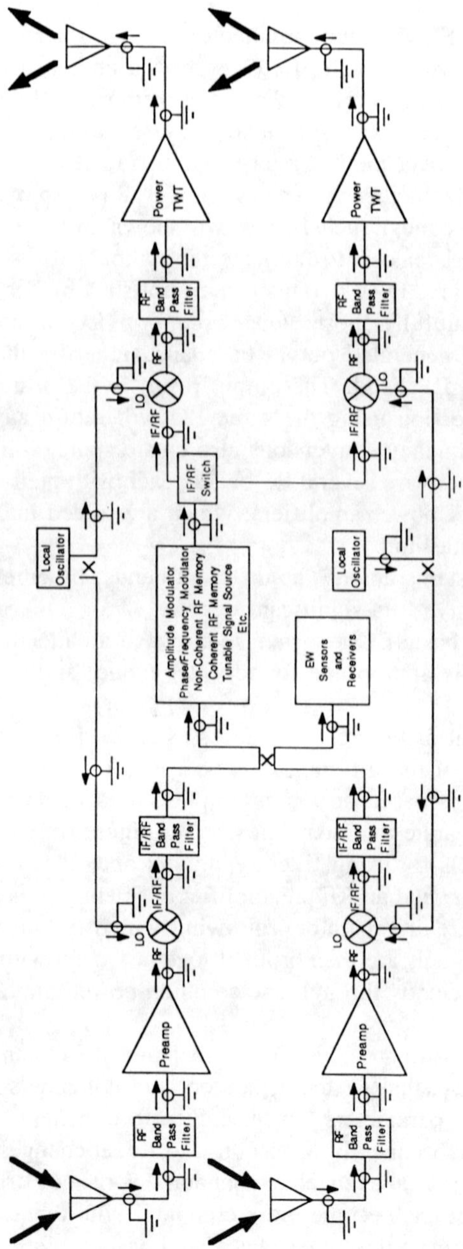

Figure 1.18 Baseband processing.

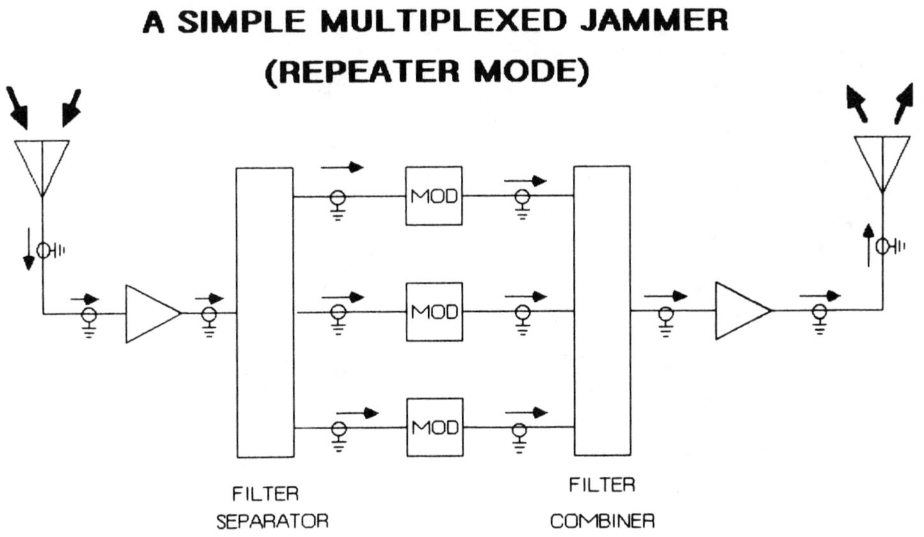

Figure 1.19 A carrier multiplexed jammer.

1.16 DIGITAL PROCESSING

The digital processing functions in an ECM system are summarized as

- signal separation,
- signal identification (ID),
- signal track,
- waveform generation, and
- central processing,

where the central processing function is a catch-all, including:

- technique determination,
- priority conflict resolution,
- resource allocation,
- parameter setup,
- data and parameter maintenance,
- gain and power control,
- slow servo functions, and
- built-in test.

Most of the data flow between the central digital processing subsystem and other subsystems or components at present is to or from the receiver, or sensor, and the signal source. In the future, the data flow ranking, in order, is expected to be:

(1) receiver, (2) signal memory, (3) antenna, and (4) signal source. In other words, we anticipate substantial changes.

A complete tabulation of the key hardware subsystems, including digital and microwave, would therefore consist of

- antennas,
- RF amplifiers,
- RF modulators,
- sensors,
- signal memory,
- signal source,
- central processing unit (CPU), and
- digital signal processing (DSP).

The distinction between the CPU and the DSP, described earlier in this chapter, is that the DSP subunits are more specialized, in order to optimize processing speed. In many cases DSP subunits have been hardwired, or use firmware, for maximum speed.

Chapter 6 describes digital processing in more detail.

1.17 THE EVOLUTION OF ECM SYSTEM STRUCTURES

Table 1.6 summarizes the development time line for ECM systems. Efficient power management concepts, along with advanced technologies like *microwave integrated circuits* (MIC), solid-state power amplifiers, and digital *large scale integrated* (LSI)

Table 1.6
The Evolution of ECM System Structures

Era	*Technology*
The 60s	Hardwired analog waveforms
	Fixed RF multiplexing
	TWTs used throughout
The 70s	Hardwired digital waveforms
	Digital integrated circuits (ICs)
	Baseband structures
	Software programmability
The 80s	Solid state RF amplifiers
	Power management with receiver and processor
	LSI digital signal processing
	Microwave integrated circuits (MICs)

circuits have made advanced sophisticated ECM systems practical, even to protect vehicles or craft with a relatively small in-band insertion volume, such as fighter aircraft. The actual deception and jamming modulations have also evolved, largely because of improved hardware and software capability.

With regard to total bandwidth coverage, there has been very little evolution. This is because the bands of operation for threat tracking radars is determined more by the laws of physics than the restraints of technology. The ECM system microwave band coverage is about 2–20 GHz. Some of the search radars, which can use large antennas, operate at lower bands, and ECM systems have operated there as best they could within the limitation of ECM antenna size. There has been speculation for several decades that ECM will soon be needed for the higher *millimeter wave* (MMW) bands; this speculation may some day prove true, especially for space applications where atmospheric propagation loss is of no concern.

Although the total bandwidth coverage has remained unchanged, the evolution of the components has been steady, with a resulting major change in the ECM system structure. In the 1960s, octave components were considered "high tech." For ECM systems, this meant at least three octave bands, each broken into numerous channels, as shown in the carrier multiplexing variant described above. Some key components proved to be exceptions to the octave capability, however; for example, microwave phase modulators generally were not available with more than a channel's bandwidth. As the components available to ECM system houses have become wider in bandwidth, the system structure has generally evolved to two bands. There is a strong cost and size incentive to pursue this trend to its ultimate limit of just one band.

Chapter 2
Components and Subsystems

2.1 INTRODUCTION

Based on the functional partitioning of ECM systems shown in Figure 1.3, the hardware can be partitioned as follows:

- functional subsystems:
- antenna,
- sensor,
- controllable elements,
- digital processors;
- additional hardware:
- components without electrical control,
- power supply, and
- structural and mechanical.

The traditional ECM antenna employed a passive broadband broadbeam EM wave transducer. Because of the need to enhance *effective radiated power* (ERP), more modern and capable ECM systems employ active aperture antennas; in such cases it is less appropriate to make a distinction between the EM transducer and the controllable elements. As stated in Chapter 1, the control of such antennas may become intimately associated with the waveforms needed for the individual jamming and deception techniques and the multiplexing waveforms. Similarly, certain sensors are intimately associated with the EM wave transducers and are used to measure certain parameters of the EM propagating waves; these measurements include

- angle measurement:
- monopulse amplitude comparison,
- monopulse phase comparison, and
- polarization measurement.

The other hardware implementation choices available to the ECM system engineer for sensing intercepted EM wave signals include

- broadband detector (using crystal diodes),
- *digital instantaneous frequency measurement* (DIFM) (using delay line phase comparison),
- narrow-instantaneous-bandwidth tunable receivers (e.g., superheterodynes, *yttrium-iron garnet* (YIG) tuned receivers),
- channelized receiver (using *surface acoustic wave* (SAW), *magnetostatic wave* (MSW) lumped element, *et cetera*),
- RF memory (using digital sampling),
- microscan and compressive receivers (using dispersive delay lines),
- AO receivers (using acousto-optic Bragg cells), and
- digital microwave receiver (using high-speed *analog-to-digital converters* (ADCs)).

There are two types of RF transmission lines in general use between components packaged with connectors: coaxial or "coax," and waveguide. Stripline is used within components and for MIC circuits. Coaxials are universally used for low power RF transmissions between components individually packaged with connectors. Coaxials have a center conductor separated from a surrounding conductor by a supporting insulator. The most common low power coaxial lines for use in systems are semi-rigid cables with either 0.141″ or 0.086″ outside diameters. Specialized tools are employed to install screw-on (SMA) connectors. Specialized tools are also employed to bend these miniature cables to the correct shape to make an RF connection from one component to another; in making these bends there usually is a minimum bend radius that must be respected, an important factor when designing the mechanical layout of the components within the system. Coaxials are characterized by very wide bandwidths: for all practical purposes, the complete microwave and millimeter wave bands; the loss per foot, however, is substantial at the higher frequencies, as illustrated in Figure 2.1. Coaxials are also characterized as being non-dispersive; that is, all frequency components travel at the same speed through the line, at approximately 70% of the speed of light in free space. Transmission losses above 20 GHz are generally considered unacceptable for many applications. The potential for increased loss and mismatch in the cable connectors also becomes high above 20 GHz, unless great care is taken in their original fabrication and in their subsequent treatment.

Whereas coaxial use predominates for low-power microwave transmission between components individually packaged with connectors, waveguide is often preferred for high-power transmission, such as between the power TWT and the transmitting antenna. Waveguide looks something like a water pipe, albeit usually rectangular. The mathematical equations describing the propagation are somewhat complicated, but the transmission, in rectangular waveguide, can be thought of as bouncing the EM wave signal from side to side, rather than following a straight path down the center. The net path length depends on the EM wavelength. As such,

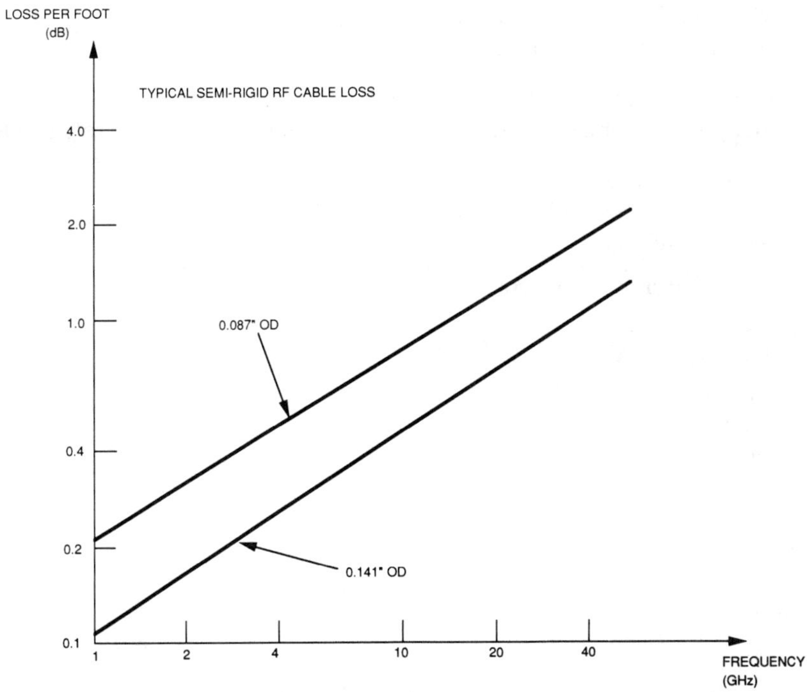

Figure 2.1 Coaxial cable loss.

waveguide does indeed exhibit at least some dispersion, the degree of which often becomes unacceptable for bandwidths greater than an octave. The dimensions of the waveguide are determined by the wavelength of the band to be transmitted, and the resulting transmission line is usually too large and too heavy to allow using waveguide for general purposes. The loss, however, at least in band, is usually superior to coaxial, and efficiency is the important consideration for the transmission line from the final output amplifier to the antenna, because the last few decibels of RF power are invariably expensive to generate.

The technologies available for electronically controllable elements include

- amplitude modulators,
- phase and frequency modulators,
- (microwave) switches,
- voltage controlled oscillators (VCOs),
- RF memories,
- programmable RF delay lines,

- TWTs and their modulators, and
- recirculating-memory loops.

Components which are considered active, because they consume power from a power supply, but which functionally are not under electrical control, include

- solid-state amplifiers,
- traveling wave tube (TWT) amplifiers, and
- fixed-RF oscillators.

The passive microwave components include

- transmission lines,
- delay lines,
- fixed attenuators (pads),
- couplers,
- circulators,
- isolators,
- filters,
- mixers,
- frequency multipliers and dividers,
- limiters, and
- frequency equalizers.

Note that transmission lines are included on this list of passive components. None of the above passive components consume power from a power supply, except those with RF diodes, which may need a bias current, including mixers, multipliers, dividers, and limiters. Active amplifiers are sometimes used for the limiting function; circulators are sometimes used for switching, which consumes transient power. All except the circulators, isolators, and perhaps the multipliers and dividers are reciprocal, in the sense that the power flow between the input and output ports can be reversed with a resultant matched attenuation. All except the mixers, multipliers, dividers, and limiters can be considered linear. Superposition applies for linear components in which multiple signals will not interfere with one another. The degree of linearity of a component is generally characterized by the degree to which the amplitude changes of the output match those of the input. By this measure, the mixer actually has quite good linearity over much of its range. Nonlinear components, such as diodes, can be used to perform linear functions, such as constant-gain frequency translation of a signal.

The functional categories of controllable elements and sensors include three main subsystems, as opposed to components, which will be described in this text:

- signal sources;
- receivers; and
- recirculating-RF memory loops.

The remainder of this chapter will describe these components and subsystems, with the exception of the sensors and receivers and the recirculating-RF memory loops which are each described in separate chapters.

2.2 PASSIVE RF COMPONENTS

2.2.1 Transmission Lines

In an ECM system, connections between components and subsystems are made via wires, busses, and transmission lines. Wires, or insulated-conductor lines, are used to carry low-frequency video signals or power-supply currents and voltages. Busses are used to carry digital-data words, that is, many bits in parallel, and are fabricated with many individual wires. Transmission lines are used to carry RF signals and sometimes higher rate video signals or long runs of a line that would otherwise be subject to *electromagnetic interference* (EMI).

Lately, industry has been pursuing alternative interconnection means. The rat's nest of semi-rigid cables between components may not be as common in the future, both because the components are integrated and because the interconnections are fabricated with a more sophisticated physical structure. The trend is towards improved serviceability, as with digital circuit boards, through the easier removal and insertion of modules of RF components.

The key functional parameters and their representative values (shown in parentheses) for low-power transmission lines are

- Loss per foot (*versus* frequency) (0.5 dB)
- *Voltage Standing Wave Ratio* (VSWR) (1.1:1)

2.2.2 Delay Lines

RF delay lines have three main roles:

- to act as a signal memory for subsequent transmission,
- to delay the signal long enough to steer an accurate measurement receiver to the approximate frequency, and
- to create a phase difference to
- steer antennas and
- measure frequency.

The delay line technologies are summarized in Table 2.1. There are other delay alternatives that employ intermediate carriers. The acoustic and magnetostatic delay lines were, at least at one time, considered exotic material delay media.

Table 2.1
Delay Line Technologies

Technology	*Approximate Propagation Speed*	*Status*
Coaxial	c	Widely used
Acoustic	$c \times 10^{-5}$	Occasionally used
Magnetostatic	$c \times 10^{-5}$	In development
Fiber optic	c	In development

The coaxial delay lines are employed in wide-band applications requiring sub-microsecond delays, such as in recirculating-memory loops (described in Chapter 5) and as a delay unit to allow an accurate receiver to be steered to the approximate frequency of an input pulse. The propagation speed in typical coaxial is about 70% of the free space speed of light. Because of the frequency *versus* loss characteristic, as shown in Figure 2.1, the delay line complete unit usually includes a frequency equalizer, so that the loss will be constant across the band. The complete delay unit often includes a heating element and thermostat to maintain the loss as a constant, because it will vary with temperature. Because coaxial does not exhibit dispersion, neither does a coaxial type of delay line.

Coaxial is quite frequently coiled into unusual shapes for insertion into a system, often with the center of the coil occupied by other components, such as a TWT.

The exotic material delay lines, although they do not employ an intermediate carrier frequency, have the signal propagate in another medium rather than as an electromagnetic wave, generally at about five orders of magnitude slower than the speed of light. These have often been referred to as *solid-state* delay lines, probably because of their small physical size compared to the bulky coaxial, although this is, of course, a misnomer. Whereas the coaxial approach is used for sub-microsecond delays, these exotic material delay lines are used for longer delays, because once the wave has been launched, it exhibits relatively low loss per unit of delay. For delays in the range of up to one microsecond, the loss budget typically has about one third the loss in the input transducer, one third in the medium, and one third in the output transducer. For delays under a microsecond, 20 dB is a typical loss value. Both surface and bulk waves are employed. Some applications, such as antenna applications, need dispersion. Other applications need electronic control of the delay, which is normally a difficult feature to achieve, but the magnetostatic approach, in principle, offers this feature.

More recently, optical fibers have been considered for RF delay lines. These consist of extremely thin transmission lines which can be coiled into quite small volumes. The transmission loss is quite low, which is why the communications industry has employed fiber optics extensively. As shown in Table 2.1, fiber optics is the only technology which uses another carrier; although the true carrier has indeed changed, the microwave carrier is still present, and is used to modulate the laser signal. At the reception end, the RF signal is detected.

The key functional parameters and representative values for delay lines are

- delay 0.2 μs
- loss 20 dB
- ripple 3 dB
- bandwidth 1 GHz
- VSWR 2:1

2.2.3 Fixed Attenuators

The purpose of fixed attenuators is to reduce an RF signal amplitude by dissipating some of the RF signal power as heat.

There is no special symbol for fixed attenuators in a system block diagram, except perhaps a rectangle with the word "pad" inside and the attenuation value. Indeed, although fixed and manually adjustable attenuators are widely used in microwave laboratories, they are generally shunned by system design engineers, who prefer the flexibility of voltage variable attenuators or gain elements. Pads are sometimes employed, however, where VSWR match is critical and isolators would be too large or expensive.

The common values for fixed attenuators produced by vendors are 3, 6, 10, and 20 dB units. The VSWR is generally excellent and the low power units small, the ripple (variation with frequency) is small, and the devices exhibit wide bandwidth and reciprocity.

The key functional parameters and representative values for pads are:

- attenuation 10 dB
- ripple 0.5 dB
- VSWR 1.2:1

2.2.4 RF Couplers

The RF signals passing through transmission lines cannot generally be summed or split as easily as signals on wires are. The purpose of RF couplers is to couple RF power from one transmission line to another.

Four representations of RF couplers are shown in Figure 2.2. This figure shows the power flow conventions for the four representations, with the solid arrows and the dashed arrows each making a complete set. Couplers are inherently four-port

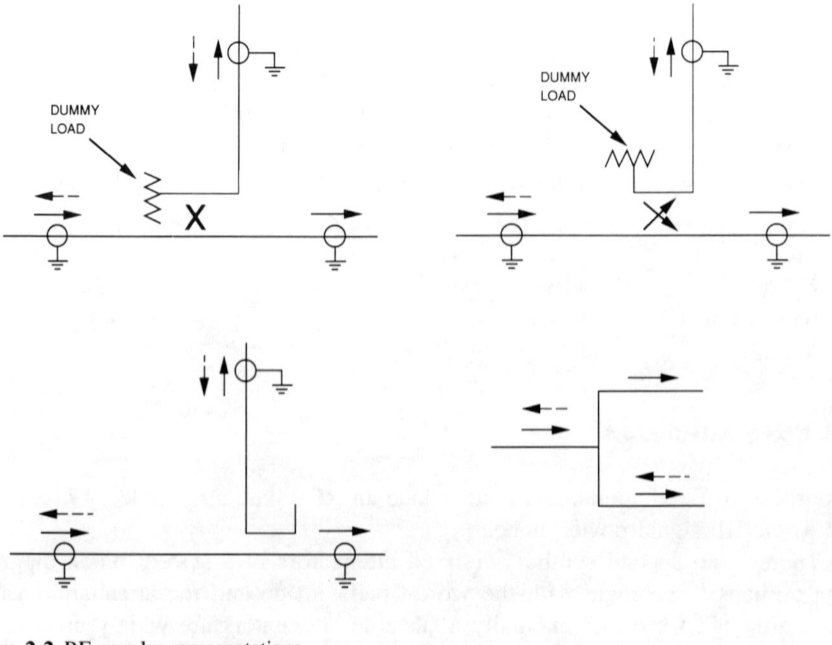

Figure 2.2 RF coupler representations.

devices, although couplers are quite often fabricated with "dummy" loads built in, effectively making them three-port components. In the third and fourth representations, the dummy load is not shown explicitly. The first representation is the one generally preferred by system engineers, and is used in the block diagrams of this text, whereas the third representation is the one generally preferred by component design engineers, and the fourth is often used by antenna design engineers.

The resistive portion of the complete insertion loss is usually quite good, generally being well under one decibel. The complete insertion loss is determined by the conservation of energy, so that the total power out of the other three ports, plus the resistive insertion loss, equals the power at the input port. Figure 2.3 shows the conservation of energy coupling loss in graphic and tabular form, that is, ignoring the resistive loss and assuming that the fourth port is perfectly isolated. For example, the outputs from a 10 dB coupler will be down 10 dB at the coupled port and down 0.46 dB at the straight-through port, both with respect to the input power. The isolation rating of a coupler determines the worst case power transfer to the unintended fourth port. For example, a 10 dB coupler with 20 dB isolation rating would have an output at the isolated port that is down at least (10 + 20 =) 30 dB. The net VSWR at the input port is determined by the impedance at all the ports, including

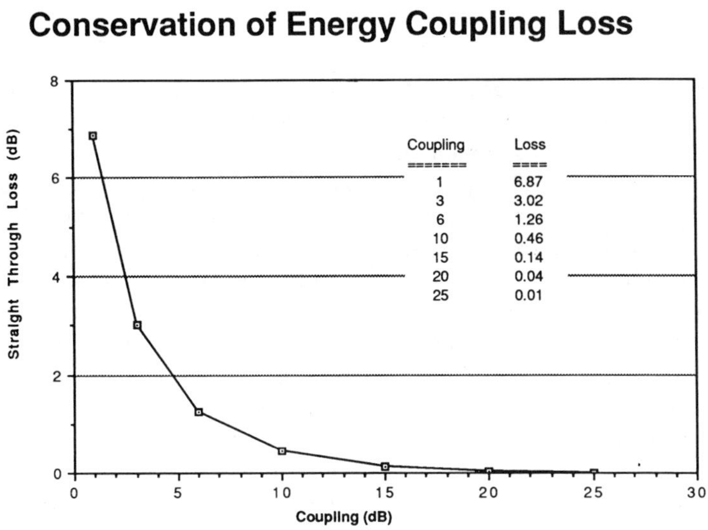

Figure 2.3 Conservation of energy-coupling loss.

the impedance of other components connected to these ports. Generally, and properly, the coupler vendors specify the input VSWR when all the other ports are perfectly matched with dummy loads. These devices exhibit reciprocity.

The commonly available coupling values are 3, 6, 10, and 20 dB. One type of coupler is called a hybrid; from the user point of view this can be considered a 3 dB coupler, that is, a coupler which splits the input power into two equal power signals. The common frequency independent phase relationships between the outputs of 3 dB couplers are 0°, 90°, and 180°.

In recent years, the trend to MIC has led to nonstandard coupling values chosen by the system user.

The key functional parameters for couplers and their representative values, shown in parentheses, are:

- resistive loss (0.5 dB)
- coupling (and ripple) (10 dB; 1 dB)
- isolation (20 dB)
- VSWR (1.2:1)
- bandwidth (4:1)

For some applications the phase and phase tolerance of hybrid couplers are key parameters.

2.2.5 Circulators and Isolators

The purpose of RF circulators and isolators is to create a nonreciprocal loss; this property is used to improve the impedance match between RF components.

An RF circulator is a nonreciprocal three-port component, with resistive insertion loss on the order of about 1 dB, and an isolation loss typically on the order of 20 dB. The isolation loss is between the unintended pairs of ports. These devices have bandwidths on the order of octaves. An isolator is a circulator with a dummy resistive load built into, or connected to, the third port. Figure 2.4 is a representation of an isolator. In such representations on block diagrams, it is important to show the direction of circulation—clockwise as in the figure, or counterclockwise. The solid and dashed arrows show the direction of RF power flow in consistent sets. In the figure, an RF signal into port 1 will emerge from port 2, whereas an RF signal into port 2 will emerge from port 3. The circulator in the figure is configured as an isolator, so an input into port 2 will not emerge from port 1 even though an RF signal into port 1 emerges from port 2. This feature isolates the source from the reflections of a mismatched load.

By casual inspection, an RF circulator would seem to be a reciprocal device; however, gyrotropic permanent magnets are built into the structure, causing Faraday rotation of the EM wave's polarization, so that the RF signals from the two paths either reinforce or cancel. The net result is that the magnetic field steers the EM wave to the appropriate port. Because of the permanent magnet, the devices are often considered to be unacceptably large for avionics applications. Some circulators use electromagnets, so that the field can be reversed; such a device acts like a switch and should be considered an electronically controllable device. Circulator switches

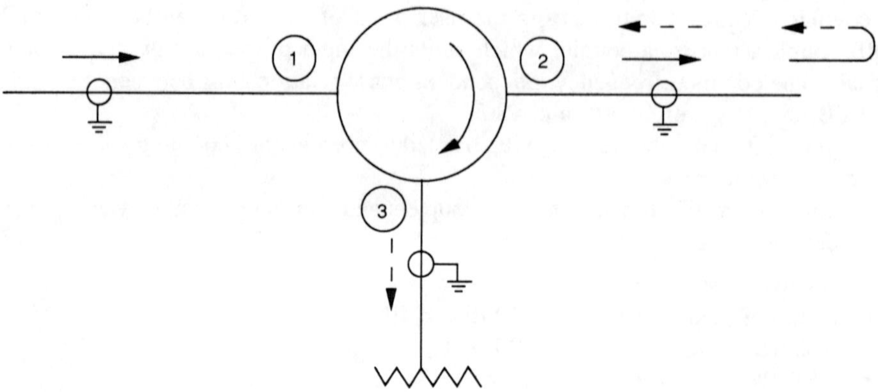

Figure 2.4 A representation of an RF isolator.

usually have a higher power rating than RF diode switches, and are therefore more appropriate for switching a high power TWT output between different antennas.

Passive circulators with permanent magnets are often used with single-antenna systems, including radar systems, to steer the transmitted power out the antenna, while simultaneously steering the impinging EM wave captured by the antenna into the receiver, instead of back to the transmitter output tube. Isolators, as the name implies, are intended to provide a good match when inserted between the source generator and the ultimate bad-match load. As will be explained in a later chapter, a bad match means that the load reflects a large percentage of the impinging signal; an isolator causes this reflection to be steered into the dummy load instead of back to the source generator. This is to be contrasted to the use of a fixed pad attenuator, which will also provide a good match in a small size and at much lower cost; unfortunately, the pad's forward loss is equal to the return loss, whereas for the isolator, the forward loss is small while the reverse loss is large, which is a much more desirable electrical performance feature.

The key functional parameters for circulators and isolators and their representative values, shown in parentheses, are:

- insertion loss (0.4 dB)
- isolation (20 dB)
- bandwidth (2:1)
- VSWR (1.3:1)

2.2.6 Filters

The purpose of filters is to create a frequency-dependent loss.

There are many types of filter designs, but from a system user's functional point of view the available types are

- bandpass,
- band notch,
- low-pass,
- high-pass,
- multiplexers and demultiplexers, and
- tunable.

Filters are passive and reciprocal components, except that complete multiplexer and demultiplexer units often include circulators, and hence are not reciprocal. When system engineers refer to a filter's bandwidth, it is assumed that they are referring to the 3 dB bandwidth, as illustrated in Figure 2.5, unless explicity stated otherwise. Filters have low insertion loss, the degree of insertion loss being approximately proportional to the number of poles on a log scale. As shown in Figure 2.5, the filter skirts asymptotically approach a slope of 6 dB per octave per pole in the limit of

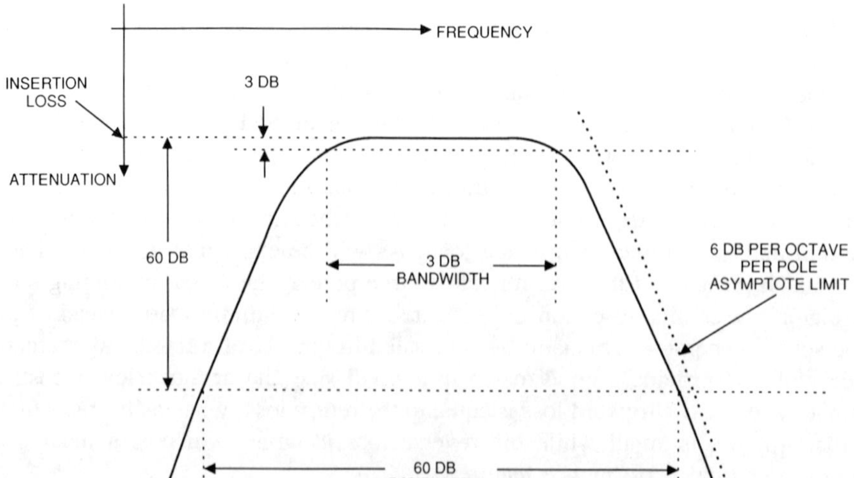

Figure 2.5 A filter's frequency response.

separation from the center. For example, if a 4-pole filter has a 3 dB bandwidth of 10 MHz, then its (4 × 6 + 3 =) 27 dB bandwidth would be about (2 × 10 =) 20 MHz, its (4 × 6 + 27 =) 51 dB bandwidth would be (2 × 20 =) 40 MHz, etc. In other words, the number of poles is a measure of how steep the skirts are, and hence a measure of the ability of the filter to reject frequency components not within the bandpass. The term poles is derived from the mathematical representation of filter responses. The number of poles a filter has is approximately proportional to the internal circuit complexity. The shape factor is also used to describe a filter. Unlike specifying the number of poles, it is an explicitly functional description, and is probably better. The shape factor is simply the ratio of two bandwidths that intersect certain attenuation values, which are the outward frequencies if the attenuation curve is not monotonic; if not explicitly stated otherwise, the shape factor is assumed to be the ratio of the 60 dB bandwidth to the 3 dB bandwidth, as illustrated in Figure 2.5.

The most common technology for tunable filters at direct RF, as opposed to superheterodyne filtering, is the *yttrium iron garnet* (YIG) filter, which uses a sphere of YIG that couples signals from one line to another at a frequency determined by an imposed magnetic field. The instantaneous bandwidth varies from 20–40 MHz. The YIG's magnetic field is generated by an electromagnet, so the control of the current in the electromagnet therefore tunes the filter's center frequency.

Filters are used in ECM systems to eliminate harmonics from RF power transmitters, to eliminate unwanted mixer images and spurs, and to separate or demultiplex threats based on their carrier frequency. Figure 2.6 illustrates one of the key problems encountered when designing a system to use frequency multiplexing and demultiplexing, also known as channelizing. ECM systems are usually required to have continuous frequency coverage, so the system designer crosses the response of adjacent channel filters at the 3 dB point or half the power for each channel. It is quite possible to use a set of channel filters with good individual responses, and with properly aligned 3 dB points, but when the RF signal on a common line is separated, routed through these channel filters, and then recombined to a common path, the response will exhibit severe ripple, especially at the crossover points, as illustrated in Figure 2.6. This phenomenon occurs because the electrical phase length at the crossover frequency is different for the paths through the individual channels; the individual signals will either reinforce or cancel each other, depending on their relative phase angles at the combining point.

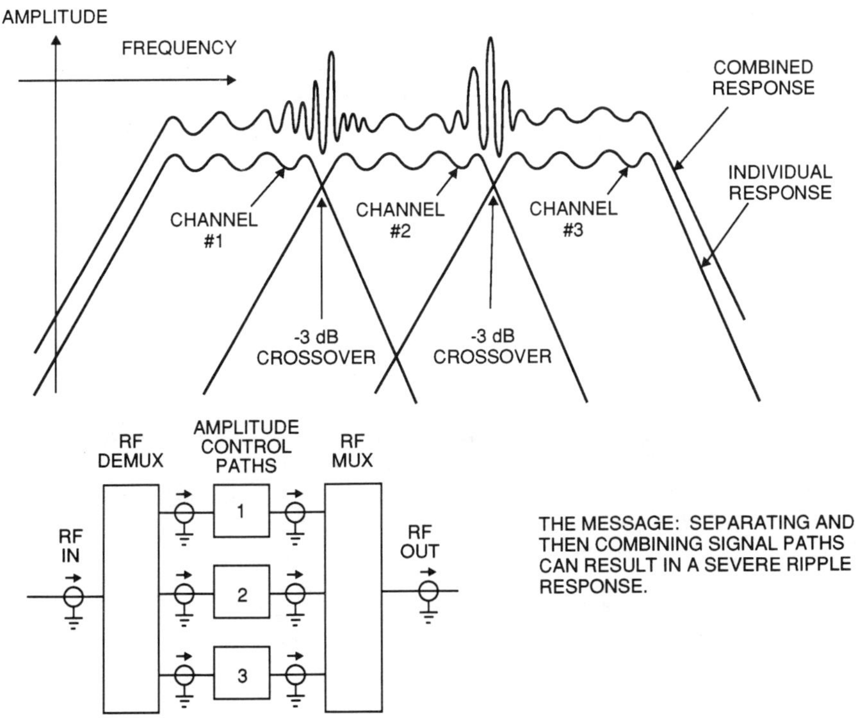

Figure 2.6 The continuous coverage problem for frequency-multiplexed repeaters.

When designing RF multiplexers, the individual filter responses can also be distorted because of match problems. Most filters do not absorb the power they prevent from passing to the output line; instead, they reflect this power. Therefore, the match is better in band, and *vice versa*. To prevent this adverse interaction, circulators are often employed as shown in Figure 2.7. Any power reflected from one channel is steered to the next channel, where it may very well be in-band and pass through. Generally, the signal is steered to the individual channels in inverse order to their band center carrier values; since losses are higher at the higher frequencies, such a sequence tends to equalize the loss of all the channels.

Some filters are characterized by high insertion loss, and in those cases the ripple is a key parameter, because the insertion loss parameter is no longer an acceptable upper limit for the ripple. In general, the key functional parameters and their representative values (shown in parentheses) for each bandpass of bandpass filters are

- center frequency (3 GHz)
- bandwidth (300 MHz)
- shape factor (5:1)
- insertion loss (1 dB)

Older ECM systems employed wide channelization filtering, with free running preset modulation on each channel. Modern systems tend to use much narrower channel filters, with programmable modulation on each channel. A modern EW receiver may

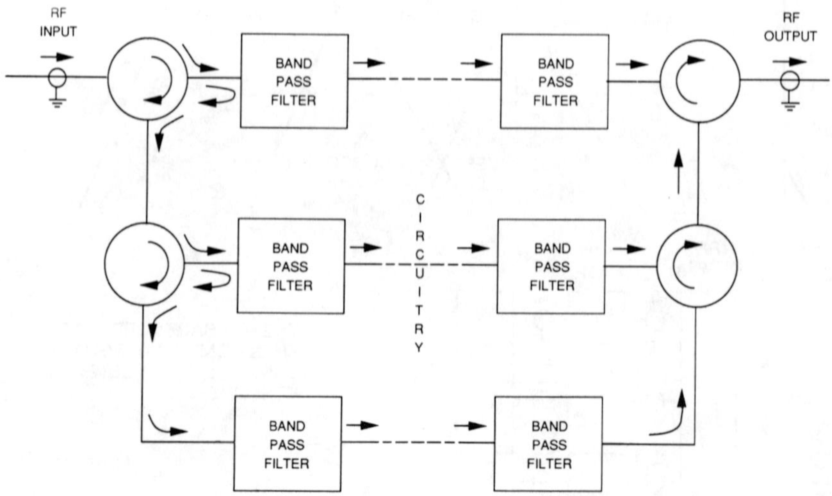

Figure 2.7 Multiplexer matching.

also use banks of narrow channels. To the extent that the channel's IBW is close to the reciprocal of the threat radar's pulse width, on the order of several MHz, the system engineer may have to specify more than the key frequency-domain parameters listed above. For the narrow channel receiver application, it may be appropriate to specify the time domain response as well.

2.2.7 Frequency Translators

The purpose of frequency translators is to convert the signal from one band to another. The mixers in Figure 1.18 are used for that function, as an example.

A representation of a mixer is shown in Figure 2.8. The convention for power flow is shown with the consistent set of solid and dashed arrows. A mixer is a three-port component: RF, LO, and IF. Two of these ports are inputs, and the third is an output, with the *local oscillator* (LO) port always being an input. The RF and *intermediate frequency* (IF) ports have quasilinear amplitude signals, while the LO port has a saturated signal. The function of a mixer is to convert frequency, and this function is achieved by multiplying signals. Multiplication cannot occur in linear networks; hence the need for a saturated signal driving nonlinear RF diodes. This saturated LO signal can be viewed as alternately multiplying the other input by +1 and then −1 at the LO carrier rate (a half cycle for each sign). It can also be viewed as switching the phase of the other input by 180 degrees at twice the LO rate. In other words, although few engineers do so, a mixer can really be thought of as a modulator. It can be shown mathematically that this simplistic model will result in two equal amplitude output tones, the carrier frequencies of which will be the sum of the input and LO carrier frequencies and the difference of the input and LO carrier

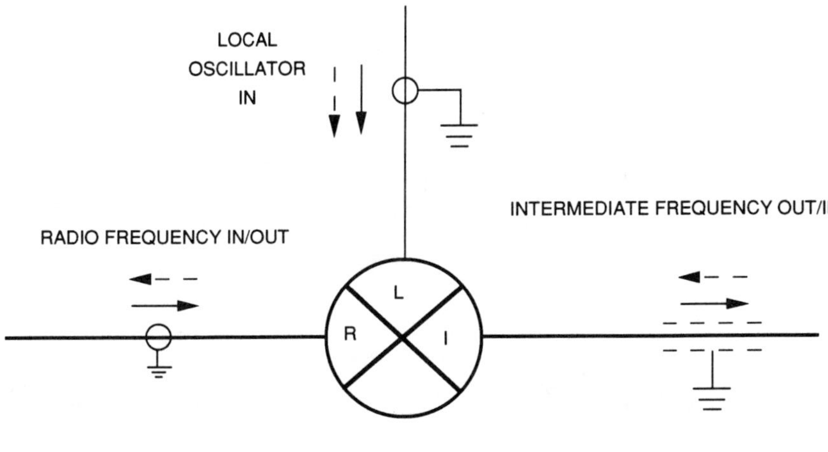

Figure 2.8 A mixer representation.

frequencies. For example, if the IF input is at 100 MHz and the LO input is at 3000 MHz, then the RF output tones will be at 2900 MHz and 3100 MHz. If an RF input is at 2900 MHz or at 3100 MHz and the LO is at 3000 MHz, then the IF output tone will be at 100 MHz. Note that a two-volt 100 MHz sinusoid is mathematically equivalent to the sum of two unity sinusoids at +100 MHz and −100 MHz. However, the multiply function is not perfect, because the LO drive does not switch the RF or IF input polarity instantaneously, and, because the LO amplitude is slightly impressed on the signal, spurious signals, or spurs are present in the output of practical mixers. This is illustrated in Figure 2.9. It can be seen that the multiplied second harmonic of the LO signal (the 2 × 1 term) tracks the input signal as the input varies, decibel for decibel. None of the other spurs shown, however, have a fixed amplitude ratio with respect to the RF or IF input signal. As illustrated, the spur level situation degrades as the RF or IF input signal approaches saturation. The multiplied second harmonic of the input (the 1 × 2 term) climbs 2 dB for each decibel of increase of the RF or IF input signal, while the multiplied third harmonic of the input (the 2 × 3 term) climbs 3 dB for each decibel increase of the input.

Mixers have very wide bandwidths, often multi-octave. Even the IF signal often has a bandwidth from close to DC to over 2 GHz. The LO power, at the level needed for saturation, is on the order of +10 dBm. A good rule of thumb is that the input power should not be within 10 dB of the LO, to avoid the spur generation phenom-

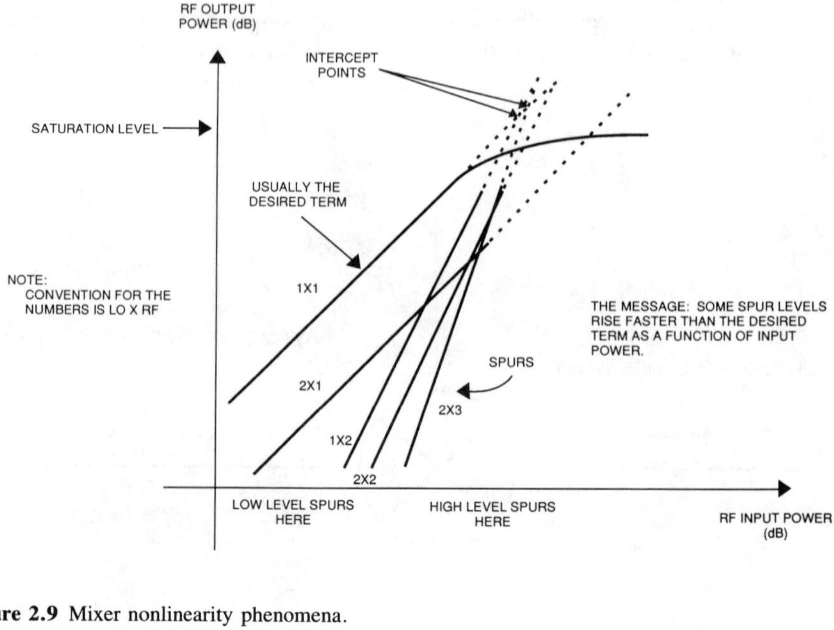

Figure 2.9 Mixer nonlinearity phenomena.

enon illustrated in Figure 2.9; the typical maximum input would therefore be about 0 dBm. The conversion loss of the wideband EW type mixers is a bit less than 10 dB, with specialized narrow-band radar-type mixers being 6 dB. The noise figure approximately equals the conversion loss. One important specification for a mixer is the LO leakage value, generally around 20 dB. The mixer VSWR is usually poor. Mixers exhibit approximate reciprocity in the sense that the power flow between the RF and IF ports can be reversed with a resultant equal conversion loss. As illustrated in Figure 2.9, mixers exhibit quite good amplitude linearity until the input is within 10 dB of the LO power. The linearity of RF components is often specified by the degree to which the output amplitude changes match the input amplitude changes. By this practical measure, mixers are linear. Superposition applies for linear components; that is, multiple signals will not interfere with one another for linear components. This practical linearity property allows mixers to be used for superheterodyne receivers. Of course, true linear systems do not alter the carrier as mixers do.

The system engineer often spends an inordinate amount of time designing around the mixer properties, especially when they are used for wide-band frequency translators. The LO value and IF range must be carefully selected, with such selection dependent on the results of calculating the amplitude and frequency domain or position of all spurs for the selected choice. A careful selection of the LO frequency and the RF and IF ranges will result in the best amplitude dynamic range.

The key functional parameters for mixers and their representative values, shown in parentheses, are

- conversion loss (10 dB)
- bandwidth (RF and IF) (4:1; 2 GHz)
- LO power (10 dBm)
- image suppression (for SSB unit) (20 dB)
- LO leakage (20 dB)
- 3rd-order intercept (15 dBm)

Mixers have been widely used in ECM systems for frequency conversion purposes. It has long been known that the frequency can be doubled by feeding an RF signal into both the RF and LO ports. Recently, both frequency multipliers and dividers have become available as another frequency conversion option for ECM system designers. These can be employed in passive microwave circuits, that is, without a phase-locked servo loop, requiring no power except for bias currents. These frequency converters usually operate over a narrow amplitude dynamic range. The output frequency is related to the input frequency in an integer proportion. There are triplers, but usually the ratios are 2:1 or 1:2; the latter is often referred to as a halver. One important point that many find confusing is that the input *versus* output time domain responses track one another. For a pulsed input, only the central spectral line frequencies are proportional to each other. For example, a 1 μs RF pulse will have a 2 MHz main spectral-lobe width at both the input and output. Therefore, only the central spectral line will have the precise double or half frequency. However, false doppler offsets created by a modulator prior to insertion into a halver or doubler will have this offset doubled or halved.

2.2.8 Limiters

The purpose of RF limiters is to limit the output power to a specified value. The means by which it does this may include dissipation, reflection, or not amplifying above a certain level.

A representation of a limiter is illustrated in Figure 2.10 as simply a rectangle. There are four components generally utilized to perform the RF limiting function: diode limiters, illustrated in Figure 2.10, solid-state amplifiers, TWT amplifiers, and *magnetostatic wave* (MSW) devices. Functionally, there are two types of limiters, the cycle-by-cycle type of limiting illustrated in Figure 2.10, and the attenuation or gain-reduction type. TWTs and some types of diodes and solid-state amplifiers represent the cycle-by-cycle limiting, whereas MSW and other types of diode and solid-state amplifers represent the attenuation or gain-reduction limiting. The distinction is critical, because the cycle-by-cycle approach creates harmonics and distorts the

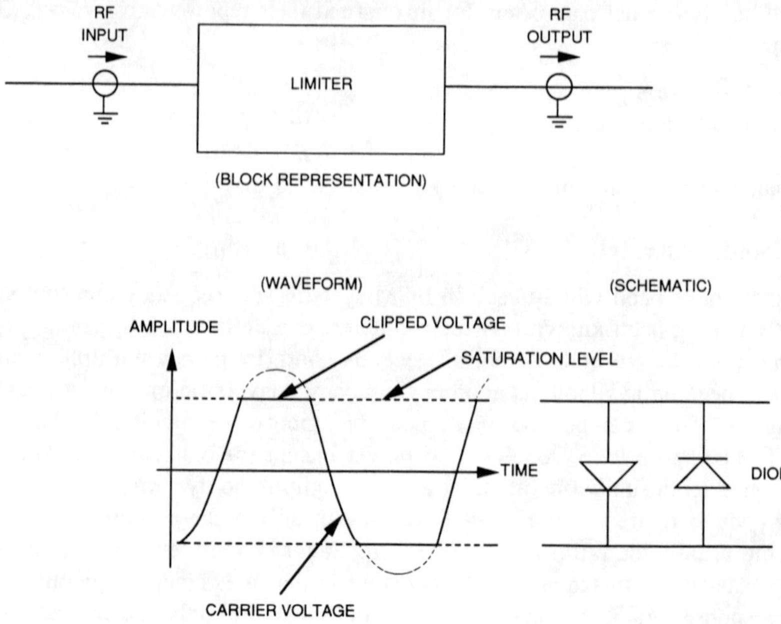

Figure 2.10 Limiters.

amplitude relationship between the saturating signal and weaker signals, and superposition does not hold. This small-signal suppression phenomena is desirable for some applications (e.g., recirculating RF memory loops), and troublesome for others (e.g., receiver protectors). Diode limiters have an insertion loss of a few decibels and exhibit poor VSWR when they go into limiting, because the excess power is reflected.

The key functional parameters and representative values for limiters are

- limiting level (+10 dBm)
- limiting dynamic range (20 dB)
- insertion loss (1 dB)
- bandwidth (2:1)

2.2.9 Frequency Equalizers

The purpose of frequency equalizers is to compensate for frequency roll off, a nominal attenuation slope *versus* frequency, and for ripple, or variation *versus* frequency, caused by other components.

The equalizer is designed as numerous tunable overlapping filter elements, often used in the low-slope skirt region. The equalizer has a bank of tuning screws, the proper adjustment of which generally requires skill and patience, because the individual tuning elements interact with one another. In some circumstances this labor-intensive tuning is more cost effective than solving the basic frequency dependence design problems of troublesome components, subsystems, or systems.

2.3 ACTIVE RF COMPONENTS

In the introduction to this chapter, the RF components were categorized. One such category was active RF components: those that consume more than just bias power from a power supply but that generally are not electrically controllable. These—the solid state RF amplifier, the TWT amplifier, and the fixed RF oscillator—are discussed in this section.

2.3.1 Solid-State Amplifiers

The primary purpose of solid-state amplifiers is to increase the signal amplitude, that is, to act as a gain element. A secondary purpose is to act as an isolator.

Solid-state amplifiers, and indeed all amplifiers, are represented on block diagrams as triangles. In the 1960s, all RF gain elements were TWTs, hence the system gain source was concentrated in one or a few components. *Solid-state amplifiers* (SSAs), although they can be fabricated with gain values as large as or larger than

TWTs, lend themselves to distributing relatively small gain values judiciously throughout the system. In other words, gain can be relatively easily subdivided.

The gain elements in RF SSAs include such wideband devices as gallium arsenide (GaAs) *field effect transistors* (FETs). The VSWR of SSAs is usually fair at the input port and poor at the output port, unless a dual channel output structure is employed to improve the impedance match and enhance the power rating. The output saturated power of wideband SSAs, suitable for ECM applications, varies from +10 dBm to a bit more than +30 dBm. These power levels are significantly lower than high power TWTs are capable of. The instantaneous bandwidth of SSAs is a bit less than two octaves. SSAs have good noise figures, often well under 6 dB, and they use low power supply voltages, such as 12 or 15 volts. SSAs have good reverse isolation, and it is becoming common to use them in lieu of isolators. They are usually quite small, especially in relation to TWTs. MIC construction, which is becoming increasingly common, can be employed to reduce the size and weight further.

For an application in a solid state antenna array, the SSA's efficiency is critical. Otherwise, the key functional parameters and representative values for SSAs are

- gain (20 dB)
- output (1 dB compression) power (+10 dBm)
- bandwidth and center frequency (2:1; 3 GHz)
- noise figure (6 dB)
- ripple (2 dB)
- 3rd order intercept (+20 dBm)

2.3.2 Traveling Wave Tube Amplifiers

The purpose of traveling wave tube amplifiers is to increase the signal amplitude to potentially high power levels, that is, to act as a high power gain element. TWT amplifiers used to be the only resource system engineers had for gain elements, but they have generally been supplanted by SSAs throughout the system, except for the output RF power amplification stages. The TWT is the key element of the transmitter section of Figure 1.5. However, sometimes TWTs are employed for their special limiting properties.

The TWT construction is illustrated in Figure 2.11. A helix is employed to create a slow wave structure, and an electron beam is directed down the center of the helix. Power applied to the cathode heats that element, which causes electrons to be emitted. High direct current (DC) voltages, on the order of several thousand volts, cause these electrons to travel down the center of the helix at about 10% of the speed of light. The RF input signal is coupled to the lead end of the helix. This RF signal has to follow the spiral path of the helix, so its linear progress down the tube is therefore only about 10% of the speed of light, even though it is indeed an EM wave propagating at about the speed of light. The key to the operation is that

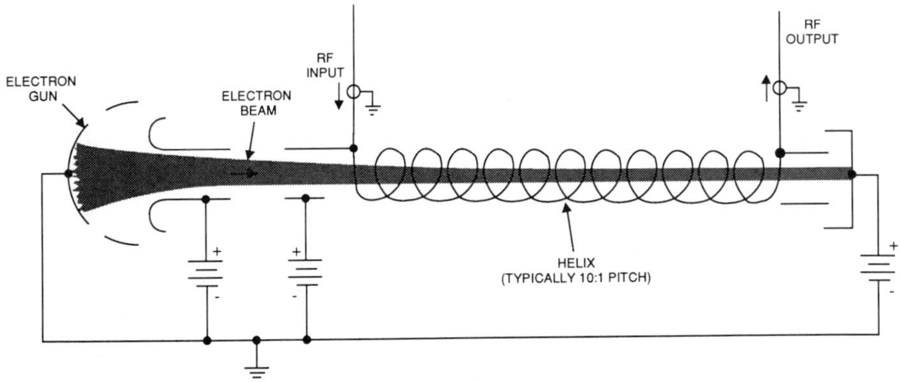

Figure 2.11 The TWT structure.

the voltage on the helix and the electron beam, both moving at about the same speed down the tube, will interact. This interaction is shown in Figure 2.12. The top of the figure shows the voltages and electron positions near the input end of the TWT, and the bottom of the figure shows the same near the output end of the TWT. The movement of individual electrons relative to the general motion tends to be toward the region where the helix is positive. As the electrons start to bunch up, the helix become even more positive. This interaction continues, and the bottom portion of the figure shows the beam bunching almost complete.

Because of this mutually reinforcing interaction between the electron density and the helix voltage, the helix voltage grows larger, and hence the signal has experienced gain. Note that the beam bunching is a nonlinear action, and, indeed, at saturation the helix voltage will not be sinusoidal; harmonics and small signal suppression are the result.

TWTs used in ECM systems have wide bandwidth, in excess of an octave, and such wideband TWTs have average power output ratings of several hundred watts in most microwave bands. Some types of tubes are designed to be pulsed, so their instantaneous power may be rated at well over a kilowatt, although the average power is not any more than *continuous wave* (CW) tubes unless cooling is used. The TWT output VSWR is usually poor.

Because of the length of the helix required to achieve reasonable gains, the precision of construction necessary and the need for high voltages, the tube and power supply combination is usually the largest, draws the most power, and is the most expensive unit in an ECM system with passive antennas.

The primary TWT use in ECM systems is as preamplifiers or drivers and as power amplifiers or output amplifiers. The key parameters for pulsed TWTs include

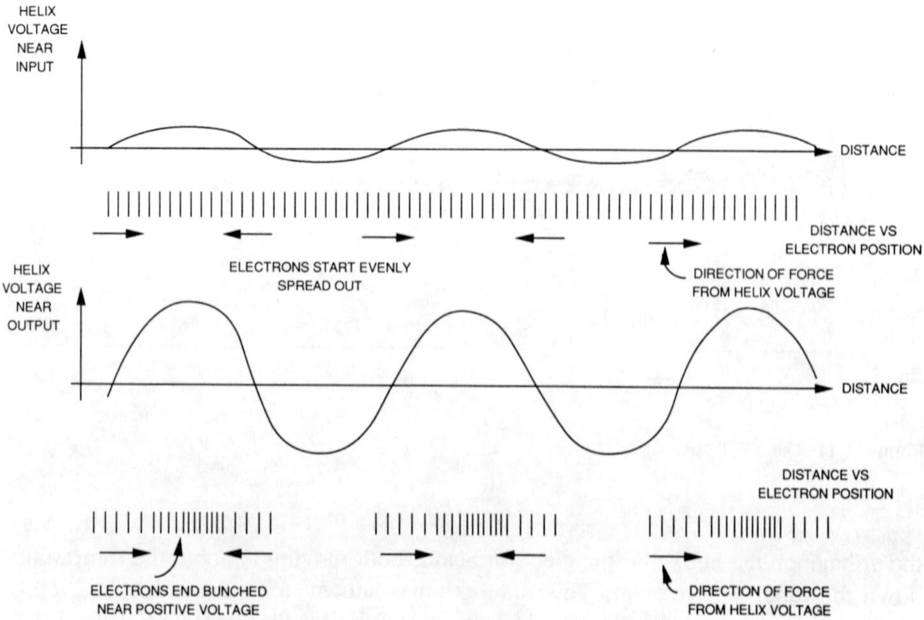

Figure 2.12 Slow wave beam bunching.

the pulse-up maximum length and the pulse-up duty cycle. A key factor in CW TWT efficacy for an ECM application used to be the ability to modulate the tube's electrical phase length; this need, however, is being supplanted by the use of external phase shifters. Otherwise the key functional parameters and representative values are

- gain (30 dB)
- bandwidth (2:1)
- power (+50 dBm)
- ripple (5 dB)
- efficiency (25%)

2.3.3 Fixed Frequency RF Oscillators

The purpose of fixed frequency RF oscillators is to convert dc input power to a pure tone RF signal. These oscillators are most commonly used as the LO signal to broadband mixers used as frequency translators.

For many ECM applications, the only key parameter for the fixed frequency RF oscillator is a power rating sufficient to drive the LO port of standard mixers.

Because of the increased importance of coherent jamming techniques, system engineers have been demanding a much cleaner spectrum and stability than formerly: a crystal stabilized oscillator in the microwave range may have a stability rated at less than 100 Hz variation. The RF output power is typically about +10 dBm, and the spurs are generally less than −60 dBc, although the harmonics, if not filtered, are considerably higher. Radar engineers usually require the oscillator's RF carrier to have such stability for the oscillator vendor to employ a temperature-stabilized oven, but most ECM requirements are not yet stringent enough to need this because the radar application is (1) more sensitive to spurs and (2) needs to integrate the LO result over many PRIs, taking a much longer time than the ECM application, which normally disposes of the signal within one individual radar PRI. The requirements for a coherent radar LO are more stringent than those for an ECM LO intended to jam or deceive that same coherent radar.

The requirements for the key parameters for a fixed frequency RF oscillator employed as an LO for a frequency translator are based on imposing no significant modulation that the radar or EW receiver can practically measure. The stability of the oscillator itself is usually dependent on the rate of temperature change, so the ECM system engineer needs to relate that to his functional needs. The key functional parameters and representative values for a fixed frequency LO are

- stability (1 kHz/min)
- spur levels (−60 dBc)
- power (+10 dBm)

2.4 CONTROLLABLE RF ELEMENTS

The class of controllable RF elements has been previously defined and their members listed. Some of these are components, while some are clearly subsystems. RF memory subsystems, and in particular the recirculating RF memory loop, will be discussed in a separate chapter. The controllable RF elements have been defined as a functional class, but the sensor functional class also includes EW receivers, many of which require considerable control. Because sensors, however, do not directly contribute to the creation of RF jamming signals, they are not included in the class of controllable elements. Sensor components and subsystems will be discussed in their own chapter; this section will describe the remaining controllable elements.

2.4.1 Amplitude Modulators

The purpose of amplitude modulators is to decrease the amplitude of the signal by dissipating or reflecting power by an electronically controlled amount. Their primary role in ECM systems is as gain control. Occasionally they are used to impose an ECM technique's amplitude modulation waveform.

The RF amplitude modulator, sometimes known as a *voltage variable attenuator* (VVA), is shown in Figure 2.13. It is often shown on block diagrams as a rectangle with the letters VVA inside. Despite its name, the amplitude modulator technology generally needs to be driven by a current. This current changes the impedance of an RF pin diode mounted either in or across a transmission line. Because current is required, the voltage source is generally buffered with a current driver, as shown. The diode characteristics are not linear, even on the log scale engineers are used to, and therefore a linearizer is needed as shown in the figure. Linearizers are themselves networks of resistors and video diodes with distributed turn-on bias points; as each diode turns on, the net impedance of the network changes. Linearizers are rather slow, with a response on the order of tens of microseconds, and are not generally compatable with high rate power management multiplexing. On the other hand, many applications for the VVA, such as repeater-mode automatic-gain setting, do not require high speed. One reason that the linearizer is shown with dashes in the figure is that the ECM system enginer has the option of dealing with the linearity problem digitally; that is, there is a one-to-one correspondence between the desired attenuation digital word, and the actual digital word needed (without a linearizer). This table can be stored in the CPU.

As shown in Figure 2.13, a latch and DAC are part of the complete RF amplitude modulator unit, as opposed to an analog-controlled RF modulator component. In modern ECM systems, the attenuation values are generated with algorithms operating in digital hardware; hence the need for the DAC. Since the data bus is multiplexed, there needs to be an addressable latch that can hold the data word until the CPU decides to change the attenuation.

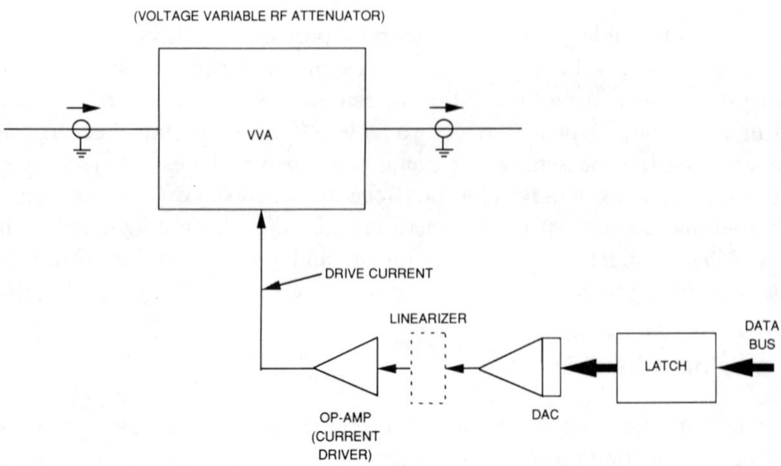

Figure 2.13 RF amplitude modulator.

These pin diode types of RF amplitude modulators generally have a dynamic range of about 40–60 dBs, and inherently about several microseconds settling if the drive current is a perfect step function. The VVA term is quite appropriate in the sense that those components are generally used more for gain alignment or dynamic gain control than as amplitude modulators.

The ripple may sometimes be a critical parameter for a VVA. In general, the key functional parameters and representative values for a linearized VVA are

- Dynamic range (60 dB)
- Linearity (±3 dB)

2.4.2 Phase and Frequency Modulators

The purpose of phase and frequency modulators in ECM systems is to create a false doppler value; that is, they are employed to impose a modulation intended to give the illusion that the craft defended by the ECM is traveling at a different relative radial velocity.

As stated previously, the difference between phase and frequency modulation is one of degree, not kind; for examle, 360°/ms of RF phase modulation is equivalent to 1000 Hz offset modulation. Depending on the situation, a certain modulation may be most appropriately considered phase modulation or frequency modulation. Hence, the following discussion does not always distinguish between the two.

In older ECM systems, the only practical resource for generating a false doppler frequency offset was to modulate the serrodyne input of the TWT tube. In modern systems, the frequency offset and similar modulations are created with either *single sideband* (SSB) mixers or with digital phase shifters, although attempts to develop practical analog diode phase shifters persist.

The requirements for phase and frequency modulators for ECM applications include a carrier offset capability of from a few Hz to up to tens, perhaps hundreds, of kilohertz. The usually linear sweep through this range takes several seconds. The sweep starts very near zero offset, so the percentage deviation required is quite large. The offset must be able to be imposed in both the positive and negative directions, that is, higher and lower than the input carrier. Since these offsets are in the kHz range, and since the carriers are in a range up to about 20,000 MHz, the deviation as a percent of the RF carrier is extremely small. As stated, the frequency controlling function, digital word or voltage or rate of voltage, is itself generally subject to modulation, such as a linear or parabolic walkout. The most important parameter requirements on the RF component is the suppression of the input carrier; the available components have carrier suppression values of 15–25 dB, with which ECM system engineers are not at all satisfied.

The TWT serrodyne voltage modulation is shown in Figure 2.14. This is an analog RF phase modulation imposed through the TWT. The TWT is driven with a

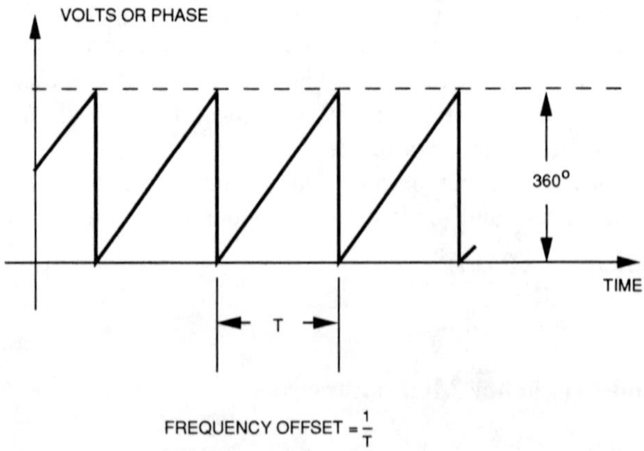

Figure 2.14 TWT serrodyne modulation.

saw-toothed waveform, the peak-to-peak voltage excursion of which causes 360° of RF phase excursion; the reciprocal of the period of this saw-toothed waveform equals the desired offset frequency. For example, if the saw-toothed ramp rate is 360°/ms, a saw-toothed period of 1 ms, the output carrier from the TWT will be offset by 1 kHz. This serrodyne capability means that extra circuitry must be included in the TWT power supply to bring the saw-toothed ramp to a high voltage bias and swing.

Unfortunately, there are practical limitations on the serrodyne performance. The peak-to-peak phase swing for the saw-toothed voltage drive input does not give precisely a 360° swing across the entire band. To the extent the swing differs from 360°, frequency domain spurs will be created, and one of these spurs will fall at the original carrier frequency, the worst possible position from an ECM jamming or deception point of view. Any frequency component at the original carrier frequency actually helps the threat radar to track, which is the exact opposite of the goal for the ECM system. In addition to the phase error, the serrodyne sawtooth also generally causes some gain modulation in parts of the band, whereas the intent of the modulation is only to change the RF phase, not the amplitude. This unintended amplitude modulation causes frequency domain spurs, one of which will also appear at the original carrier frequency.

The beating of an input signal and the LO in a simple mixer generally results in mirror-image spectral lines. The SSB mixer, sometimes called a quadrature mixer, and occasionally called an I and Q mixer, is shown in Figure 2.15. The SSB mixer includes an RF power splitter, two identical *double sideband* (DSB) mixers, and an RF power combiner. The two couplers must have proper phase shifts. Each IF port of each DSB mixer is driven by a signal which has been saturated to cause the nonlinear multiplication phenomena; the saturated signal may as well be a square wave. The frequency of the IF square wave function is the desired false doppler offset value, just as the TWT serrodyne frequency is the offset value. Figure 2.15 shows the input spectrum, the spectrum out of each individual DSB mixer, and the final SSB output spectrum. As this figure shows, each single DSB mixer generates an equal power image signal at the opposite side of the original carrier, in addition to an infinity of other frequency domain spurs, hence the name double sideband mixer. When these two mixer outputs are combined by the output coupler, the image and half the spurs are deleted, their power going to the desired offset signal and the remaining spurs, as shown. The figure shows all frequency domain components at their theoretical relative amplitude, except for the original carrier leakage signal, whose amplitude is determined by the hardware tolerances. To achieve this SSB improvement over the operation of a single DSB mixer, the saturated, square wave function, IF signal needs to be generated as a paired set, with 90° phase shift, by the function generator.

The action of each of the DSB mixers in Figure 2.15 can be thought of as multiplying the RF signal alternately by +1 and −1 at the function rate; it can also be thought of as switching in 180° phase shift steps at the desired offset rate, twice the function rate. As such, each individual DSB mixer theoretically functions the same as a one-bit digital phase shifter, and a SSB mixer functions the same as a two-bit digital phase shifter. The advantages the SSB mixer has with respect to the digital phase shifter are the good RF bandwidth and the higher offset frequency values possible.

Digital phase shifters, sometimes called digilators, have been the most common means to generate false doppler RF offsets, although the instantaneous bandwidth limitations have been troublesome. The fact that they have a digital interface makes them quite compatable with the digital processing and CPU functions in modern ECM systems. An example of a three-bit phase shifter modulation is shown in Figure 2.16. The CPU generates a series of successive digital words, each of which calls up an RF phase value successively larger or smaller than that from the previous digital word. The phase steps between successive digital words are supposed to be equal, as shown in Figure 2.16. For example, for a three-bit phase shifter, if the clock rate is 8 kHz, with each step lasting 125 μs, it will take 1 ms to step through all eight steps, and the false doppler offset frequency will be 1 kHz. Of course, it is possible to use this component with more complex phase modulation patterns, but the stepped phase ramp, approximating the intent of the serrodyne of a TWT, is best for creating the most pure offset frequency. The amplitude of the digilator is not supposed to change as the digital words are changed, and to the extent there is inadvertant amplitude modulation, there will be frequency domain spurs, one of which will fall at the original carrier frequency. To the extent that the phase step changes are not equal, or do not sum to 360°, there will be frequency domain spurs, one of which will again appear at the dreaded original carrier frequency. The digilator has the reputation of being more accurate than the TWT serrodyne, but a fair comparison

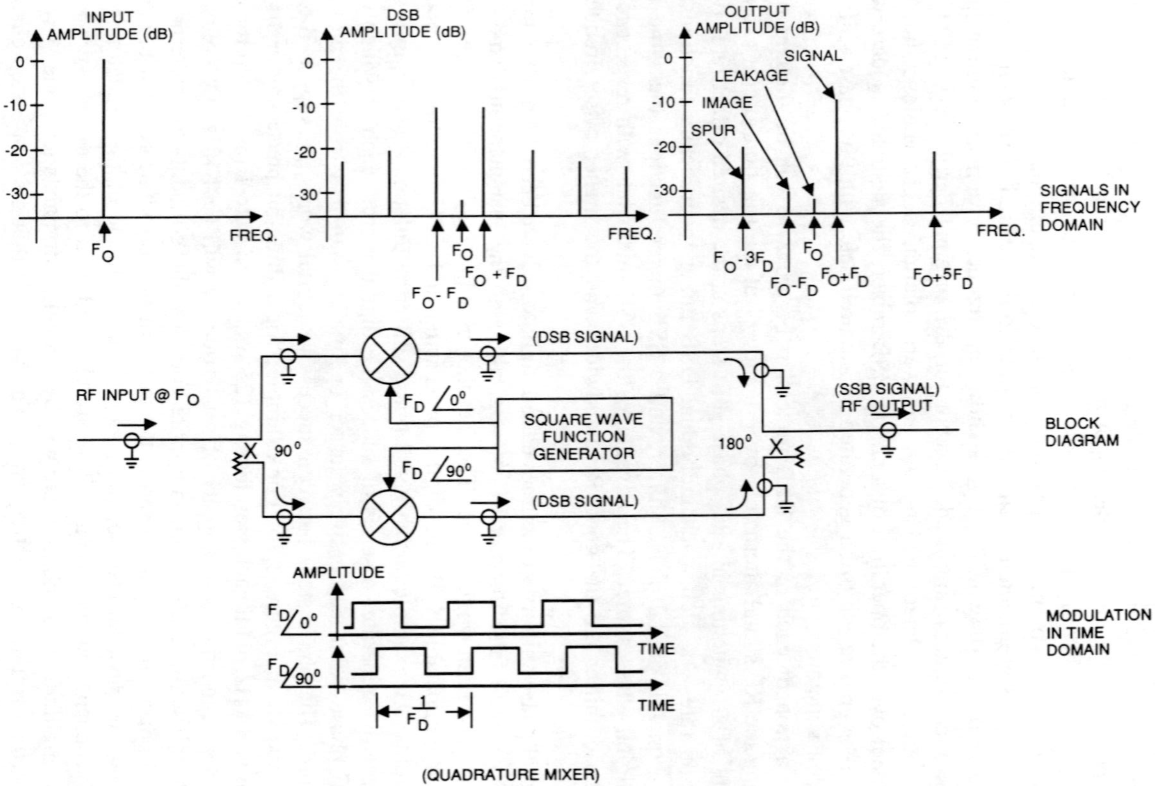

Figure 2.15 The single sideband mixer.

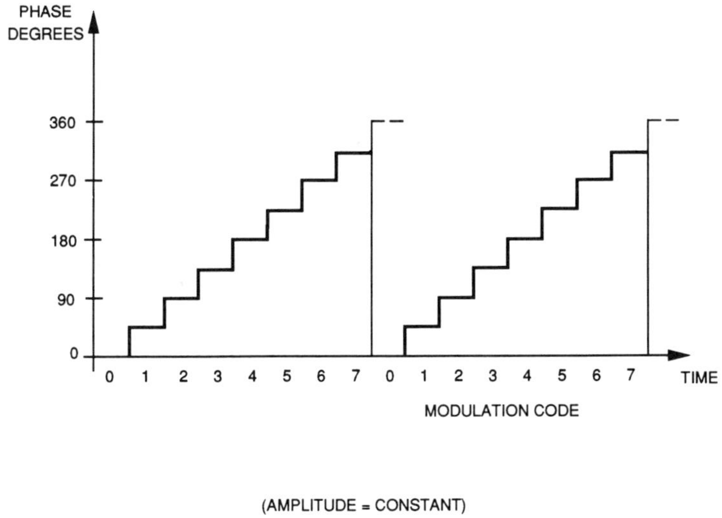

Figure 2.16 A three-bit phase shifter modulation.

must allow the TWT serrodyne to be optimized for the same instantaneous bandwidth as the digilator.

The digilator was developed as a solid-state alternative to the TWT serrodyne. It is classified by the number of bits of resolution, with one bit for 180° phase shifts, two bits for 90° phase shifts, three bits for 45°, *et cetera*. Digilators with less than three bits are generally not used, because mixers are functionally equivalent and have better bandwidths. Digilators with more than five bits of resolution are quite difficult to fabricate. Digilators have insertion losses of a few decibel, depending on the number of bits of resolution. The shortest clock period is typically several microseconds.

There is often confusion about what increasing the number of resolution bits of a digilator actually achieves and what it does not, and that subject will be clarified in this text. Figure 2.17, Table 2.2, and Table 2.3 show the theoretical digilator spectrum as a function of the number of resolution bits. The number of bits of resolution of a digilator does not determine the carrier suppression; indeed, more bits of resolution make it more difficult to achieve carrier suppression because of the added complexity. The number of bits of resolution does not determine the purity of the desired, or frequency domain, main spectral line. The number of bits of resolution does not really determine the insertion loss; in practice, the theoretical improvement for more bits of resolution shown in Table 2.3 will be offset by the resistive and VSWR losses of the more complex microwave circuitry. However, an increased number of bits of resolution does alter the amplitude of the spurs, as shown

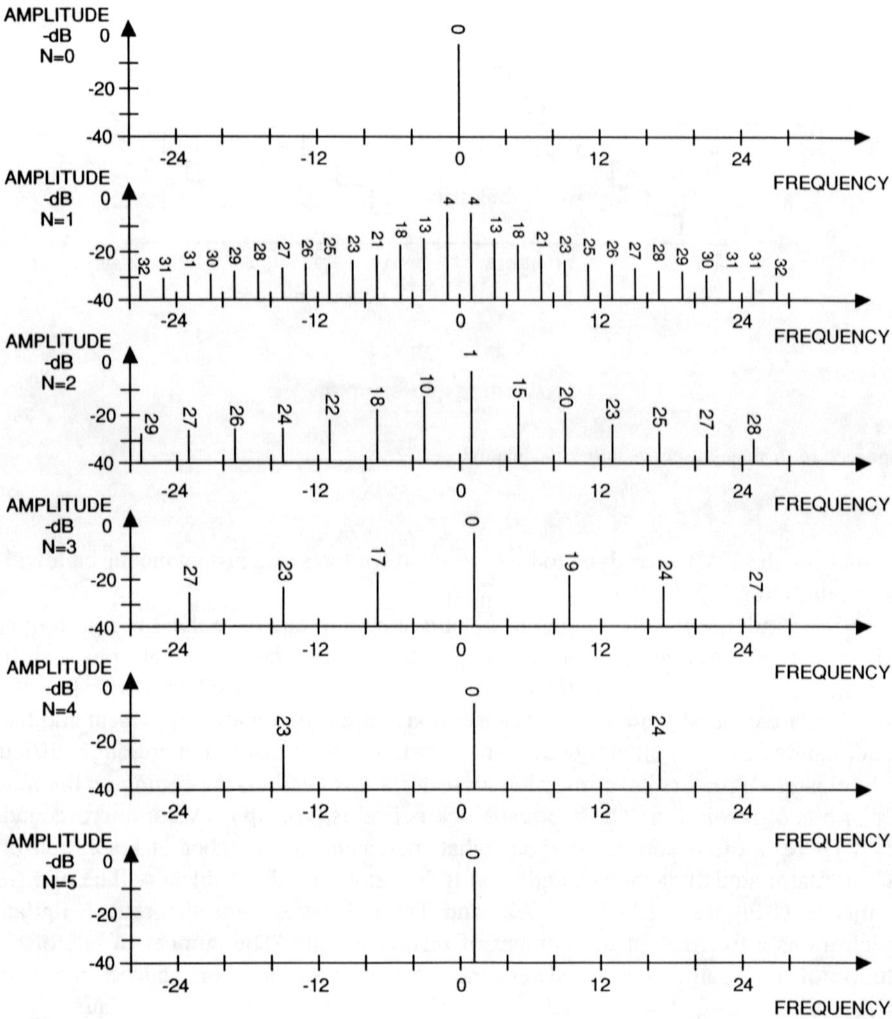

Figure 2.17 RF digilator spectra.

Table 2.2
Digilator Frequency Components

			Resolution and Harmonic Sign									
	Relative		*Negative*					*Positive*				
+/−	*Amplitude*	*Harmonic:*										
Harmonic Number	*dBc*	*Number of Bits:*	*5*	*4*	*3*	*2*	*1*	*1*	*2*	*3*	*4*	*5*
1	0.00						*	*	*	*	*	*
3	−9.54					*	*	*				
5	−13.98						*	*	*			
7	−16.90				*	*	*	*				
9	−19.80						*	*	*	*		
11	−20.83					*	*	*				
13	−22.28						*	*	*			
15	−23.52			*	*	*	*	*				
17	−24.61						*	*	*	*	*	
19	−25.58					*	*	*				
21	−26.44						*	*	*			
23	−27.23				*	*	*	*				
25	−27.96						*	*	*	*		

Note: a "*" indicates that a spectral-line frequency component is present at the indicated amplitude and +/− frequency offset.

Table 2.3
Digilator Loss

Resolution Bits	*Delta Phase Degree*	*Absolute Amplitude dBc*	*Signal to Distortion Ratio dBc*
1	180.00	−3.92	−1.66
2	90.00	−0.91	6.32
3	45.00	−0.22	12.76
4	22.50	−0.05	18.87
5	11.25	−0.01	24.94

in Figure 2.17 and Table 2.2. The frequency scale in the figure is normalized so that the desired offset is unity, and the relative amplitude in the table is normalized so that the amplitude at that desired offset frequency is zero decibel. The figure and table can be interpreted as showing that the amplitude relationship between the desired frequency component and the spurs is a constant, but, as each bit of resolution is added, half the spurs drop out. For example, Table 2.2 shows that the worst spur for a three-bit digilator is −16.90 dBc at offset harmonic −7, while a four-bit has −23.52 dBc at −15. Figure 2.17 shows the absolute amplitude of all the spectral lines; as each bit of resolution is added, the power that used to be in the spectral lines that dropped out is distributed among the other spectral components. Table 2.2 gives the relative amplitudes between the frequency components, and Table 2.3 gives the uniform frequency component multiplier to obtain the absolute amplitudes for the various resolution values in bits. For example, for a three-bit resolution unit subtract 0.22 dB from all the relative amplitudes in Table 2.2 to get the absolute amplitudes. The sum of all the absolute-amplitude spectral components sums to unity (zero dB). Table 2.3 also gives the phase step size in degrees, and the *signal-to-distortion ratio* (SDR) in decibel; the latter is similar to a *signal-to-noise ratio* (SNR) and is defined as the ratio of the desired main spectral line to the remaining power. Tables 2.2 and 2.3 are useful for a number of other applications; for example, they can be used to predict the spectrum of a DSB or SSB mixer as described above.

Figure 2.17 and Tables 2.2 and 2.3 are only appropriate for a linearly stepped ramp phase program. For all types of phase modulators, the resultant frequency offset of a ramp function is simply the smoothed phase ramp rate in degrees per second divided by 360°.

There are a number of ways to fabricate digilators; one approach is to use microwave switches to switch between paths that have certain fixed broadband phase shifts.

The key functional parameters and representative values for frequency offset modulators are

- frequency offset maximum (10 kHz)
- carrier suppression (20 dB)
- spur (including image) amplitudes (−20 dB)
- insertion loss (8 dB)
- bandwidth (2 GHz)

For phase shifters employed in active aperture antennas, as opposed to being used to generate specialized ECM deception waveforms, the additional key parameters would be the absolute phase across the band and the power level capability, including IM products; the insertion loss would be very critical.

2.4.3 Switches

The purpose of microwave switches is to gate RF signals. Microwave switches are widely used in active ECM systems, as exemplified by Figure 1.5. Microwave switches are used in active ECM systems to impose the relatively low-rate angle-deception modulation and the relatively high-rate false-range pulse (position and width) modulation, and to steer and switch the RF signal paths.

Figure 2.18 shows representations of an on-off switch and a switch used for time multiplexing. The convention of SPST, for *single-pole single-throw,* and SPTT, for *single-pole triple-throw,* is often employed, and is a terminology usage derived from mechanical switches; it is really a misnomer for this hardware, although functionally quite appropriate.

Microwave switches can have in excess of 60 dB on-off ratio, with a few decibel insertion loss. They usually have good bandwidth, well in excess of an octave. The switching time of microwave switches available from RF switch vendors, undoubtedly a result of performance insistance by ECM system houses, is typically on the order of tens of nanoseconds. These switches are usually fabricated so that the control-line inputs are *transistor-transistor logic* (TTL) or *emitter-coupled logic*

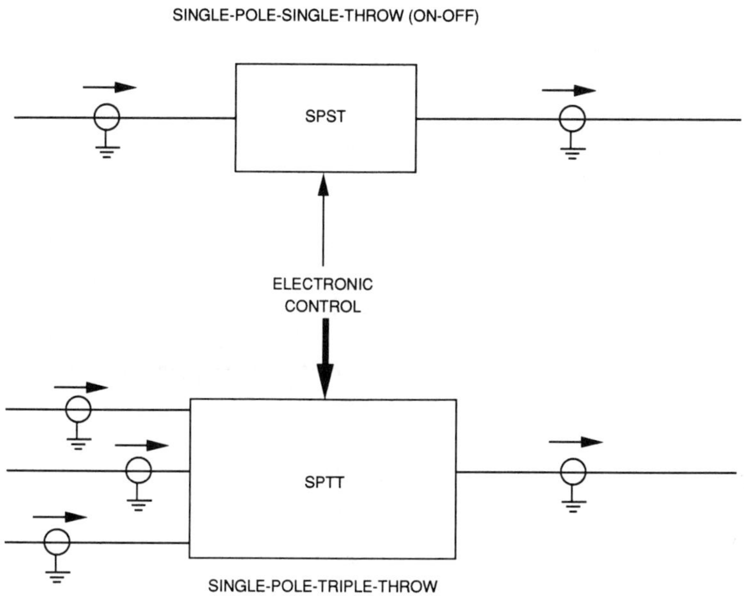

Figure 2.18 Microwave switches.

(ECL) compatable; that is, they are made to be compatable with the common digital logic families. In the on or low-loss, state, the switch VSWR is fair to poor, and in the off state the VSWR is terrible, but, because the path is blocked in the off state anyway, the bad VSWR often does not matter. To improve the VSWR, couplers can be used to combine two parallel switches in such a manner as to mutually cancel the reflections, but that is an expensive and space-consuming solution.

The diodes in the microwave circuitry are generally driven by currents rather than voltages, just as they are for RF amplitude modulators. To make the switch unit compatable with logic families for the ECM system houses, the vendor supplies the unit with a built-in current driver, and often impulses the leading edge with a spike of current to speed the RF switching characteristics. Multiple switching units, an example of which is shown in Figure 2.18, are becoming increasingly common. As pointed out in the previous chapter, switching, both on-off and between paths, is used in ECM systems to impose amplitude modulation.

One issue that the ECM system user has to address when specifying a microwave switch is whether the multi-state switches should be controlled by a coded bus or by a set of individual control lines; for example, an 8:1 switch could either have a three-wire control bus or eight individual control wires, one for each path. This author recommends the latter, so that fine tuning of the timing can be acomplished; such fine tuning includes starting to turn one path on before the previously on path has been shut off. By specifying a microwave switch with a bus control, the user has lost the chance to optimize the timing in such a manner.

In a steady-state condition only one path of multi-state switches should be closed at a time; that is, all other paths should be off, otherwise the nominal impedance will be cut at least in half, resulting in extremely poor VSWR. However, having all paths off is an allowable and common state.

These high-speed switches are more expensive than conventional RF switches, which easily take microseconds to switch. Because of the driving force of power management, however, the switches employed in ECM systems are almost exclusively these fast ones. The microwave switch is one of the most common controllable RF units in an ECM system.

The key functional parameters and representative values for fast RF switches are

- switching speed (10 ns)
- on-off ratio (50 dB)
- insertion loss (2 dB)
- number of poles and throws (3; 2)
- bandwidth (4:1)

For ease of measurement, the switching speed is often specified as the time to reach the 3 dB point when turning on, and the time to reach the 90% point of full attenuation on a log scale when turning off.

2.4.4 Voltage-Controlled Oscillators

The purpose of the RF voltage controlled oscillator (VCO) is to generate an electronically controlled RF carrier. The VCO is a key component in an ECM system, because it is used to generate the noise in the noise mode of jamming. The VCO is also used to generate the carrier for the transponder, or false, pulses.

VCOs can be fabricated to have good linearity, octave bandwidths, and +10 and +30 dBm power by using the YIG-tuned oscillator approach, but such a technical approach is not consistent with the requirements needed for noise-mode jamming signal generation. The analog high speed VCOs are sub-octave units, or at least meet an octave tuning bandwidth with difficulty. The power levels are in the range of +10 dBm. The required tuning response time to a step-control voltage may be as short as 10 ns. The dwell at each such frequency may be a small fraction of a microsecond. The combination of these stringent specifications makes it difficult even to define or measure the frequency accuracy. As discussed in Chapter 1, these requirements are not arbitrary, but can be traced to the general application and physical laws, with the typical craft size being a primary driver.

In addition to the stringent requirements on the RF oscillator component, the complete VCO subsystem unit includes a drive circuit which usually is quite complex. One potential implementation is shown in Figure 2.19, which addresses the high-speed accurate stepping problem. This block diagram includes a pair of high-speed DACs, since the conversion and settling time of each digital-to-analog converter (DAC) is too slow to create the complete waveform, with the analog switch being much faster than the DAC. In operation the high-speed data bus loads data words alternately into the upper and lower digital latches shown in Figure 2.19. Each high-speed DAC starts making the conversion as soon as its new word is available. By using the analog single-pole double-throw (SPDT) switch, each DAC has time to settle while the voltage from the other DAC is being employed by the RF VCO. The waveforms at points *A* and *B* are shown in Figure 2.19 with solid and dashed lines respectively.

As stated, the noise-mode function is the driving factor determining the requirements for the VCO unit: the drive circuitry and the RF VCO itself. Figure 1.9 showed an example of such noise generation. The VCO unit must be capable of large jumps in frequency so as to generate noise or transponder mode signals against several radar threats simultaneously, as in Figure 1.14. Although such requirements result in making the VCO unit one of the most expensive units in an ECM system, the power RF output tube is even more expensive, in terms of generating the last incremental decibel of power. Hence, the requirements burden severely stresses the most current trends in VCO unit and subsystem technology, both RF and drive circuitry, to optimize the spectrum from the transmitter.

The VCO unit design stresses that have to be dealt with include thermal and other past history effects, such as hysteresis. The RF VCO component typically

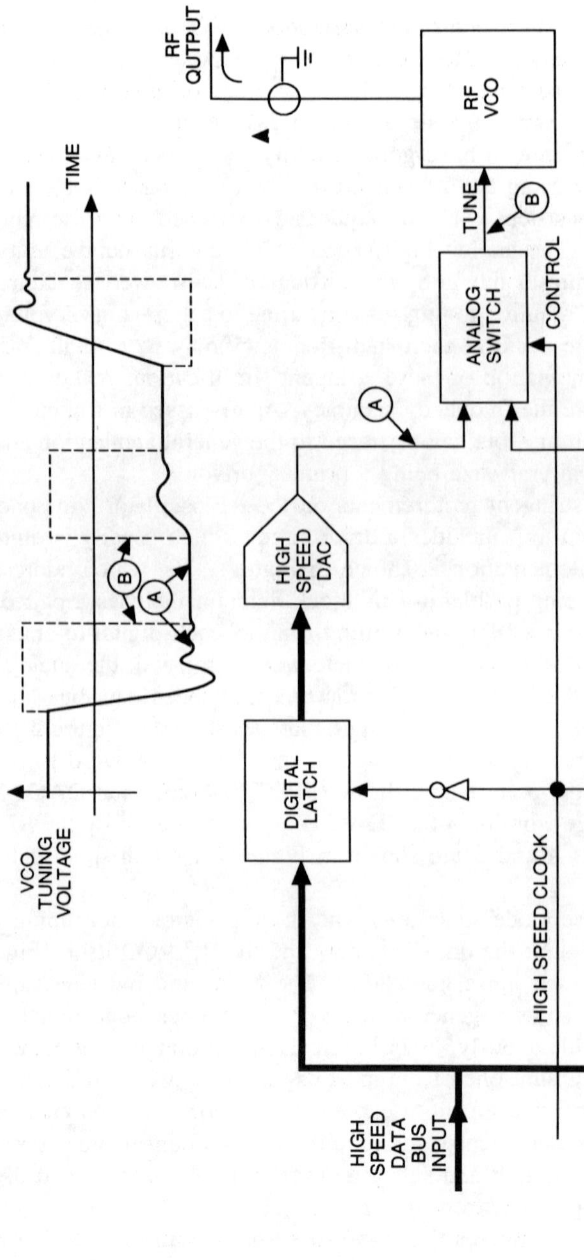

Figure 2.19 RF noise generator.

dissipates more heat at one end of the band then the other, and the temperature influences the tuning transfer function. Therefore, if the system CPU begins to start jamming new threats elsewhere in the band, the result is a temperature change, and a resulting mistuning error against the threats currently being jammed. Even without new threats, the pseudorandom quality of the noise jamming tends to make the temperature erratic. The VCO unit design must take into account these stresses to achieve the necessary accurate, repeatable, and fast tuning.

The VCO unit will be commanded to make large jumps in frequency in very short periods of time, hence the RF VCO and drive circuitry settling and ringing characteristics are significant problems. The complex waveforms characteristic of noise mode operation exacerbate the repeatability problem. One specification on the RF VCO, however, that the ECM system designer can potentially relax is the absolute tuning accuracy; a servo system can use a frequency sensor to tune the RF VCO to the threat's frequency, without the system being required to know the absolute frequency accurately.

To solve the VCO unit problems, complex mechanizations have evolved; Figure 2.19 shows just one example of several aspects of these complex and costly mechanizations. It is quite possible that, in the near future, radically different technology, such as the digital RF synthesizer approach, may replace these analog RF components. One type of RF synthesizer is fabricated with low frequency but accurate and programmable counters, which drive multiple stages of frequency conversion to achieve an accurate digitally programmable RF. Another type of synthesizer uses an RF oscillator that is phase-locked to a stable counter, but these are too slow for this ECM noise application.

The key functional parameters and representative values for VCO subsystems are

- tuning range bandwidth (1 GHz)
- response time to a step command (40 ns)
- power level and ripple (+10 dBm; 3 dB)
- repeatability and stability (0.5 MHz; 0.1 MHz/min)

The response time includes not just the time for the gross frequency change, but the time to settle to a quite fine tolerance. For some applications, linearity and absolute frequency accuracy may be key parameters.

2.4.5 Programmable Delay Lines

The purpose of the electronically-programmable RF-delay line is to delay in a coherent manner radar pulses for the transponder mode described in the previous chapter. (See Figures 1.4 and 1.8.) Both close-in moving false-range and far-out fixed-range pulse delays are needed.

One approach to meet the need for electronically-controlled coherent delay is the tapped delay line and switch matrix combination, as shown in Figure 2.20. A tapped delay line is the purest way to generate coherent delay, because the RF signal must propagate for an additional range delay time just as if the vehicle protected by the ECM system were further away. In the configuration shown in Figure 2.20, multiple range targets, or pulses can be programmed only if the switch matrix has a special structure, because, as explained, most switches cannot have more than one path closed at a time. Aside from the switch matrix problem, a multiple pulse output per input pulse is quite compatable with a tapped delay line. Regardless of whether multiple pulses are needed, the structure of Figure 2.20 is quite compatible with pulse-to-pulse operation against many simultaneous threats. All the CPU has to do is to steer those RF threat bands to this programmable delay line, and the multitude of pulses will simply pass through, with the deceptive range delay imposed. The CPU-DSP does not have to track the pulses. However, the desired false range positions, or tap positions on the delay line, have to be designed into the hardware. This limits the flexibility unless a tap will be fabricated for every possible range delay. As explained in Chapter 1, it is often desired to simulate relative range motion; such a requirement results in a large number of taps. Generally the overall cost increases substantially for a large number of taps, so range rate motion may be dif-

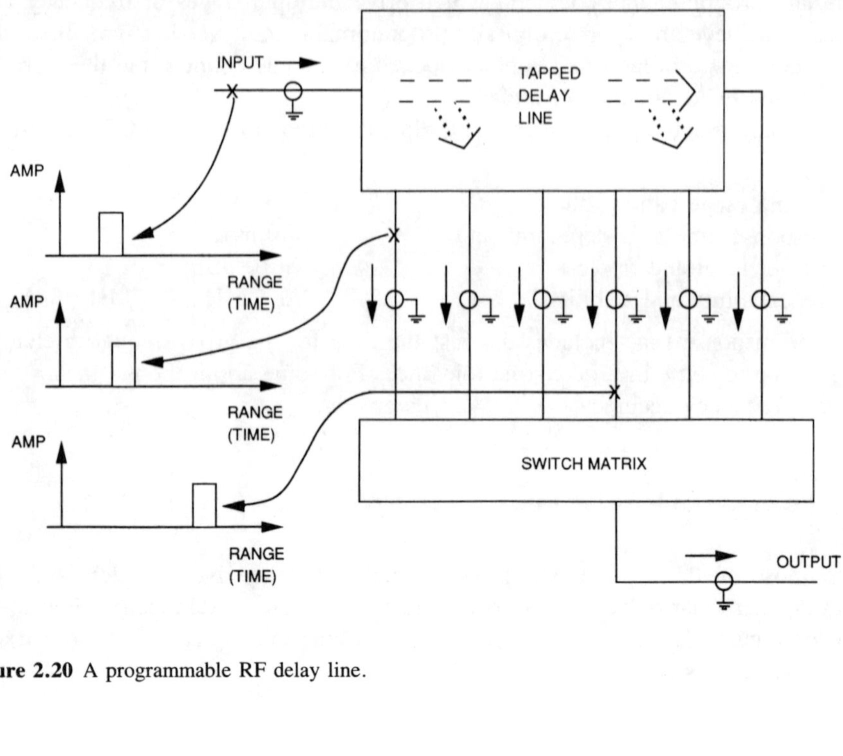

Figure 2.20 A programmable RF delay line.

ficult to implement with a tapped delay line. Another problem is that the SNR degrades as the signal progresses down the delay line; in microwave systems SNR degradation is prevented by inserting amplifiers so that the signal does not drop too close to the base noise level, but this is generally not possible within the delay medium. For the longer delays, the attenuation could be considerable, and, if so, the SNR will be a problem. Mitigating this somewhat, some delay line mediums have relatively low loss once the wave is launched, as described previously.

The instantaneous bandwidth is limited for most delay media, except for coaxial which has other problems. The key functional parameters and representative values for a coherent fixed false-delay unit are

- bandwidth (2 GHz)
- number of taps (10)
- maximum and minimum delay (100/10) μs
- ripple and amplitude tracking between taps (3 dB)

The insertion loss is not a key parameter unless the loss becomes so great that practical solid-state amplifiers inserted just before and just after the delay line cannot maintain the system noise figure and dynamic range. The key functional parameters for a tapped delay line capable of being controlled to give the appearance of range motion, in addition to the above, are

- tap spacing (50 ns)
- number of taps (100)
- tap switching speed (1 μs)

This tap spacing to simulate range motion is often in equal increments, and needs to be a small percentage of the delayed pulsewidth, because the pulsewidth is indicative of the radar's range resolution. The taps need to extend to the maximum range resulting from the range motion.

2.5 ANTENNAS

2.5.1 Passive Antennas

Because ECM antennas are so intimately tied to jamming efficacy, the relationship between the ECM function and the key antenna parameters will be discussed at the system level in other chapters. This section on passive antennas includes a brief introduction to the theory of antenna gain and directivity, and the frequency dependence of these parameters, in a manner intended to give the reader a conceptualization of the function. The reader may also wish to refer to the traditional derivation of the radar range equation, as, for example, presented by Skolnik or Chrzanowski. (See bibliography.)

The purpose of the antenna is to couple between free space and transmission lines within the system; that is, the purpose of the antenna is to radiate or intercept EM waves.

Figure 2.21 illustrates the principles of antenna directivity and frequency dependence. In the upper left of the figure, rays are seen impinging on the antenna aperture. The antenna includes means to bring or focus all the ray voltages to a single summing point, not shown in the figure. The vector diagram in the upper right of the figure provides a phasor representation of each of the voltages resulting from each impinging ray. Because the rays impinge at an off-boresight angle, for the example shown, some of the rays take longer to strike the aperture, and hence their phasor has a different phase angle. If, for example, a 10 GHz EM wave ray took an additional 0.05 ns to reach point C compared to the ray striking point A, then that ray's voltage would become a component with a 180° phase shift at the summing point, because each RF cycle period takes 0.10 ns. In the illustration shown, all the phasors sum up to zero; this occurs at the antenna null. The antenna aperture acts effectively as a spatial or angular filter. This filter response is shown in Figure 2.22,

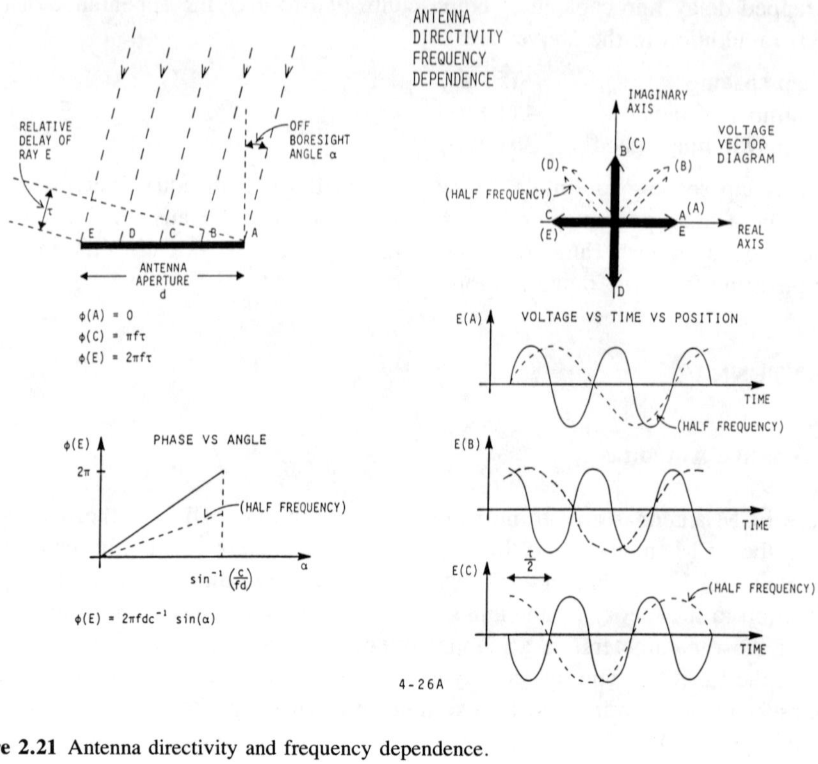

Figure 2.21 Antenna directivity and frequency dependence.

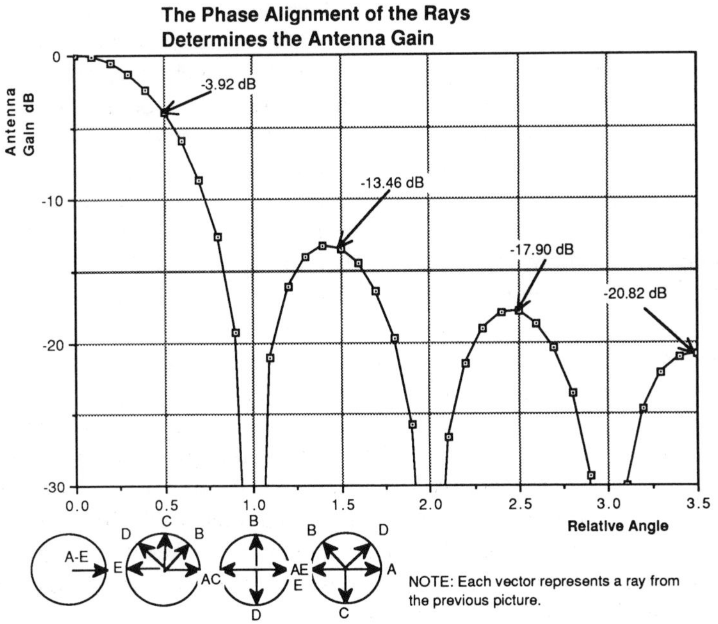

Figure 2.22 An angular filter.

for a uniform weighted aperture, again with vector diagrams. The third vector diagram in Figure 2.22 is for the antenna null, and is the same as the upper right diagram of Figure 2.21. Compare the vector diagram between the main beam and the first sidelobe, which has 13 dB less gain at that angle. The gain is less because the phasors do not align. The further out into the sidelobes, that is, the further away from boresight, the more random the phase angles will appear, and therefore the summation will on average result in less net amplitude, and the antenna will have less gain. The angle-filter shape of Figure 2.22 is described mathematically as a $\sin(x)/x$ function, which is a transformation of a rectangular uniform non-zero value function—in this case, the aperture—with zero value of the function elsewhere. (The transformation of a rectangular pulse in the time domain to the frequency domain also results in a $\sin(x)/x$ spectral pattern.) Antenna designers can make the sidelobes be lower than this classic $\sin(x)/x$ shape by non-uniformly weighting the aperture.

The time domain representation of the phasors is shown in the lower right of Figure 2.21. The frequency dependence of antennas is also illustrated in Figure 2.21, with the specific example of a wave at half the frequency shown consistently with dashed lines. Because the time delay to the far edge (point E) on the aperture represents less of a phase shift at half the frequency, phasors A through E do not sum to nil at half the frequency; the illustrated boresight has the wavefront angle well within the main beam at half the frequency, whereas the wavefront is at the antenna null at the original frequency. Figures 2.21 and 2.22, therefore, reveal how the antenna's directivity is a function of both the physical size of the aperture and the frequency of the intercepted or emitted EM wave.

The directivity of an antenna is a quantitative measure of its ability to accept only incoming EM waves at the boresight angle at which it is pointed. As shown in Figure 2.21, the directivity is better at higher frequencies because the ray phase angles change much more rapidly as a function of angle at the higher frequencies. Therefore, a physically larger antenna is needed at lower frequencies to achieve a given directivity. The quantitative value for directivity, at least for high-gain antennas, is roughly the solid angle of a sphere divided by the solid angle of the main beam. For example, a 3.6° by 3.6° main beamwidth antenna subtends about $[(3.6/360 \times 2 \times 3.14)^2]/(4 \times 3.14)$ 0.03% of a spherical surface, for a directivity of 35 dB. The difference between gain and directivity is that gain includes manifold feed losses. The gain of an antenna is the power or radiation intensity transmitted or intercepted in the boresight direction relative to the power of a lossless isotropic, or omni-directional, antenna. The ERP of an antenna is the combination of the power fed to the antenna and the gain of the antenna, or else the power radiated from the antenna and the directivity of the antenna.

To reiterate, an antenna is a transducer between free space and transmission lines, inasmuch as it sums voltages across an aperture, each derived from the components or rays of the impinging electromagnetic wave; it acts as a spatial filter with directivity because the delay the wavefront experiences in reaching different portions of the aperture causes the voltages to have phase angles proportional to both the boresight angle and the carrier frequency.

Passive antennas are reciprocal, that is, they respond in the same way whether they receive or transmit, provided that the power is not excessive enough to damage the structure. An appropriate passive EW antenna, suitable for radiating ECM power levels at both horizontal and vertical polarizations simultaneously, is the horn antenna. Other passive EW antennas include linear and dual polarized log periodic antennas and right- and left-handed circularly polarized planar spiral antennas. The key to designing EW antennas, in addition to installation considerations and power handling capability for those that must radiate, is achieving a good match over a wide range of wavelengths. The placement of the antenna on the asset, for example, a craft to be protected by the ECM, is critical, since it is desirable to have the leading edge of the reflected radar signal have jamming superimposed on it.

The key functional parameters and representative values for passive antennas are

• directivity	+6 dB
• loss	0.5 dB
• bandwidth	4:1

2.5.2 Antenna Control

In older ECM systems, antennas were generally passive. The purpose of electronic control of antennas is to increase the ERP, to facilitate the function of certain techniques, and to incorporate more efficiently the power management function.

Figure 2.23 shows two examples of antenna control schemes that rely on microwave switches. Many horn antennas are fabricated with two input ports so that RF signals into these ports result in vertically and horizontally polarized propagating EM waves respectively. If equal magnitude signals are inserted into the two ports simultaneously, then the result will be a 45° polarization with respect to both horizontal and vertical. Therefore, as shown in Figure 2.23a, one way to control polarization is simply to switch between the two ports, assuming intermediate polarizations are not needed.

In the second example, Figure 2.23b, a SPTT RF multiplexing switch is used to steer the antenna beam. In this example, there are several antennas, each with a

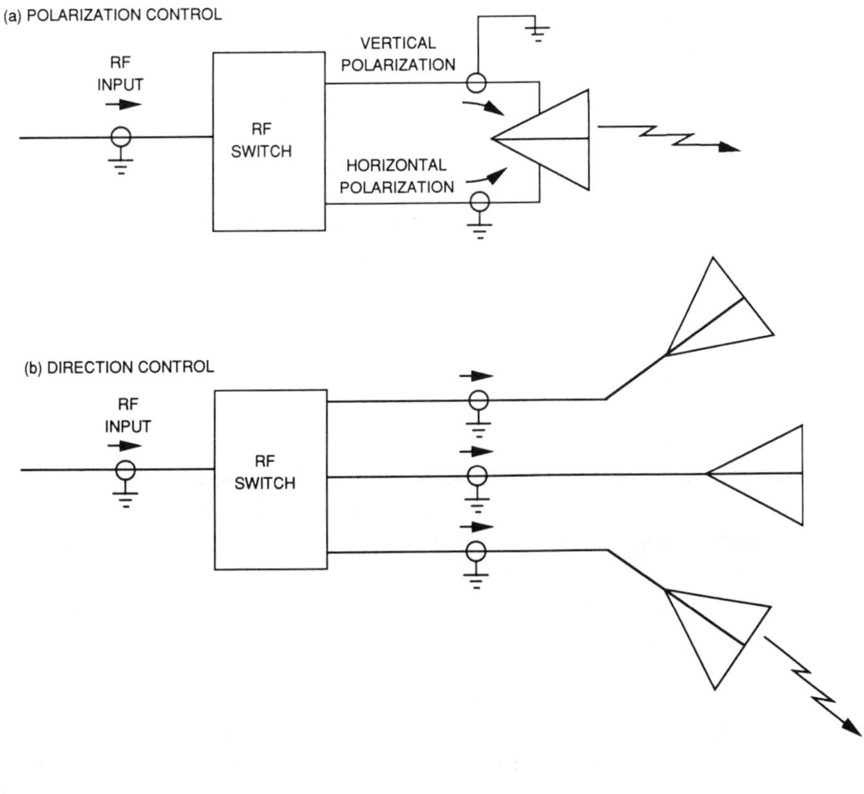

Figure 2.23 Antenna control using switches.

beamwidth about one third of the total angular coverage. To transmit in a given direction, the RF switch is toggled so as to steer the RF input to the proper antenna. Since all the power, less the switch insertion loss, is going into one of the individual antennas, the transmitted power density is three times (4.8 dB) more than it would be if a single passive antenna were employed.

There are several problems with these switched antenna approaches. One problem is that presumably the high power RF TWT amplifier is driving the RF switch; RF switches, since they are fabricated with microwave diodes, generally cannot take the very high RF power levels. Even if the switch could take the power level, the switch insertion loss is quite burdensome. The RF insertion loss of switches elsewhere within the ECM system, as, for instance, in any of the numerous low-level switches generally found, can be compensated for with relatively inexpensive gain elements. However, the insertion loss of the switches shown in Figure 2.23 directly reduces the RF output power, the last few dB of which may be critical for effective jamming performance.

Figure 2.24 shows alternative ECM antenna direction control schemes using RF phase shifters instead of RF switches. Figure 2.24a is again a multiple antenna system, similar to Figure 2.23b, in which the control electronics steers all of the RF power to one of the individual antennas. The transmission direction control using RF phase control works as follows: the low level RF input is split into several paths, four in the example of Figure 2.24a, using a coupling arrangement with certain necessary phase relationships. After the signal in each of these paths passes through the phase shifter, experiencing the ensuing insertion loss and phase shift, it enters the power amplifier, usually a TWT. The output of each TWT power amplifier then enters another coupling network. This coupling network is either a traditional set of couplers having high power capability, or a spatial combiner, such as a Rotman lens. Depending on the phase relationships of the coupling, or splitting and combining, arrangements, which are fixed, and the phase shifters, which are under electronic control, all of the power will impinge on one particular individual antenna. All of the other output ports from the second coupling network have the multiple signals combine in such a way as to mutually cancel.

The selected antenna in Figure 2.24a has the combined power of multiple TWT power amplifiers; this means four times or 6 dB more than the power from one individual power amplifier. The extra power to the selected antenna, plus the extra antenna gain derived from each individual antenna having a narrower beam width than the total required angular coverage, one fourth, or 6 dB, for the example of Figure 2.24a, both enhance the ERP. Another advantage of the configuration of Figure 2.24a is that the output coupling network is purely passive, unlike the switch of Figure 2.23b, which contains diodes, and is therefore more easily fabricated to withstand the stress of the high RF power levels. Furthermore, there is no coupling loss in the output combiner, unless one of the driving TWTs goes bad, only the resistive loss, which is considerably less than the insertion loss of a microwave switch;

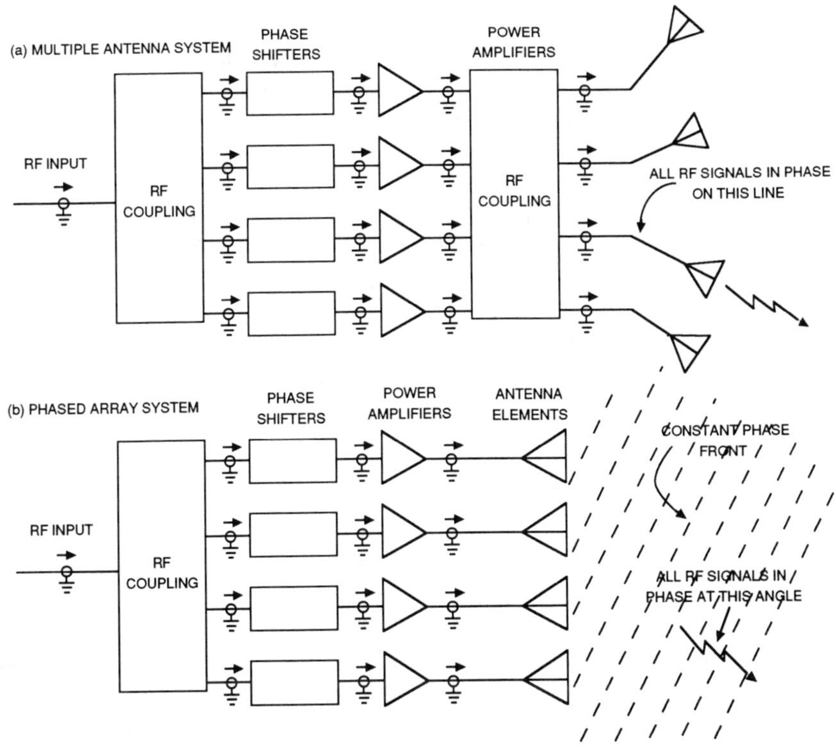

Figure 2.24 Antenna control using phase shifters.

hence the expensively generated RF power is not wasted. The electronic control of the phase, via the RF phase shifter, which typically does require diodes, is positioned ahead of the TWT power amplifier, and hence experiences relatively low-level RF power. Finally, if additional RF power is required, it is usually easier to build several power amplifier tubes than to solve the design problem of dissipating the internally generated heat of a higher power tube, especially if a CW capability is required. For all these reasons, the configuration of Figure 2.24a is electrically superior to that of Figure 2.23b. The configuration of Figure 2.23b is simpler, however, in that it does not require line matching. The electrical lengths in the configuration of Figure 2.24a must be carefully matched, with the appropriate phase offsets, including the power amplifiers, which will be driven into their nonlinear saturated region. The phase matching needs to be on the order of 10°; this tolerance is not easily achieved over the very wide bandwidths, which ECM systems must operate across.

The electrically steerable antennas in Figure 2.23b and 2.24a have a disadvantage not yet mentioned, and that relates to aperture. The problem is that a narrower beam antenna, that is, a high-gain antenna, must have a wider aperture than a broad beam antenna. Hence, switching between several high-gain antennas, instead of using one low-gain wide-beam antenna, requires considerably more aperture on the surface of the vehicle. That is, each and every one of the multiple high-gain antennas of the switched approach is larger than the alternate single low-gain antenna. However, many vehicles do not have one single location on the surface that can cover all the required angles, so a multiple antenna implementation is not a drawback in such situations.

Figure 2.24b is a phased-array antenna system, often referred to as an active aperture. The principle of operation is quite similar to that shown in Figure 2.24a, except that (1) the combining is done in space, and (2) each individual aperture should be considered an element of an antenna, instead of a separate antenna, with each such element having a wide full-coverage beam width. The array of all these antenna elements makes up the full antenna aperture; the net result of such a large net aperture is a high-gain narrow-beam antenna, made up of small elements each with low-gain wide-beam patterns. The system in Figure 2.24b is thus even more efficient than that in Figure 2.24a, because (1) not even the small resistive loss of the combining network is experienced, and (2) the individual antenna elements sum to a larger net aperture. However, if the vehicle shape is not conducive to pointing in all the required directions from one location on the vehicle surface, either the configuration of Figure 2.24a or a combination of both configurations is needed.

The active aperture structure of Figure 2.24b is used for direction control, for polarization control and for radiation pattern control, the latter two requiring more complex configurations than shown in the figure. With regard to pattern control, it is highly desirable to switch between high-gain operation and low-gain operation, because the CPU-DSP may not be able to track all of the threats, for example in angle, RF, or PRF, by switching to low gain; the net beam width becomes as wide as one of the elements, and the RF power is then transmitted equally in all directions within that wide beamwidth. The beam can be made wide by deliberately mistuning the phase shifter RF phases.

The phased array usually has equally spaced elements. The array can be in one dimension or in two dimensions. Usually the array is wider in azimuth because angular resolution is more important in azimuth than in elevation. Most distant threats are somewhere near the horizon, which means that the elevation angle uncertainty is much less than the azimuth angle uncertainty. For an antenna used just for ECM jamming transmissions, the gain in the direction of the threat resulting from a sufficient aperture is of primary importance, whether achieved with a larger azimuth or achieved with a larger elevation dimension. For an antenna employed for both ESM and ECM purposes, however, including ESM to support ECM, a larger azimuth dimension is more useful than a larger elevation dimension. Just like any aperture with constant power density across its surface, the resultant phased array antenna has the "sin(x)/x" antenna-pattern shape, with the first sidelobe at −13.5 dB, the second sidelobe at −17.9 dB, the third at −20.8 dB, *et cetera,* just as shown for the passive antenna in Figure 2.22. It is possible, however, at the expense of some ERP in the main beam, to weight the power levels to the individual antenna elements unequally so as to suppress the sidelobe power relative to the main beam. This is accomplished by putting appropriate attenuators at the input to certain antenna elements.

The economics of active aperture antennas will be described in another chapter. An even more complex array structure than that of Figure 2.24b can be used both to receive and transmit from an electronically controllable antenna; this has the advantage of ensuring that the transmitted signal is indeed going back toward the source of the radar signal. The key functional parameters and representative values for an electronically steerable antenna are

- gain (directivity and efficiency) (25 dB)
- power (1 kW)
- angular coverage at a min gain (90°)
- bandwidth (2:1)
- steering and focusing speed (1 μs)
- gain and beamwidth control range (20 dB)

Chapter 3
System Engineering Principles

3.1 SPECIALIZED KNOWLEDGE

This chapter describes aspects of ECM systems design expertise. Before such a subject can be properly presented, a certain knowledge base is required. This knowledge base includes

- general technical engineering background,
- general system overview knowledge,
- familiarity with components and subsystems, and
- certain specialized knowledge.

It is assumed that the reader has the general technical background. The system overview was presented in Chapter 1. The majority of components and subsystems available to ECM system engineers were presented in Chapter 2, with the major exception of sensor and transponder subsystems, the discussion of which will be postponed to subsequent chapters. The needed specialized knowledge includes VSWR, noise figure, receiver sensitivity, and component saturation; these subjects will be briefly presented in this section.

3.1.1 Voltage Standing Wave Ratio

Figure 3.1 shows a standard video circuit. Several lumped element components are shown in parallel, connected at points *P1*, *P2*, *P3*, and *P4*. These points represent a common node, and such circuits are analyzed as if there was a voltage on this node. That is, it is assumed that points connected by a wire have the same voltage. For microwave circuits, such an assumption is not valid; points connected by a wire do not, in general, have the same voltage in microwave circuits if such points are physically separated, and, indeed, the voltages may be radically different. The same laws of physics apply for both video circuits and microwave circuits, but the periods

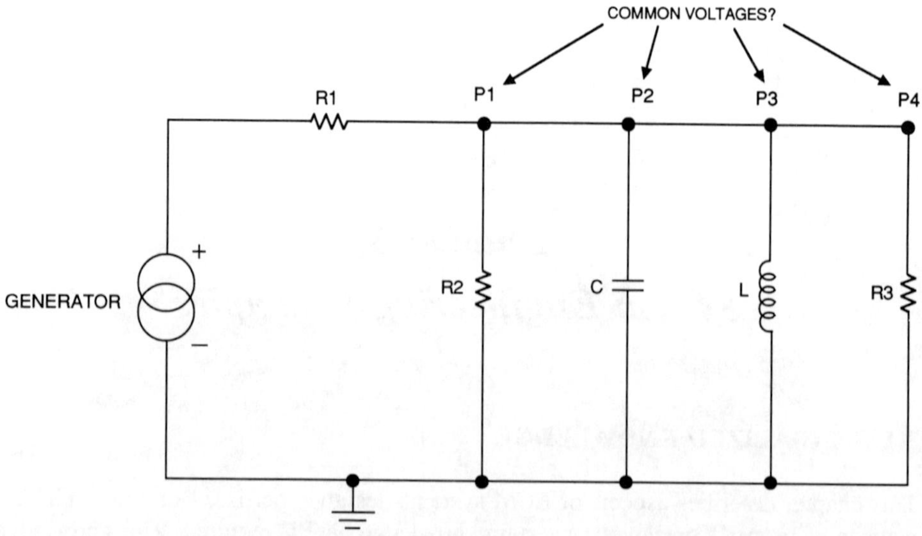

Figure 3.1 A standard video circuit.

of the highest video frequency AC signal are so long, relative to typical propagation times, that it is a very good approximation to assume that separated points or nodes connected by a wire have the same voltage.

In frequency bands where the carrier period is relatively short compared to the time needed for the signal to propagate between various points in the circuit, some complex phenomena occur, usually classified under the heading of VSWR phenomena. The speed of EM wave propagation in free space is about 186,000 mi/s, that is, 186 mi/ms. At 5280 ft/mi, this is 982,080 ft/ms, or 982 ft/μs, or 0.98 ft/ns; that is, the speed of EM wave propagation is about 1 ft/ns. This speed is the same for all bands, RF through optical bands. In a transmission line, such as coaxial, the speed is about 70% of the free space propagation speed. The consequence of (1) this magnitude of EM wave propagation speed, (2) the typical size of microwave circuits, and (3) the nominal period of microwave signals, is that, in general, the propagation time, which is the distance divided by the speed, exceeds the carrier period. This means that when a microwave signal leaves a generator and propagates down a transmission line toward a load, the voltage at the generator port will actually reverse sign before the signal voltage reaches the load. The typical microwave transmission line has radically different voltages, including opposite polarities, connected by a wire. Figure 3.2 illustrates this. This is the reason why all those complex VSWR phenomena occur.

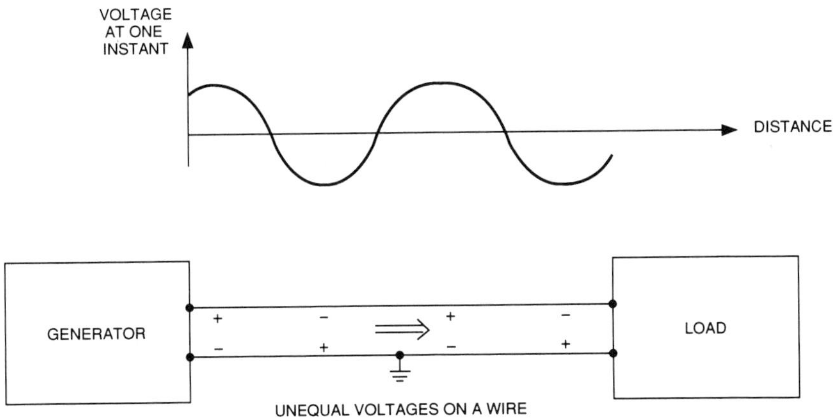

Figure 3.2 The significance of short wavelengths.

For an example of such relatively slow propagation in relation to the carrier period, or, alternatively, of short wavelength in relation to the distance to be transversed, consider the case of a 5 GHz carrier propagating 2 ft from the generator to the load in coaxial. The signal in this example will take about 3 ns to reach the load. Since the carrier period is 0.2 ns, the generator signal will then go through 15 cycles of its AC sinusoidal pattern before the first such cycle reaches the load.

A similar phenomenon occurs for video non-AC signals, although for most applications the transient phenomena is not noticed. An example of a transient response of a video signal on a transmission line, with unmatched load and generator, is shown in Figure 3.3. The transmission line is four units of distance long, the speed of EM wave propagation is one unit of speed, and the frames illustrated are separated by two units of time, with the first frame, at the top, at time equals one (i.e., time values are 1, 3, 5, 7, *et cetera*). For the first two frames the one volt signal is propagating from left to right. By frame three, the signal has reached the far end, and the mismatch is such that only half the voltage can appear across the load. To achieve this voltage, a reflected voltage of −0.5 V needs to be added to the incoming +1.0 V. This voltage difference is the result of an impedance difference between the transmission line and the load, and creates a −0.5 V wave propagating right to left, which subtracts from the incident voltage still propagating from left to right. When this reflected wave reaches the generator, a similar result occurs. It can be seen that after the multiple-reflection waves propagate back and forth several times, the transient conditions will end, and the voltages will have stabilized. Propagation of these waves should not be confused with propagation of power; once the voltages have settled out, and there are no more waves, power is still propagating from the source to the load.

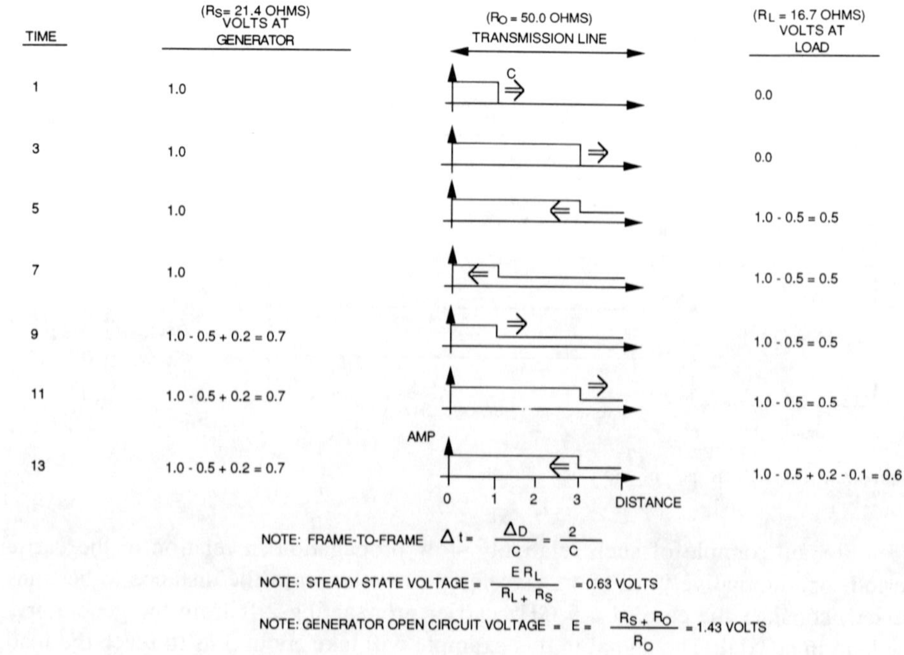

Figure 3.3 Video mismatch and propagation example.

In summary, it takes voltage and power time to propagate finite distances, and if the load and transmission line impedances are not properly matched, the load voltage will also not match the voltage on the incident signal; the unmatched net voltage on the load will then be the sum of the incident signal and the reflected signal. In other words, an unmatched load will cause a usually partial reflection of the incident wave, and the unmatched net voltage can be viewed as caused either by the mismatch in impedance or the subtraction of the reflected signal. Finally, the original generator in turn acts as a load for the reflected signal, which may or may not be matched, and so the reflected signal now becomes the incident signal to the generator, about which similar statements can be made.

Figure 3.4 illustrates the relative power transmitted to a load as a function of the degree of match of the generator and load resistances. The abscissa is the load resistance normalized to the generator or source resistance, while the ordinate scale is the power delivered to the load normalized to the maximum possible. We can see that the maximum power delivery is achieved when the load resistance equals the generator source resistance. If the load resistance is reduced, then the voltage will drop faster than the current increases, so the power, which is the product of current

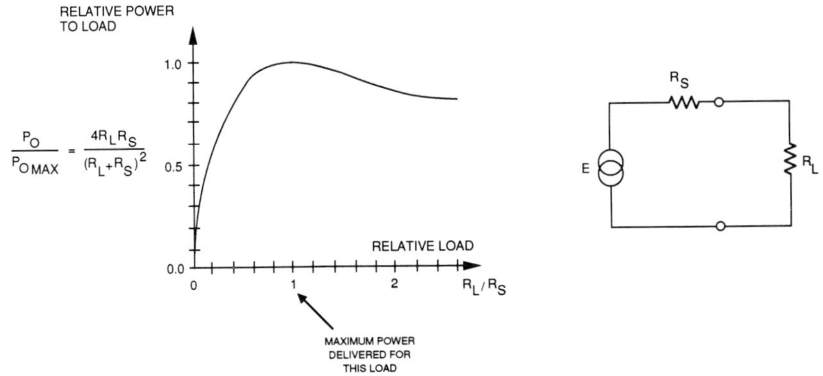

Figure 3.4 Power delivered *versus* load resistance.

and voltage, decreases. If the load resistance is increased, the current drops faster than the voltage increases, and again the power delivered decreases. The graph in Figure 3.4 is for pure resistances, but a similar result occurs for complex impedances. A complex-impedance load is a load that changes the phase of an AC signal as well as the amplitude. By convention, a matched component or transmission line in microwave RF systems has 50 Ω impedance. Any other value of impedance is a mismatch. A matched transmission line is supposed to provide a perfect 50 Ω load to a generator, at least until a signal is reflected back from the far end terminal, although the transmission line does not dissipate the power itself. In other words, a matched load, including transmission lines, is a load that does not reflect a signal back to the source.

Practical transmission lines are close to being ideal matches. The typical RF component in an ECM system, however, has a noticeable mismatch. Therefore, in the typical case, there is an incident wave and a reflected wave on the transmission line. Figure 3.5 shows such a situation. The reader should understand, however, that these two waves, the incident and reflected waves, are just the steady-state solution to an infinity of reverbations, as was illustrated in the video example of Figure 3.3. Since the steady-state incident and reflected waves are coherent for the distances and signal bandwidths of interest, the two voltages of the two waves will, in general, reinforce or subtract from each other, and a standing wave will result, as shown. The standing wave example shown in Figure 3.5 is for a badly matched load, since the cusp-shaped nulls are quite deep. In the early days of the microwave industry it was conventional practice to measure the impedance match of a load by using a special transmission line, between the generator and the load, which had a movable probe. This probe was moved up and down the line in search of the maximum and minimum voltages, which were duly noted. The ratio of these voltages were then

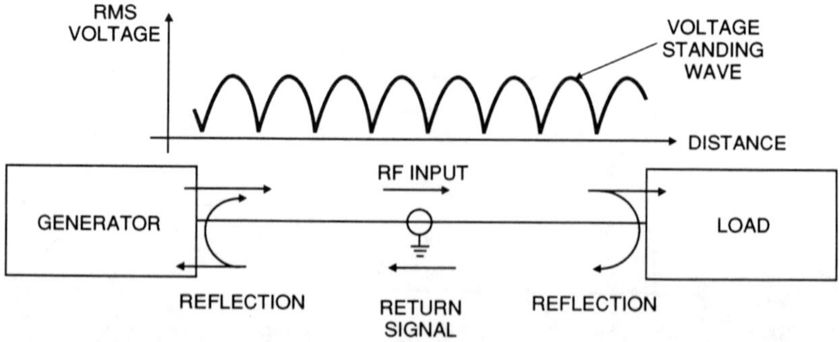

Figure 3.5 A standing wave.

computed to normalize the result, and hence the name: voltage standing wave ratio. The most commonly used RF test equipment now, however, is often automated, and generally measures the incident and reflected power directly, using these measurement results to compute the traditional VSWR value. Figure 3.6 relates the load interface VSWR to the reflection return loss in decibel. Figure 3.6 can also be used with Figure 2.3 to relate the VSWR and the return loss to the power passed to the load; simply consider the coupling axis of Figure 2.3 also to represent the relative return loss resulting from a mismatch. For example, a component with a VSWR of 4:1 will reflect a signal 4.4 dB less than the incident RF power, and the power delivered to the load will be 1.9 dB less than the incident RF power. By convention, the match of components is usually expressed in VSWR, sometimes it is expressed in terms of reflection loss, and it is almost never expressed in terms of impedance.

Some general comments regarding VSWR phenomena are in order. The instantaneous voltage on a transmission line is dependent on distance along that transmission line; in other words, it is dependent on the physical location, even for steady state conditions, because of the relatively slow propagation rate compared to the carrier periods, as already noted. The net *root mean square* (rms) voltage on a transmission line is also dependent on distance, as shown in Figure 3.5, if the load is not perfectly matched. The power flow on the transmission line, however, is not dependent on distance; there is just as much power flowing to the load at the VSWR nulls as there is at the VSWR peaks, and likewise for the reflected signal. The incident and reflected waves have a fixed phase relationship for a given distance down the transmission line and a given CW carrier frequency. The incident and reflected waves are 180° out of phase at the rms nulls and in phase at the rms peaks. What determines which way the RF power is flowing? The phase relationship between the voltage and current determine the direction of power flow for both the incident wave and the reflected wave. The load impedance is, in general, a complex numerical value, not just a magnitude.

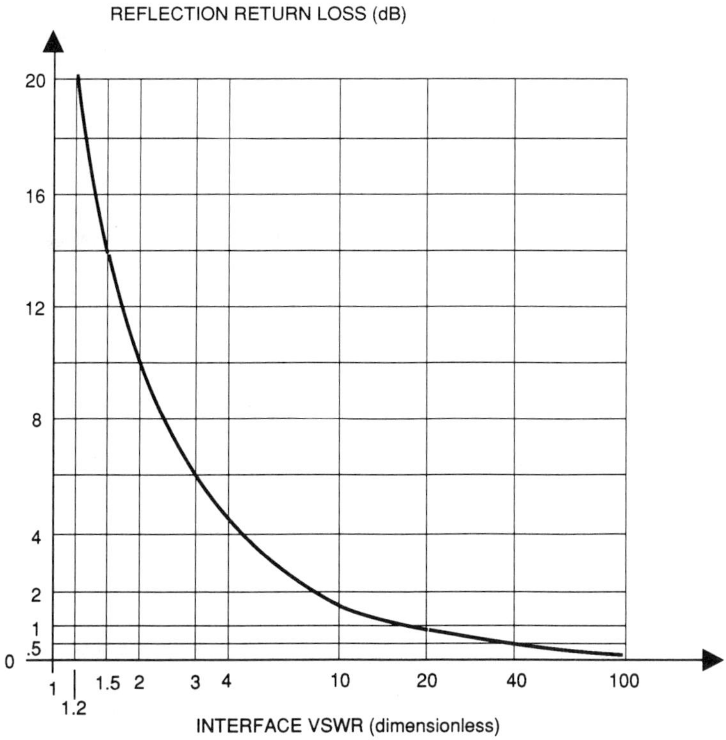

Figure 3.6 Reflection loss *versus* VSWR.

As shown in Figure 3.5 and explained previously, the VSWR phenomena can be viewed as a recirculation result: the signal propagates from the generator down the transmission line to the load where a portion is reflected; this reflected signal then propagates back to the generator, where a portion again is reflected. This signal reflected off the generator can be conceptualized as either (1) continuing its journey around this loop an infinity of times, or (2) combining coherently with the original incident signal from the generator to give a net incident wave. The second conceptualization is more appropriate for steady state conditions, while the first is more appropriate for dynamic or transient situations. It is traditional to teach VSWR calculation methodologies based on the use of Smith charts. However, the Smith chart is not appropriate for the typical bandwidths and dimensions of ECM systems. This text will, by integrating various concepts, exploit the fact that VSWR phenomena is a recirculation phenomenon, as illustrated in Figure 3.5, and will therefore explain how to make precise VSWR calculations based on the recirculating RF memory loop theory to be presented in another chapter. Although the computation methodology will be explained elsewhere, this section will qualitatively discuss VSWR phenomena further.

The VSWR phenomena results from the carrier period being short in comparison to the time it takes the EM wave to propagate between various points in a microwave system or circuit. Because of this, voltage, including RMS voltage, is characterized by standing wave patterns, and is therefore a less appropriate parameter, whereas power flow, since it is uniform along the transmission line length, is a more appropriate parameter to use in describing the functioning of microwave systems. The VSWR value of a component is intended to define a component's impedance referenced to its rated impedance; by convention, the rated impedance for microwave components is usually 50 Ω. Note that even generator output impedances are commonly defined with a VSWR value, which is definitely a misnomer, even though it accurately defines the generator's reflection coefficient, because only the load's VSWR impedance determines the standing wave ratio on the line. It is common industry practice to specify component impedances with VSWR magnitude values, with no mention of phase. This practice makes it impossible to calculate the loss through a string or network of microwave components at any particular frequency, although ripple and average losses or gains can be calculated for these systems based on such information. A component's true insertion loss or gain is defined and measured with perfectly matched generators and loads. This point is especially important to keep in mind regarding couplers and low-loss components. Although the voltage standing wave ratio on a transmission line is only dependent on the load's VSWR impedance, the ripple caused by the interaction of the load and generator is dependent on both the load and generator VSWR impedances. Ripple is the variation of transmission power delivered to the load across the frequency band of interest. If the load and generator are both mismatched, then the net delivery of RF power across a given band is most commonly less power than if they were matched, but the average power is more than if they were matched. That is, the typical and average power-delivery values have opposite errors with respect to the matched case.

The ripple, resulting from both the load and generator being mismatched, is reduced by inserting an isolator as shown in Figure 3.7. (The reader may wish to review the description of isolators given in Chapter 2.) Notice that the voltage standing wave ratio just in front of the load has not been changed by the isolator; the voltage standing wave ratio on the line in front of a component will always be precisely the VSWR expression of that component's impedance. The reader should understand the distinction between VSWR and ripple. The isolator reduces the ripple. As shown in Figure 3.7, the rms voltage between the generator and isolator is independent of distance; hence the power delivered is independent of distance and frequency. There is no reflected wave on the transmission line between the generator and isolator; all of the reflected power from the load is dissipated in the isolator's dummy load. The percentage of power the load reflects is approximately constant,

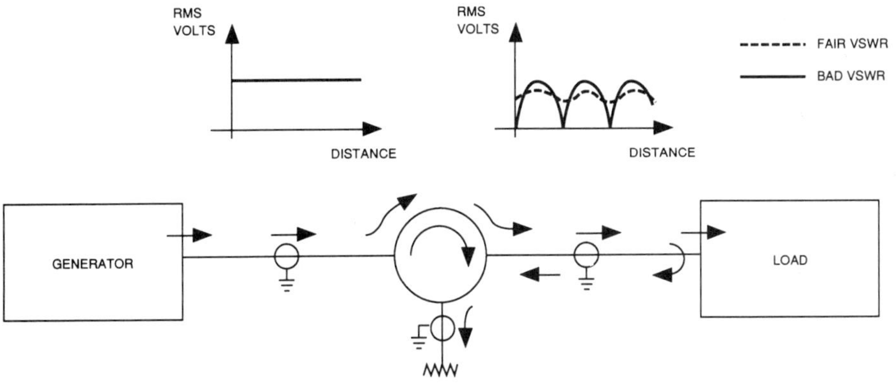

Figure 3.7 RF isolator example.

so the power delivered to the load will not change if the frequency is changed or the cable lengths are changed. In other words, there is no ripple for the situation in Figure 3.7 because the RF signal cannot make a complete round trip up and down the transmission lines between the generator and load.

One final VSWR example is shown in Figure 3.8. This figure shows a block diagram with a repeater path, the main path, from receiving antenna to transmitting antenna. A coupler is used to couple some of the signal and deliver this signal to a *crystal video receiver* (CVR) through an RF switch, switch number 2. The question is, does turning this switch number 2 on and off influence the gain of the main path? Because switch number 2 is not in the main path, it would seem as though the state of switch number 2 could not influence the gain of the main path, but such a cursory conclusion ignores the influence of the VSWR phenomena. If switch number 2 is shut off, all of the incident power to switch number 2 will be reflected, that is, the switch's off state has high VSWR. This reflection will propagate back to the coupler where the signal will be partially dissipated in the dummy load and partially coupled across to the main path to continue to the preamplifier output. If the preamplifier output impedance is poorly matched, that is, in common terminology, it has a poor VSWR, some of this reflected signal impinging on the preamp will be reflected again. At this point, the reflected signal will be at a certain phase angle with respect to the original signal traveling down the main path; these two signals will either reinforce or subtract from one another. So, indeed, because of VSWR phenomena, the control of RF components off the main path can influence the gain of the main path.

The above text has followed convention in using the term VSWR to refer to both an impedance and an actual voltage standing wave ratio. To summarize some key points, as commonly employed, the term VSWR

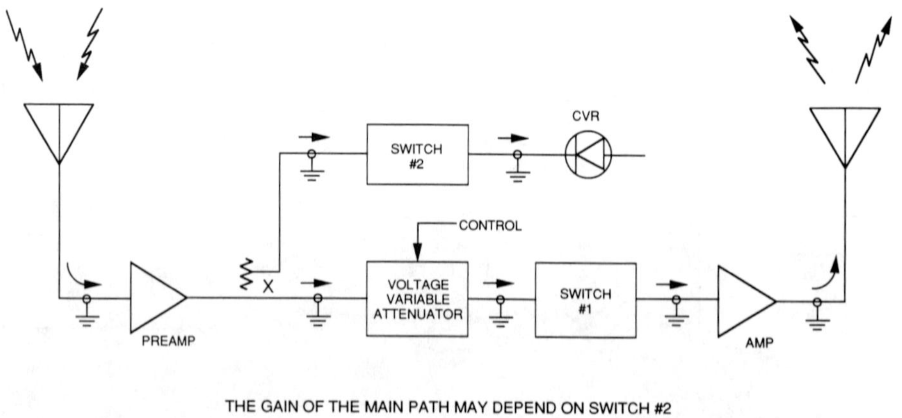

Figure 3.8 VSWR example.

- defines a component's input or output impedance magnitude,
- does not necessarily define a voltage standing wave ratio, and
- does not define ripple.

3.1.2 Noise Figure

A transmission line at room temperature, terminated at both ends, will exhibit a noise power level spectral density in the microwave band of kT, approximately -114 dBm/MHz, where k = Boltzmann's constant = 1.38×10^{-23} J/K (or W · s/K or W/Hz · K), and T is temperature in K. This microwave power is created by thermally induced means. Just as much noise power is flowing downward on the line as upward on the line, so there is no net power flow, assuming the temperature is constant along the line. As a general rule, a component with a given gain or loss will not have a ratio of output noise to input noise equal to that gain or loss respectively. For example, a 10 dB attenuator pad will have the same (-114 dBm/MHz) noise level at both the input and output ports. As a more interesting example, a 24 dB gain 1000 MHz (30 dB) bandwidth amplifier will not have an output noise power that is 24 dB higher than the kT level; that is, the output noise power is not equal to $-114 + 24 = -90$ dBm/MHz, or $-114 + 24 + 30 = -60$ dBm total; the noise power will instead be higher than this simplistically determined noise level. The difference on a log scale between this simplistic determination and the actual noise level is known as the *noise figure,* as shown in Figure 3.9. Figure 3.9 summarizes the definition of noise figure and derives a conveniently usable equation for the output noise; the equation for the net noise figure of a series string of components is also shown. For example, if the amplifier has a noise figure F = 6 dB, a gain G =

DEFINITION: $F = \dfrac{S_{IN} / N_{IN}}{S_{OUT} / N_{OUT}}$

(WHEN T=290°K)

SUBSTITUTING THESE: $G = \dfrac{S_{OUT}}{S_{IN}}$

$N_{IN} = kTB$

GIVES: $N_{OUT} = kTBFG$

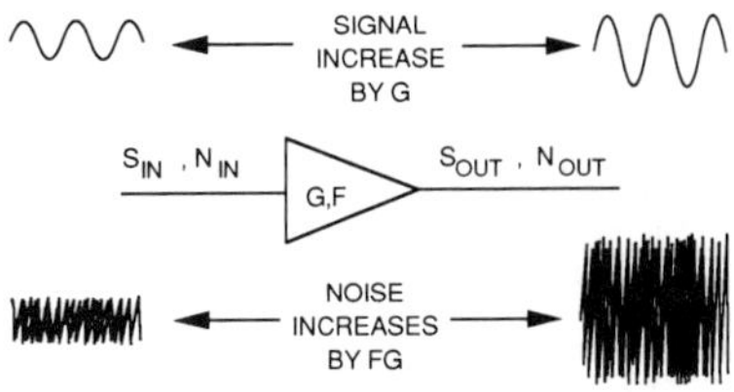

FOR A SERIES STRING OF COMPONENTS:

$$F_0 = F_1 + \frac{F_2 - 1}{G_1} + \frac{F_3 - 1}{G_1 G_2} + \ldots + \frac{F_M - 1}{G_1 G_2 \ldots G_{M-1}}$$

WHERE:

F = NOISE FIGURE
S = SIGNAL POWER
N = NOISE POWER
G = GAIN

B = BANDWIDTH (THE NARROWEST IN A STRING OF COMPONENTS)
T = ABSOLUTE TEMPERATURE (KELVIN)
k = BOLTZMANN'S CONSTANT = 1.38×10^{-23}
M = NUMBER OF COMPONENTS IN SERIES

NOTE: $kT \simeq$ -114 dBm/MHz AT ROOM TEMPERATURE

Figure 3.9 Noise figure definition.

24 dB, a bandwidth B = 1000 MHz (30 dB), and a thermally induced input RF noise power of $kT = -114$ dBm in each MHz, then the output noise level will not be $kTBG$, it will instead be $(kT)BFG = -114 + 30 + 6 + 24 = -54$ dBm.

Although the equation used to define noise figure in Figure 3.9 seems simple, confusion often arises because of the meaning of the input noise, and the reader should be aware of the cause of this error. The input noise is from the thermally generated power at the input port; this power spectral density equals kT. The rated noise figure of a component is defined for T = 290° K (17° C or 63° F), so the input noise power in the noise figure equation of Figure 3.9 is defined to be about −114 dBm/MHz. If noise is created by a preceding component, for example, that noise must not be substituted as the input noise in the equation defining noise figure in Figure 3.9. Such noise, to the extent that it is above the kT level, experiences only the normal gain or loss of that component, like any other signal; to this output noise must be added each component's output noise as determined by the equation for output-noise power derived in Figure 3.9. The equation for the net noise figure of a series string of components, although not suitable for conceptualization purposes, is quite convenient for use with a programmable calculator to compute a net noise figure.

Figure 3.10 shows an example of calculating noise levels using two different methodologies. A block diagram is shown, with a loss followed by an amplification,

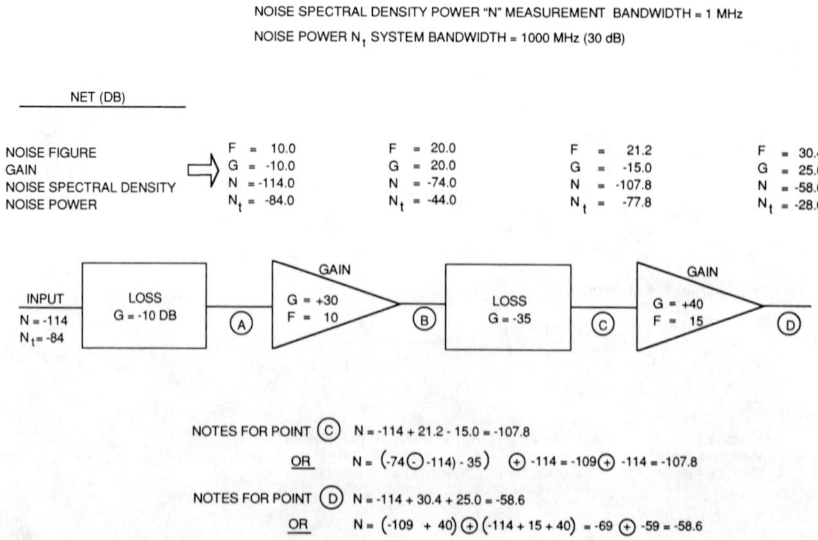

Figure 3.10 Sample noise level calculation.

followed by a loss, followed by an amplification. The gain and noise figure of each component is tabulated. The net noise figure to that point, the net gain to that point, the noise level per MHz at that point and the total noise level at points A, B, C, and D, that is, after each component, is also tabulated. The net noise figure and gain are referenced to the input. Two methods for calculating the noise level spectral density at points C and D are shown. The first method is the traditional approach using the equation derived in Figure 3.9. The reader may wish to study the second method, which gives the same answer, since it may be more helpful with the conceptualization of the phenomenon, when the use of the equation in the first method is not satisfying. The second method also conveniently lends itself to making noise level running value calculations at each point in the block diagram, based on the values of the previous calculated point, without explicit regard to the noise figure referenced back to the original input. The circle-minus and circle-plus signs represent an operation on two values expressed in decibel; these operators take the log of the subtraction or sum, as the case may be, of the antilogs of the values being operated on.

As shown in Figure 3.10, the noise density at point C is the noise density at point B, less the -114 dBm/MHz thermal noise at point B, through a loss of 35 dB, with that noise power density result summed with the -114 dBm/MHz thermal noise otherwise present at point C. Note that for the example of calculating the noise density at point D, the noise density at point C, the input to the amplifier, less the

−114 dBm/MHz inherent thermal noise, has the straight 40 dB gain enhancement imposed, while the −114 dBm/MHz inherent thermal noise has both the 40 dB gain enhancement plus the 15 dB noise figure imposed; these two output components (−69 dBm and −59 dBm) are then summed to determine the net output noise density. As stated before, the common error is to apply both the noise figure and gain to the −109 dBm portion of the input to the amplifier, which would be incorrect; noise generated by a previous amplifier will experience straight gain just like a signal.

Noise level calculations are usually made on a power spectral density basis, with 1 MHz measurement bandwidth commonly, and conveniently, employed as shown in Figure 3.10. For purposes of assessing system performance, such as repeater mode dynamic range, the noise power a particular threat radar will see is then equal to the output power spectral density altered by the ratio of the radar bandwidth to 1 MHz. However, ECM systems cover very wide bandwidths, so the total integrated noise power will be much higher. The primary purpose for calculating the total integrated noise power is to ascertain that no component saturates on the thermal noise power.

The typical noise figure for a TWT falls in the range of 10–20 dB. A solid state amplifier will often have a noise figure less than 6 dB. An attenuator, or other passive loss, will have a noise figure equal to its attenuation, since a signal will drop in amplitude by the attenuation loss, whereas the thermal noise floor will remain at $kT = -114$ dBm/MHz on both sides of the passive component. To a good approximation, the conversion loss of a mixer can be treated as a signal attenuation, so that the noise figure of a mixer is approximately its conversion loss. However, unlike a truly passive device, this is just an approximation, and, indeed, a mixer's noise figure in decibel may actually be slightly less than the conversion loss in decibel.

3.1.3 Receiver Sensitivity

Receiver sensitivity is a measure of the ability of a receiver to sense a signal, especially a weak signal, in the presence of noise. The noise referred to includes the background EM radiation noise, but generally the predominant inherent noise contribution is the thermally generated broadband noise floor (kT, approximately −114 dBm/MHz) that propagates in both directions in transmission lines. Of particular interest is the receiver's *minimum detectable signal* (MDS) input level that reliably results in an output digital logic bit or word with an acceptable false alarm rate. However, the response of the receiver to intercepted deliberately radiated noise or electronic equipment inadvertent noise should also be considered an important issue, especially for ECM equipment.

Figure 3.11 illustrates the detection of a pulsed signal in the presence of noise. The receiver consists of an RF preamplifier with gain G, some loss L, a crystal

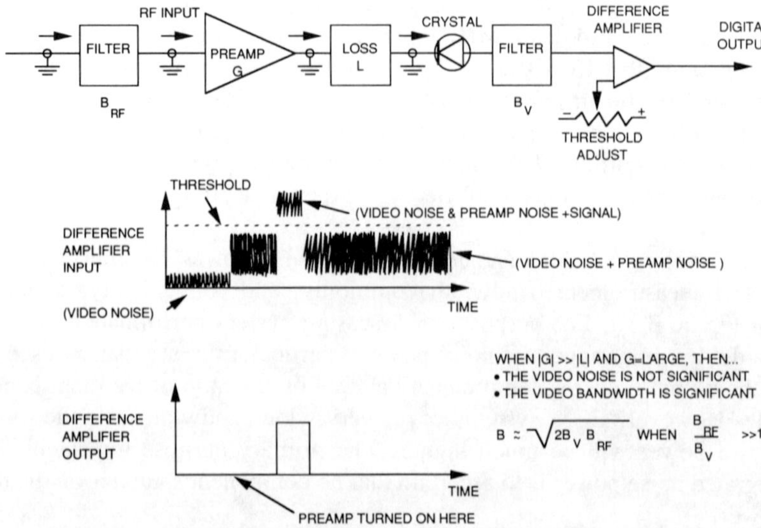

Figure 3.11 Detecting a signal with noise present.

detector and a difference amplifier with adjustable threshold. The crystal detector converts the RF AC signal to a rectified unipolar video signal. The operation of this most simple of EW receivers is treated more fully in Chapter 4, which describes EW receivers. At the left of the waveform portion of the figure the RF preamplifier is not on, so the difference amplifier input is just the video thermal noise coming from the crystal. When the RF preamplifier is turned on, its noise, as detected by the crystal, is summed with the thermal video noise. In this example, the absolute value of G is much larger than the absolute value of L, so the RF noise generated by the RF preamplifier swamps out the crystal's thermally generated noise. As can be seen, the difference amplifier's threshold is set high enough so that the noise generally does not cross the threshold. When the pulse signal is present, the sum total of the RF signal and the RF noise, both converted to video by the crystal, and the crystal's own video noise all sum together to give a voltage, which for the example shown in Figure 3.11, is sufficient to cross the threshold and generate a logic signal.

The situation shown in Figure 3.11, where the RF preamplifier noise swamps out the crystal's own noise, is appropriate, because such a situation is a design rule advocated later in this chapter (Section 3.4). When this rule is followed, the more complex equations that describe the combination of the RF preamplifier noise and the crystal's own noise can be ignored. This design rule will always give the best sensitivity. Although this design rule eliminates consideration of video noise amplitude, the video bandwidth is still significant. The net bandwidth, for SNR crystal receiver sensitivity calculation purposes, is a little more than the geometric mean of the RF and video bandwidths; this bandwidth rule applies when, as is usually the case in ECM systems, the RF bandwidth is much wider than the video bandwidth.

When the RF pulse enters the receiver system, the RF noise can be in phase, out of phase or in quadrature phase, with respect to the input phase. Therefore, the noise may help or hinder the signal in causing the crystal output video voltage to cross the threshold. Two key questions that should be addressed are (1) how best to set the threshold, and (2) what the resulting MDS sensitivity is.

The setting of the receiver threshold, such as the receiver in Figure 3.11, should be set by considering these difference-amplifier related factors:

- noise level,
- time constant or bandwidth,
- acceptable false alarm rate, and
- desired probability of detection for each input pulse.

The performance of the receiver with regard to all these factors is shown in Figure 3.12, based on calculations made by the author. The abscissa is the threshold to noise ratio, that is, the precise trip point, as measured by the power of a pure noise-free signal *versus* the average noise power. The left ordinate scale gives the false

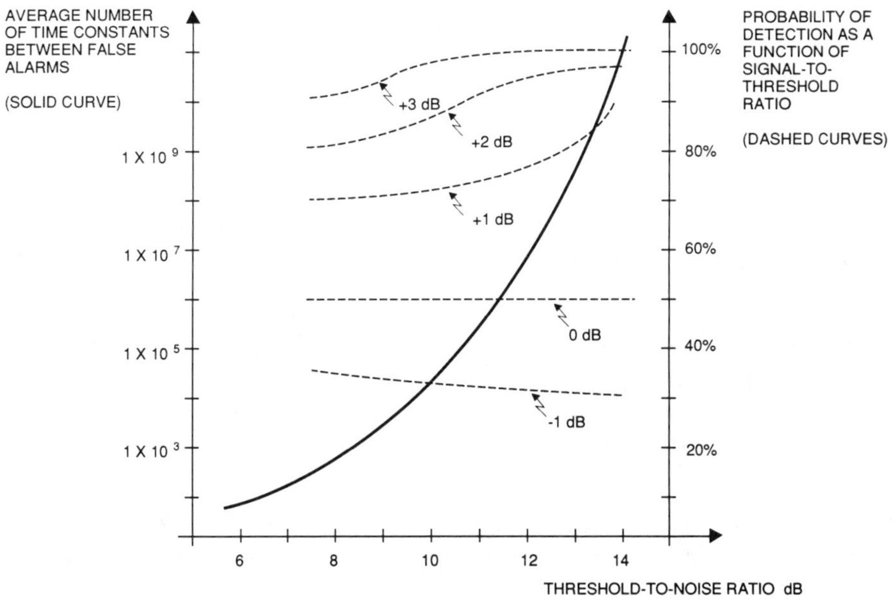

Figure 3.12 Sensitivity design graph.

alarm rate, normalized to the receiver's time constant. The time constant is approximately the reciprocal of the video bandwidth when the RF bandwidth is much wider than the video bandwidth. The right ordinate scale gives the probability of detection for a family of input signal power curves. For the overall receiver of Figure 3.11, the combination of false alarm rate and probability of detection for given signal input levels is determined by the

- video bandwidth,
- RF bandwidth,
- front-end noise figure, and
- threshold level.

Therefore the above factors determine the receiver MDS. The front-end gain is indirectly normalized out of consideration, assuming the advocated design rule is followed and the threshold-level value is normalized by that gain. In using the graph of Figure 3.12 to determine receiver MDS sensitivity, the usual starting point is the false alarm rate, which, when converted to normalized units of time, is used to read the threshold-to-noise ratio. Given this relative threshold value, the noise level, and the probability of successful detection (the right ordinate scale), the family of dashed curves gives the MDS sensitivity. Alternatively, given the relative threshold, noise level, and input amplitude, the graph shows the probability of detection.

A receiver with a threshold-to-noise ratio within the span shown in Figure 3.12, which covers typical EW values, is almost guaranteed to detect successfully any input that is more than 3 dB above that threshold value; that is, the probability of detection is close to 100% for signals more than 3 dB over the threshold, as shown in the figure. For signals right at the threshold, there is a 50–50 chance of detection, because it is equally likely that the noise will reinforce or reduce the signal.

We emphasize that a value for receiver sensitivity is meaningless unless the false alarm rate and probability of detection are both specified. To reiterate, assuming that the RF preamplifier is the main contributor to the video noise from the crystal, the MDS sensitivity of a system employing an idealized CVR can be determined by using Figure 3.12, given the

- front-end noise figure,
- RF bandwidth,
- video bandwidth,
- acceptable false alarm rate, and
- acceptable probability of detection.

Figure 3.13 gives an example of such an MDS calculation. The MDS calculation includes intermediate calculations of noise levels, and is therefore another example of that also.

With regard to the above five parameters that determine an EW system's MDS, some comments are appropriate. The front-end noise figure is generally determined

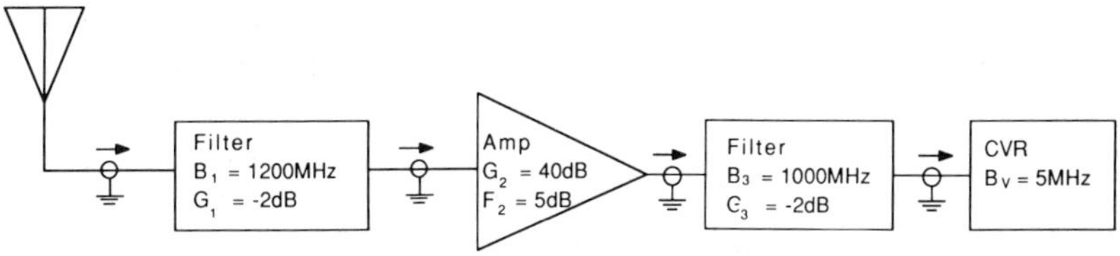

GIVEN:
(1) Above block diagram values
(2) Required false alarm rate = 1 millisecond
(3) required probability of detection = 90%

QUESTION: What is the sensitivity at the antenna (for pulses of 0.1 microsecond)

METHOD: Effective noise power into CVR is:

$N_e = KTB_e F_2 G_2/G_3 + KTB_c = KTB_e F G_2/G_3$
$= -114 + 20 + 5 + 40 - 2 = -51\text{dBm}$

where $B_e = (2B_r B_v)^{1/2} = 100\text{MHz}$
$B_r = 1000\text{MHz}$

The number of time constants between false alarms is determined by the video bandwidth and required false alarm rate and equals $1\times10^{-3} \times 2\times5\times10^{6} = 1\times10^{4}$. The graph of time constants VS threshold to noise gives 10dB for the threshold-to-noise ratio, and the signal-to-threshold ratio needs to be (between 2 and) 3dB. Therefore the threshold needs to be set at -51+10 = -41dBm, and the minimum signal for 90% probability of detection needs to be -41+3 = -38dBm. The net gain from the antenna to the CVR is -2+40-2 = 36dB. Therefore the sensitivity at the antenna is -38-36 = -74dBm. ANSWER

Figure 3.13 Sample CVR sensitivity calculation.

by the noise figure of the RF preamplifier and the losses between it and the receiver antenna. The RF bandwidth is determined by the coverage desired or needed. The video bandwidth is determined by the expected pulse widths. The acceptable false-alarm rate and probability of detection are generally determined by digital-processing requirements, so as to take the appropriate action reliably.

3.1.4 Overdrive and Saturation

Except for phased array systems, traveling-wave tubes are generally used in the ECM output transmitters because they have the best combination of power and bandwidth. It is tempting to set the drive levels so that the tubes are driven hard into saturation, squeezing the last few dB of power out of the transmitter. However, when this is done, the tube acts in a highly non-linear manner. Indeed, if it is driven too hard into saturation, the output power may actually drop. Furthermore, saturated TWTs, and saturated amplifiers in general, create spurious signals. In theory, the TWT will create the 3rd, 5th, 7th, *et cetera,* harmonics of the driving signal, which is the frequency domain description of an unbiased square wave; in practice, however, the construction tolerance of the TWT assembly is such that even harmonics will be present, too. For an octave tube, these harmonics can either be ignored or filtered out. Usually, the luxury of ignoring such harmonics is not permitted for a variety of reasons, often associated with the viability of the other on-board electronics; that is, the EMI is unacceptable. If a bandpass filter is used at the transmitter output, that will cause some insertion loss, as well as take up a not insignificant volume because of the power levels involved. Because of such considerations, care must be taken not to drive the TWT too hard.

Saturated TWTs, and some types of solid state amplifiers and RF limiters, are characterized by strong signal capture. This is a phenomenon that is exploited in the design of recirculating RF memory loops; a detailed description is given in the section on *recirculating memory loop* (RML) transponder storage time. Suffice it to say that simultaneous weak signals are suppressed. Therefore, if full power noise mode operation is being conducted in one part of the band, normal repeater mode operation may not be viable in the other parts, because the gain will be suppressed by the saturation of the TWT transmitter on the noise jamming. This may be another reason to moderate the TWT drive.

Although weak simultaneous signals tend to be strongly suppressed in a TWT transmitter by high level signals, the *intermodulation products* (IMPs) are not suppressed, nor are the harmonics. In this regard it should be noted that if a single pure CW tone is passed through a phase or frequency modulator before entering the TWT transmitter, and this modulator creates a complex spectrum with many spectral lines, some of which are small, theory says that the TWT will not suppress such lines even if driven hard, except for the in-band overdrive fall-off already discussed.

3.2 SYSTEM ENGINEERING METHODOLOGY

Given the system overview description, the description of components available for use, and the specialized knowledge needed regarding VSWR, noise figure, sensitivity, and saturation, the subject of systems expertise and systems design methodology can be described.

The elements to system engineering can be itemized in steps as follows.

1. requirements and constraints:
   - define and document, and
   - establish traceability;
2. structured design:
   - functional analysis,
   - functional block and flow diagrams,
   - partitioning analysis,
   - straw-man formulation,
   - requirements allocation,

  - optimization, and
  - cost-effectiveness analysis;
3. design selection:
  - reiterate for several designs,
  - straw-man presentation,
    - summary description, and
    - summarize advantages and disadvantages,
  - discuss tradeoffs with:
    - management, and
    - engineering,
  - select design, and
  - final optimization;
4. documentation:
  - design,
  - parameters,
  - issues, and
  - traceability.

Steps 1 and 4 tend to be avoided or minimized by many, but experience teaches that they are very important. In particular, to document the source of the requirements is very important. In the future, certain design changes may be considered, and there will be great difficulty in making the decision regarding these design changes if there is no certainty about why the original decisions were made. However, if the reason for the requirement value is indeed documented, future design changes can be made intelligently, quickly, and prudently. The traceability aspect means that the various requirement parametric values and technique descriptions are documented as coming from certain pages of certain technical memos, kept in a certain library or filing system. Alternatively, but less preferable, the traceability documentation may state that a particular value came from the verbally expressed statement of a certain expert. Very frequently, the requirement origination is simply the customer specification. Often, because of time pressures or other constraints, the system engineer has to make an educated guess about the requirement value; in such a situation, it is especially important to document the reasoning, otherwise in the future that value may become "gospel," although there is no good justification. It is advisable for the traceability documentation to include the confidence factor for all values or technique descriptions, as well as the source.

System engineering is, almost by definition, an iterative process. Components and subsystems can be assembled in an infinite variety of forms. In step 2, after the functional analysis, with any associated diagrams, and the partitioning analysis are completed, a "straw man" formulation is proposed. This proposed approach must have the requirements applied to it and be optimized, and then its cost effectiveness must be analyzed. Based on this experience, other approaches should be formulated and compared. Many of the straw man approaches will be quickly labelled as clearly less viable. Quite often, after several iterations, the one clear choice will emerge. More often, the trade-offs between the different approaches will need to be discussed with management and other engineering personnel. Finally, one approach is selected. Soon after the selection is made, the key system engineers must immediately perform step 4. A tabulation of the more important system parameters will be found to be quite useful.

In many programs there will be strong incentives to skimp on these four system engineering steps so that the actual detail design can commence. However, properly conducting the system design function almost always results in a better design and a more successful program.

System engineering is similar in many ways to component and other types of engineering, but it also has several different attributes. Some of the unique system engineering methodology elements include the following considerations:

- diverse expert-knowledge synthesis is required;
- issues should be explicitly identified and addressed;
- partitioning is often based on time scale analysis;
- determination of cost per unit volume is a key trade-off tool;
- identification of limiting parameters is necessary;
- critical items and functions should be identified;
- the goal is a flexible and generic approach; and
- realism within the present current trends.

The system engineering process should bring together the expert knowledge regarding the constraints associated with each of the key system elements. As part of this process, the design issues should be tabulated, and each described with at least a short paragraph in a technical memo, at the beginning, mid-point and completion of the design process.

Proper partitioning is crucial to achieving a successful design, although unfortunately it is often not considered to be part of the system design process. Partitioning is the process of taking the design structure and drawing boxes around components, making elements, so that specific engineers or vendors can be assigned responsibility for each such element. In other words, partitioning divides the system into distinct elements and correspondingly allots the detailed specialized design tasks. This must be accomplished in the most sensible manner. Partitioning should meet three goals: (1) the interfaces are clean and simple, (2) the engineering specialties required for each are not diverse, and (3) the magnitude of the design task for each is appropriate. A 3–12 person-month specialized design task for each such in-house-designed element is an appropriate and efficient magnitude. This author recommends that the partitioning be based on a time scale analysis, which will be discussed in this chapter.

Determination of the average cost per unit volume, at least for systems where cost and size are critical, is a key trade-off tool. To use this tool, after assembling the initial allocation budgets, element designs, and quotes, the system engineer simply sums both the costs and volumes of all the system elements and divides. Each element is then compared to the average to see if it is in line with the other elements. What usually becomes apparent is that the volume budget can be reallocated to give an element for which the designer or vendor has a difficult packaging task a little bit more volume, by taking volume from another or several elements that are not as badly constrained. This will bring all the elements closer to the average cost-per-unit volume, thereby lowering that average and saving expense.

Another attribute of good system engineering is the explicit identification of the key parameters that limit the performance or cost effectiveness of the design. Examples of such limiting parameters might include the speed of tuned oscillators, the number of filter channels per unit volume, or the output power of the transmitter TWT. The engineers associated with a program need to be cognizant of these limiting parameters, especially if dramatic improvements are to be realized. Although this author claims that dramatic improvements are much less likely to occur if these parameters are not so publicized, doing so in actual practice does not normally result in an immediate solution, so this aspect has a different longer term role to play in the system engineering process. In other words, such explicit written statements may not lead to a solution until the next system design, perhaps a few years later; some percentage of the problems will be solved quickly, however, to the benefit of the present system. A similar comment can be made about explicitly identifying the critical electrical functions and other critical items.

The next item on the above list of good system engineering attributes is especially appropriate to ECM system engineering: the goal should be to synthesize a design that is flexible and generic. As stated in the introductory chapter, electronic warfare is a rapidly evolving field. Hardwiring a technical approach to achieve successful jamming of a particular threat or class of threats means that it is very likely that the system will soon be outdated. The EW development process, in particular, has been continually characterized by new designs, both patches and system structures, intended to counter particular functions; that is true of both radar ECCM and ECM system designs. Designing a power-managed system that can flexibly adapt or be reprogrammed to alter truly generic output parameters and to generate a wide range of modulations should be an important design goal leading to a generic ECM system that will not be as quickly outdated.

The most important aspect of the system design process is realism. Once the designer starts depending on advancing current technology to gain volume and to reduce cost, system performance becomes unpredictable, and no amount of repartitioning will solve the problem. Trade-offs must solve problems, not create new ones. There is a time to challenge the current technology and a time to build practical systems. The engineers must be fully cognizant of current technology, and management must assume the responsibility for making decisions that will permit the timely building of practical systems. If the system design goals cannot be achieved in a timely, cost-effective manner, management must be given the necessary facts and alternatives to permit good decisions to achieve satisfactory performance. The goal, the technology, and the resources must all be realistic.

The system design process entails many decisions that should not be made solely on technical bases, such as "make or buy" decisions about subsystems. The system design process, if properly accomplished, is a task effort that judiciously integrates the considerations related to

- customer requirements and user needs,
- current technology,
- laws of physics,
- laws of economics,
- risk, and
- economics and schedule of the design process itself.

3.3 SYSTEM QUALITY

What are the criteria for judging the quality of military electronics equipment, such as avionics, ground vehicle, or naval electronic equipment? The criteria are summarized in the following list:

- cost, affordability, and producibility,
- electrical performance,
- electrical interface compatibility,
- operational compatibility (EMI-EMC),
- packaging fit, volume, and weight,
- power consumption,
- operating temperature and heat dissipation,
- human engineering, and
- the "ilities":
  - maintainability,
  - reliability,
  - availability,
  - survivability,
  - transportability,
  - environmental vulnerability, and
  - safe usability.

The cost of an ECM system should not be too large a percentage of the asset to be protected (5–20% may be appropriate) while at the same time significantly increasing the protected asset's effective lifetime. The operational compatibility is very important for ECM systems, much more so than other types of military electronic equipment, because, since it is a jammer, the other on-board electronic systems may be inadvertently jammed.

The ratings of a military electronic system by the above list determines the quality of that system. Hence, system engineers need to rate their straw-man candidates against this list.

One of the rating items on the list is the electrical performance. What determines the performance of an ECM system? Performance is rated by the improved survivability of the asset protected by the ECM. This survivability in turn is determined by the ability to inhibit the threat from firing, or if the threat fires, performance is often quantified by such measures as miss-distance or probability-of-kill. These performance criteria are in turn generally determined by the electrical performance criteria summarized in Table 3.1.

As revealed in Table 3.1, there are five critical specifications for an ECM system, the first of which is operating bandwidth. Bandwidth is critical if the ECM system is to handle a variety of threats, each operating in its own part of the spectrum. Quite often, the next specification mentioned is RF power capability, or sometimes ERP, although we have chosen to list the effective ECM power (EEP). The system determinants of EEP are

- output amplifier, or tube, power rating,
- transmission losses,
- antenna gain,
- antenna polarization,
- signal frequency center, and
- signal bandwidth.

The last four of these determinants are with respect to a corresponding value for the radar. The conventional, antenna ERP is the product of the antenna gain in the direction of the threat and the transmitter's power drive after transmission losses. It is appropriate to consider the EEP as being the product of the transmitter power, the antenna gain, the percentage of the radiated power that is accepted by the radar's antenna polarization dependency, the percentage of the transmitter's spectral power accepted by the radar bandpass, and the percentage of the power not lost through miscellaneous transmission losses, including impedance-mismatch-related losses. By this measure, the antenna ERP and the EEP would be equal in repeater mode, except for polarization effects, because the ECM system uses the radar's own signal in repeater mode, which is approximately matched to its own receiver. Our opinion is that this use of EEP, or effective ECM power, would avoid confusion about the quality of ECM transmissions. At this time, however, the use of EEP is not a standard convention in the EW community; only the use of antenna ERP is a standard convention, and it only takes into account the total RF power, the miscellaneous transmission losses, and the antenna gain.

Table 3.1
ECM System Electrical Performance Criteria

Criteria	*Examples*
Critical Specifications	
Operating bandwidth	
Effective ECM power	
Antenna coverage	
Propagation delay	
Raw power consumption	
Power Management Parameter Control	
Modes	Repeater, transponder and noise
Parameter ranges	Dynamic range, noise bandwidth, *et cetera*
Instantaneous bandwidth	
Techniques	
Repertoire	Swept square wave, AM, RGPO, VGPO, *et cetera*
Parameter ranges	Maximum and minimum doppler, maximum and minimum range delay, *et cetera*
Programmability & flexibility	
Compatibility with other electronics	
Multiple Simultaneous Threat Handling	
Power, time, and angle sharing	Jump time between frequencies, *et cetera*
Sensor capability*	IBW, dynamic range, selectivity, *et cetera*
Signal processing capability*	Number of PRI trackers, MOPS, *et cetera*

*A means to an end.

In theory, the output tube power rating should not be considered a critical specification; as such, it is not included in Table 3.1. However, if the case is made that the DSP may not be able to cope with the signal environment to steer the antenna or tune the VCO properly, then the predominant determinant of ERP is the output tube rating. If a fixed passive antenna is used, with a given wide beamwidth, the tube rating is the only practical variable to influence ERP. For such justifications, and because of its brute force simplicity, it is quite common for the EW community to consider the raw RF-power-generating capacity a critical specification, regardless of how sophisticated the power management actions are.

To reiterate: ignoring the issue of whether the output power tube rating is critical, the EEP consists of the transmitter tube power rating, the antenna gain, the polarization match loss between the ECM and radar antennas, the percentage of power that falls in the threat radar's bandpass, and any miscellaneous losses. For example, if an ECM system operates against a 5 MHz IBW radar with a 200 W TWT transmitter, with a 10 dB gain antenna, with 1 dB of polarization loss, transmitting a centered barrage noise spectrum with bandwidth of 100 MHz (10 × the radar IBW) through 1 dB of miscellaneous losses, the EEP is 53 + 10 − 1 − 10 − 1 = 51 dBm. If the same ECM system now transmits a centered 20 MHz noise strobe spectrum (4 × the radar IBW), the effective ECM power is 53 + 10 − 1 − 6 − 1 = 55 dBm. If the same ECM system transmits a saturated repeater mode signal, the effective ECM power is 53 + 10 − 1 − 0 − 1 = 61 dBm. If the same ECM system operating in repeater mode has the antenna boresight misdirected so that the threat is really at the first sidelobe angle, with the uniform-aperture pattern of Figure 2.22, the effective ECM power is 53 + (10 − 13) − 1 − 0 − 1 = 48 dBm.

The full antenna gain is only realized if the threat is in the main beam. In other words, a high-gain antenna means a narrow beam antenna, so a high-gain antenna could result in a disastrous effective ECM power value if the beam is pointed at a slightly different angle from one aimed directly at the threat. Therefore, unless a phased array or other steerable antenna is employed, it is necessary to use a broad-beam antenna for full angular coverage, although it lowers a simplistic ERP calculation. Indeed, antenna coverage is listed as a key specification in Table 3.1.

Whereas the repeater mode automatically optimizes the transmitted spectrum bandwidth, care needs to be taken in the centering and bandwidth of the transmitted spectrum in noise or transponder modes, or else the effective ECM power will be reduced. The effective ECM power does not continue to increase as the transmitted bandwidth is reduced. Once the transmitted bandwidth approaches the radar receiver's bandwidth, no further improvement in effective ECM power is possible. For example, a properly centered 40 MHz wide noise strobe jamming a radar receiver with 10 MHz IBW can have a 3 dB EEP improvement by lowering the noise bandwidth from 40–20 MHz; however, lowering the noise bandwidth from 40 MHz to 40 kHz will not improve the effective ECM power by 30 dB. For noise jamming in particular, the quality of the noise will degrade if the transmitted bandwidth is made less than about twice the radar receiver's bandwidth. In other words, for a signal to be noisy, it cannot have as much effective ECM power as transponder or repeater modes.

The critical specifications most often mentioned in a short statement about an ECM system have just been addressed, but Table 3.1 also includes propagation delay and raw power consumption. Propagation delay is critical for repeater mode. The rule is that the propagation delay should be a small percentage of the shortest pulse-width expected; otherwise, a radar operator will see the true skin return "peek out" at the leading edge of the pulse. This rule generally requires the propagation delay to be a small percentage of a microsecond.

The next criterion for electrical performance, in Table 3.1, is power management parameter control ranges and the modes of operation. The IBW is shown separately because of its importance, although it is but one of a number of parameters. The narrower the IBW transmitted, as a percentage of the band covered, the more the effective ECM power becomes dependent on the *digital signal processor* (DSP) accuracy. Another good example would be the steering of a phased array; generally the commands to steer to a particular angle are dependent on the carrier frequency. Wide IBW, such as in repeater or transponder modes, can be important for the viability of a free running operation, although the trend to power management dependent on DSP control mitigates that somewhat. It is prudent, however, to allow a fall back to a free-running capability.

The power management parameter ranges include depth of amplitude modulation, frequency offset extremes, noise bandwidth, repeater bandwidth, noise multiplexing rates, *et cetera*. The techniques criteria include a set of parameters of a different nature. These parameters include the programmable minimum and maximum doppler ranges and range delay, the frequency sweep of an on-off square wave, *et cetera*. The distinction is that the parameters within the power management parameter control are concerned with the parameter ranges of the microwave components being controlled, whereas the parameters within the category of techniques are concerned with the range of the source of the control signals.

The effective performance of an ECM system is definitely dependent on the repertoire set of techniques available for jamming. High RF power and wide bandwidth are not going to be effective if certain particular techniques needed to counter a particular radar are not available. The system design approach advocated by this text, and which the current state of technology almost allows now, is to be able to vary every potential parameter of an RF transmission ultimately under computer software control. This will allow the system to generate the jamming signals needed for particular jamming techniques yet to be invented, possibly for use against radar systems which have not yet been deployed.

As shown in Table 3.1, the jamming techniques need to be programmable in a flexible manner. It must be kept in mind that the key system engineers, including the hardware, software, and technique engineers, may not be available in the future to reprogram the system. The software, therefore, needs to be structured with appropriate menus and other user-friendly features. For ECM systems in particular, the quality of the ECM system is quite dependent on the flexibility and quality of the controlling software. Although the trend is to self-contained ECM systems, the electrical performance is also dependent on its compatibility with other on-board electronics equipment. The two most common electrical systems that need to be interfaced with compatibly are the on-board ESM set and the status display set; the display set is for an EW officer, or for the single seat vehicle driver or aircraft pilot.

The last category in Table 3.1 is simultaneous threat handling. To a purist this category should have been included in the power management parameter control category; however, it is such a critical and distinct issue that it is listed separately from all the other system parameter ranges. An important example of multiple simultaneous threat handling performance is the frequency multiplexing between different distinct carrier frequencies, as shown in Figure 1.14. To generate noise against two threats simultaneously, as explained in the first chapter, the multiplexing rate must be sufficiently higher than the threat radar bandpass widths. The question is, when this criteria is met, will the RF VCO generating the microwave signal spend too much of its duty cycle slewing between the threat frequencies, or ringing before settling? For a phased array or other steerable antenna system, it likewise must not spend too large a duty cycle slewing from one angle to another. Actually, for a phased array, it may be more appropriate to think of the process as focusing a good beam at the new angle, rather than as slewing a beam. In any event, the multiplexing rate for steerable antennas needs to equal the highest expected average pulse reception rate from all the threats being jammed with transponder or repeater pulsed modes or gated narrowband noise mode. Normally broadband noise jamming is used with a broad beam antenna, so the steering multiplexing rate for that mode is generally not an issue. The multiplexing rate requirements are summarized in Table 3.2.

With regard to time-sharing in general, the performance is dependent on the capability of the built in DSP, with its PRI trackers. For most power-managed ECM systems, the DSP needs to track the threat signals of interest, although we recommend there always be a free-running back-up. This defines the number of PRI trackers that must operate simultaneously and harmoniously. The number of trackers, therefore, and their jitter or tracking accuracy when confronted with a complex signal environment, is an important performance criteria.

The last two items within the simultaneous threat handling category of Table 3.1, sensor and signal processing capability, are marked with an asterisk, indicating they are means to an end. These criteria do not strictly belong in Table 3.1, because the table is intended to rate only the electrical performance by characterizing the output signals. It is indeed true that a "black box" specification for a complete ECM system should not include these two criteria, only the response under various signal conditions. Such a puristic concept is not normally followed, however, because it

Table 3.2
Multiplexing Rate Requirements

Multiplexed Parameter	*Required Rate*
Frequency	Faster than the reciprocal of the threat's instantaneous bandwidth
Angle	Higher than the maximum average pulse reception rate
Time (switching between unique modulations)	Higher than the maximum average pulse reception rate or Faster than the reciprocal of the threat's instantaneous bandwidth

is much simpler to understand and to test the system performance if the sensor and processor capability are described or tested in addition to the system output. Therefore, from a practical point of view, the sensor capability is indeed a performance criteria. Examples of this would include an *instantaneous frequency measurement* (IFM) receiver's frequency resolution, the sensitivity of a CVR, and the data buffer size. The issue with regard to the DSP, is the question of whether the DSP will choke on the incoming data, or go into overload. There is a maximum average pulse rate expected, and although the DSP can buffer the incoming pulses, the average pulse rate must still be handled. Of course, large numbers of pulses belonging to signals that do not need to be tracked can simply be dumped into a special processing bin; the capability to do this successfully is vital to ensure system effectiveness but poses a difficult processing problem in itself.

Finally, it should be pointed out that every ECM system should be designed to counter the threat to the specific craft to be protected, not every radar in the general area.

3.4 MICROWAVE SYSTEM DESIGN

The above text has addressed system engineering methodology in general, the issue of determining military electronic system quality, and especially the issue of determining the quality of an ECM system's electrical performance. The system design methodology included the generation and then the evaluation of straw-man candidates. This section now describes how the microwave aspects of these straw-man candidates are so evaluated and some guidelines for eventual redesign. Before this evaluation method is described, some design rules of thumb should be noted:

- the repeater mode noise figure should be almost equal to the first amplifier (the preamplifier) noise figure, plus loss in front of that amplifier, in decibels;
- the receiver's inherent internal noise level should be swamped out by the noise from the preamplifier;
- the ripple between any two component interfaces should be much less than the desired system ripple; and
- the signal power must not approach closer than 10 decibels of the LO power in a mixer.

If confronted with the task of evaluating a microwave system block diagram, especially for an ECM system, how does one know what constitutes a good and complete evaluation? The answer is that the evaluation has been properly accomplished if one has correctly computed and checked all of these evaluation items:

- calculate signal ripple and rate (amplitude *versus* frequency) (probabilistic on large systems);
- calculate noise levels;

- calculate gain;
- calculate intermodulation levels (e.g., third-order intercept value);
- calculate propagation delay;
- check the power handling capability of every component (especially in receivers);
- check if VSWR will hurt tubes;
- examine (amplitude) dynamic range for variable attenuator extremes;
- examine drive and saturation levels for variable attenuator extremes, especially the preamplifier, and
- examine output noise level for worst case variable attenuator value.

The calculation of signal ripple and rate, that is, the amplitude variation as a function of frequency, is based on the VSWR theory presented in this and other chapters. The noise calculations are based on the noise figure theory presented in this chapter. When making the calculations for ripple and noise, the calculations are dependent on the character of the components. The components fall in several classes for calculation purposes, with identical groupings for both ripple and noise calculations. These classes are illustrated in Figure 3.14. For example, the default value for the noise figure for passive linear reciprocal components is the attenuation value.

The gain calculation is made by simply summing the gain values of the components, expressed in decibels, with negative values for those components with a

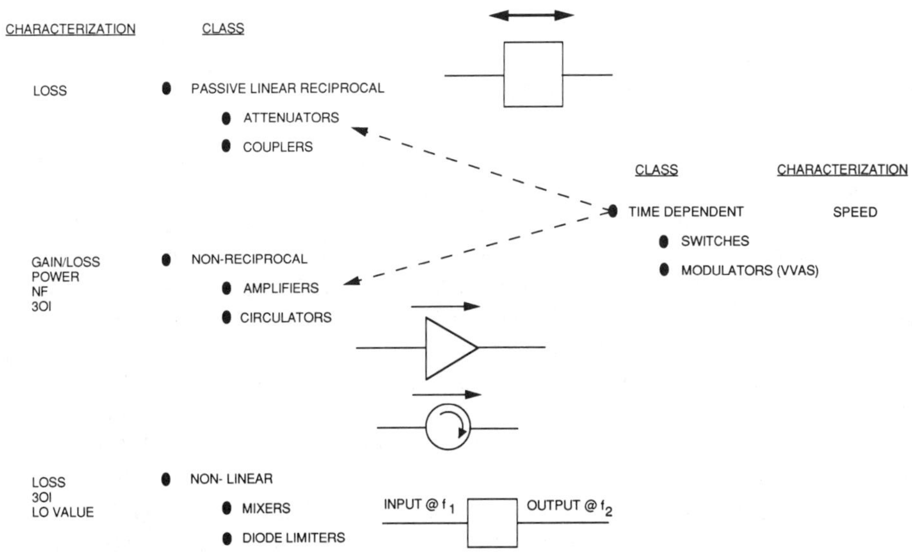

Figure 3.14 Classes of components.

loss. The intermodulation level calculations are based on using the intercept rating model, such as that illustrated in Figure 2.9.

The propagation delay is calculated by simply summing the propagation delays of the individual components and adding the propagation delay of the transmission lines. Mechanical sketches or drawings are therefore needed in addition to an electrical block diagram. The propagation delay of a non-dispersive transmission line is usually the length of the cable divided by the velocity of the EM wave in the line, typically about 70% of the free-space EM wave velocity.

The majority of the system propagation delays generally results from

- RF cable routing,
- transmission through slow-wave helix structures in TWTs, and
- RF filtering.

Most other broadband components do not contribute much delay unless there is a physically long signal path from input to output. The next iteration straw-man system design candidate can have reduced propagation delay by

- carefully locating the components and optimally orientating them,
- using solid-state amplifiers instead of TWTs,
- mounting several individual components on a common surface within one chassis, instead of each component on a separate chassis with its own set of connectors, and
- using MIC technology.

As shown in the preceding list of evaluation items, it is important to ascertain that, under all conditions, no component will burn out. One such potentially troublesome condition is when the load is inadvertently disconnected from the output power TWT. In this case, all the RF power will be reflected back, which could damage the tube. A variation on this, for a system that may seem to be operating satisfactorily, is a VSWR-caused frequency dependence. Although the tube and load may have been previously tested at maximum power, a signal at a particular frequency may result in the standing wave peak's occurring at the tube's output port and causing a tube burn-out. A sufficient safety margin must be included in the microwave design to ensure that this does not happen, or protection circuits must be included to ensure that no damage results.

Sensor components are the next most vulnerable to burn-out. It is therefore important to have one component near the front end that will limit the RF power safely, at a level which will not overstress the other components. This limiter is usually the RF preamplifier, but it may be necessary to include a diode limiter in the system design. It should be pointed out that the maximum signal strength entering the receiver antenna can be quite large under certain conditions, so the system design must be capable of withstanding RF power levels considerably above the dynamic range of the signals of interest. For example, an ECM system mounted in an aircraft could be illuminated by a nearby radar while the aircraft taxis on the runway. Therefore, the burn-out point should be at least 40 dB more than the maximum signal of interest, and possibly much more.

The typical ECM RF system design includes electronically controllable RF variable attenuators. As indicated in the preceding list of evaluation items to compute and check, the dynamic range, the drive and saturation levels, and the final output noise level all need to be checked for the minimum and maximum extremes of these variable attenuators.

Figure 3.15 shows an example of such microwave system calculations for a particular set of variable attenuator values. For illustrative purposes, this example shows the case for a system with insufficient drive for the final output tube. Note the organization of the chart, and how both the individual values and the net values are tabulated. Note how the paths from the receiving antenna and the VCO signal source are both shown. Calculations are made for both small signal and large signal conditions. The calculation for dynamic range, which is very important for repeater mode, is based on a 10 MHz IBW and a 10 dB SNR. Note how the noise figure

Sample RF System Calculation

Small Signal Gain	dB	0	-6	30	-10	20	30	0
Noise Figure	dB	x	6	6	10	6	10	x
Propagation Delay	ns	1	10	6	1	4	6	1
Saturation Rating	dBm	x	x	10	x	20	60	x
Third Order Intercept	dBm	x	x	30	x	30	75	x
Net Small Signal Gain	dB	0	-6	24	14	34	64	64
Net Noise Figure	dB	x	6	12	12	12	12	12
Output Noise Level	dBm	-114	-114	-78	-88	-68	-38	-38
Large-Signal Level	dB	-10	-16	10	0	20	50	50
Net Large Signal Gain	dB	0	-6	20	10	30	60	60
Dynamic Range	dB	94	88	78	78	78	78	78
Depth of Modulation	dB	0	0	0	60	60	60	60
Net Propagation Delay	ns	1	11	17	18	22	28	29
Spur Level	dBc	x	x	28	28	20	20	20

Note: RF switch on/off ratio equals 60 dB.

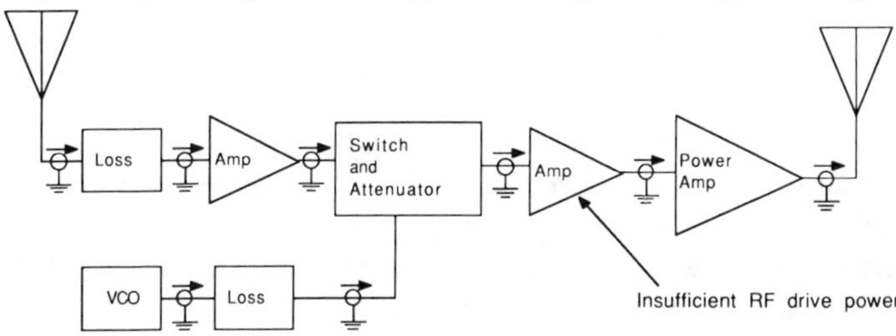

Small Signal Gain	dB	x	-3	-10	20	30
Noise Figure	dB	x	3	10	6	10
Net Small Signal Gain	dB	x	-3	-13	7	37
Net Noise Figure	dB	x	3	13	19	19
Output Noise Level	dBm	-114	-114	-114	-88	-57
Large-Signal Level	dBm	20	17	7	20	50
Net Large Signal Gain	dB	x	-3	-13	0	30
Depth of Modulation	dB	124	121	111	98	97

Figure 3.15 Sample RF system calculations.

does not change after the preamplifier. This text provides the reader with sufficient background to make similar calculations.

The key parameters needed for each component to make system calculations like those illustrated in Figure 3.15 are

• noise figure	dB
• small-signal gain (SSG)	dB
• large-signal gain (LSG)	dB
• large signal output power	dBm
• propagation delay	ns
• third- or second-order intercept	dBm.

The large signal output power and gain are usually defined for amplifiers and limiters as the 1 dB compression point power, which means that the LSG is 1 dB less than the SSG. If the RF system designer intends to drive the amplifier harder, then more appropriate values should be used in the calculations. It is good design practice to ensure that only amplifiers and limiters go into saturation. In this regard, it is important to check that the large signal power into each mixer does not exceed 10 dB below the LO power of that mixer.

The running or net parameters to be calculated are illustrated in Figure 3.15. These should include

• net SSG	dB
• net noise figure	dB
• output noise level	dBm/MHz
• power	dBm
• net LSG	dB
• dynamic range	dB
• depth of modulation	dB
• net propagation delay	ns
• spurious levels	dBm.

For the large-signal RF power calculations, start with the largest expected input signal level and compare the LSG = SSG determined output value with the power rating of the component at each point in the system. If the power rating is exceeded, then it is necessary to adjust the LSG value appropriately. Following the above guidelines of doing these calculations at the extremes of the electronically controllable variable attenuators may well result in the discovery that at some point in the system there is insufficient drive for the next stage to reach saturation.

To find the system dynamic range, the difference in decibels is calculated between the large signal output power and the output noise level, within an appropriate bandwidth, for no input signal. However, there is no universally accepted convention for the thermal noise-limiting bottom end of the dynamic range; some may choose to define the bottom end of the dynamic range differently. In any case this dynamic range is dependent on the chosen bandwidth. It is most appropriate that this bandwidth equal either the widest threat bandwidth expected or the average bandwidth of all the threats expected. A value of 10 MHz can be used as representative of non-coherent threats to determine the dynamic range. Since coherent threats have much narrower bandwidths, the calculation for them can probably be ignored, since the resulting dynamic range will be much better. The depth of modulation calculation gives the modulation result a radar would detect if the on-off microwave switch shown in the diagram had been toggled. This depth of modulation may not equal the dynamic range. The VCO signal depth of modulation shown in the figure is based on the VCO being directly gated on and off—not a common practice. For a good quality ECM system, the depth of modulation should exceed the sidelobe depth of the threat radar's antenna; hence, depth of modulation values in excess of 40 dB are recommended for good performance. The impact of dynamic range is somewhat more subtle, but values in excess of 40 dB are desirable in order to mimic normal amplitude changes as a function of range.

The net propagation delay is simply the net running total of the component delay values, just like the net SSG, except that the units are nanoseconds. Although not shown in Figure 3.15, for simplicity, the losses and propagation delays in the transmission lines between components should also be included in the calculations. The description of transmission lines in this text gives instructions for doing this. It is advisable to include a 0.05–0.1 dB allowance for loss per connector, depending on the band. Also shown in Figure 3.15 is the third-order intercept point, which defines the spur value for two equal power signals. The *third-order intercept* (3OI) value specified for each component is used to calculate the spur level or, alternatively, to calculate the worst 3OI up through each component's output. The 3OI at each point in the block diagram should be equal to or less than that component's 3OI rating. Finally, it is important to note that only the repeater path needs a propagation delay calculation. The delay through other paths for other modes is usually not important, except for the minimum RGPO delay of transponder mode.

With the preceding qualifications, the calculations typified by Figure 3.15 include time dependence to the extent that depth of modulation was calculated. To take into account the time dependence of Figure 3.15, the RF values are calculated for the different conditions that occur at different times. The range of time scales is so great that this method of "frozen time" calculation is quite appropriate. In particular, the preceding text has emphasized making calculations similar to those in Figure 3.15 for the extreme values of the variable attenuators. Hence, the variable attenuator is simply an attenuator pad for the purposes of such calculations; that is, there are no explicitly dynamic aspects to the calculations.

As part of the RF design process, the RF components need to be specified. There are certain parameters that are often inadvertently left out of such specifications. Here is a list of component specifications that must be checked:

- both minima and maxima,
- ripple (amplitude variation across band),
- unit to unit variation, and
- propagation delay.

As an example, the minimum output power of an RF amplifier, but not always the maximum power, is specified; we recommend that both be specified. The VSWR minimum is often not specified; this may be one case where it is inappropriate to specify both maximum and minimum, in order not to unduly burden the vendor, but such a decision should be conscious and explicit, and not occur by default. It should be noted that the above list includes both ripple and unit-to-unit variation; the insertion loss or gain is usually the most important parameter where unit-to-unit variation needs to be quantified. For example, a component could be specified to have a maximum 3 dB ripple, no specified minimum ripple, and 2 dB unit-to-unit variation of the nominal, or statistically determined average loss.

The propagation delay is often left out of component specifications; we recommend that the minimum and maximum both be specified if the maximum is greater than 2 ns.

In addition to the above specifications to check, there is one parameter that is always specified but not often properly defined: the receiver sensitivity. If the specification requires, for instance, a receiver with a −60 dBm sensitivity, the result may be that, although the receiver generally does indeed give an output trigger at −60 dBm input, there are too many false alarms. The moral is that MDS sensitivity is a

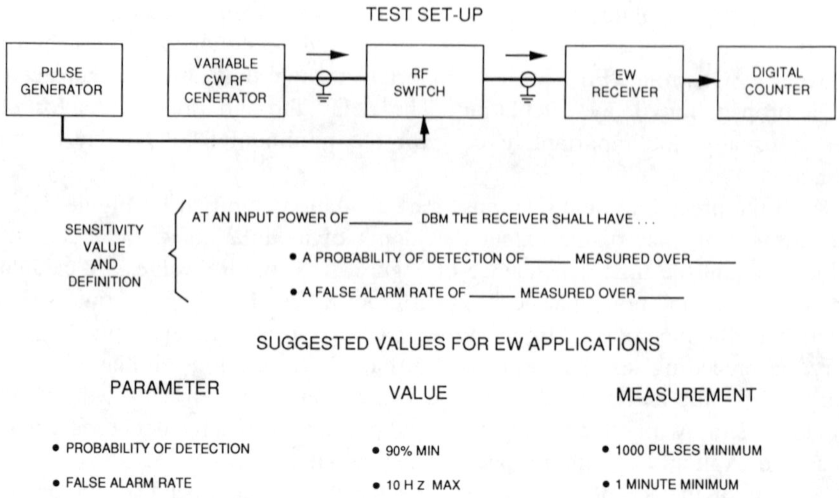

Figure 3.16 Specifying receiver sensitivity.

combination of probability of detection and probability of false alarm. Figure 3.16 illustrates how to specify sensitivity, and gives suggested parameters.

3.5 VIDEO AND DIGITAL DESIGN

One of the most important system design tasks is to partition the system properly. Doing this

- minimizes mutual interference,
- allows specialized engineers to be assigned clear responsibility,
- ensures minimized interface complexity,
- provides for a logical hierarchical testing and debug sequence,
- aids in mechanical packaging, and
- aids in modularity-based maintenance.

In most modern ECM systems with extensive digital signal processing capability, it is common and good practice to minimize video transmission length, unless shielding steps are taken. Therefore, a good rule for a partitioning design that minimizes the potential for mutual interference is to enclose video circuits in such a way that the enclosed unit's interfaces are RF, digital, and power only. For example, a CVR would include the crystal, the video or difference amplifier, and the *analog-to-digital converter* (ADC) within a shielded unit. In other words, the partitioning will group an appropriate set of components into a unit so that a specialized engineer can be responsible for the unique properties and interfaces of the components, and for shielding the whole unit. All circuitry is subject to interference problems, but digital circuitry is somewhat resistant to interference because of the binary nature of the signals. For long runs on long busses, digital signals should have line drivers and matched receivers. The RF microwave circuits are also resistant to interference because of the very nature of the circuitry; microwave circuits need to be well shielded to be viable in the first place. Therefore, the video circuitry that buffers the RF and digital is the most sensitive EMI weak link in the system structure, and the partitioning needs to take that into account. Among all these links, the receiver detection and the signal source drive are the most sensitive points in terms of impedance matching, interference, and rise time.

In addition to this guideline regarding video shielding, what are the overall governing principles regarding system partitioning? We recommend a set of system-partitioning principles based on time scale analysis. The range of time scales, and hence operating frequencies, used in an ECM system was illustrated in Figure 1.13. Time scale analysis leads to these partitioning principles:

- high level algorithms are performed only in the CPU-DSP;
- high-speed quasi-redundant functions are performed in specialized hardware elements;

- hardware-element partitioning is based on the association with an analog process;
- hardwired circuitry should have as few assumptions built into it as possible; and
- data flow is minimized at the interfaces.

Four examples of applying these principles to specific instances are given in the following four paragraphs:

(1) Sensed data caused by short noncoherent pulses is reduced within the sensor subsystem to a stream of digital words passed on to the system data bus. The digital-word reduction includes components, or fields, describing nominal amplitude, carrier frequency, pulse width, time of arrival, *et cetera*.

(2) The tuning of a narrow-band, swept superheterodyne receiver is done directly by the CPU, via a DAC included in the superheterodyne unit, from system data bus words.

(3) Noise generation from a VCO signal source is generated by free-running high-speed specialized circuits included in the signal source unit, once the parameters have been set up by the CPU, with DACs included for system data bus interface.

(4) Amplitude modulation intended for angle deception, generally under 1 kHz, is created directly by the CPU, applied to the modulator through a DAC included in the modulator unit, via the system data bus.

Notice how, by applying the above principles, examples 2 and 4 show direct control by the CPU, while example 3 reveals indirect control necessitated by the required speed of the circuitry. Notice also, in example 1, that the information regarding the entire pulse was reduced to one word; the reduction did not occur outside the receiver subsystem.

Just as RF microwave has VSWR impedance considerations, and video has impedance considerations, so too the interfaces on digital data busses need to be considered and designed for best operation. Here are some considerations that need to be addressed:

- bus size,
- clock rate,
- logic family,
- tri-state capability,
- impedance and load,
- transmission length,
- noise immunity,
- differential drivers and receivers,
- bus format:
  - data word size,
  - address,
  - clock,

- strobe,
- miscellaneous commands,
- interrupt capability, and
- data priority and collision avoidance.

As part of the system design process, a document is usually generated that defines the word formats on the data bus, specifies the addresses of all receivers, and specifies the protocol for control of the bus for information transfer. For a large ECM system, a competent system engineer needs to be assigned this duty, with responsibility for the generation and revision of the bus description document. This person needs to be a system engineer because the definition process requires meeting with all the specialized engineers responsible for the individual elements. Furthermore, such discussions will inevitably reveal misunderstandings and discrepancies regarding the transferral of needed information. These problems need to be resolved from a system perspective. The definition of the interfaces, including data bus information flow, is a key system engineering function.

Once the system has been partitioned, and once the digital interfaces have been defined, what other aspects of the digital design should be considered? Following is a list of considerations that can be used to evaluate the straw-man structure, or which could be used for potential improvements:

- placing special purpose hardware functions under CPU control;
- clock rate operations per second;
- memory size and speed;
- memory sharing;
- memory buffering;
- RAM *versus* nonvolatile RAM *versus* ROM;
- overflow, overload, priority, and conflict resolution;
- built-in test;
- interrupt capability;
- use of specialized hardware elements *versus* software-driven general-purpose processing;
- pipeline *versus* parallel *versus* distributed processing; and
- software-derived considerations.

The design of such digital structures often requires that the specialized digital hardware elements have their parameters and processing paths set by more general CPU units. This is consistent with the partitioning principles cited previously.

The software requirements need to be an important driver for the digital hardware design. The considerations for the software design itself should include

- language,
- architecture,
- modularity, including subroutines and macros,

- maintainability,
- flexibility,
- documentation,
- interface:
  - menu driven,
  - graphics,
  - queuing, and
  - interrupt alarms.

Finally, the partitioning and use of the memory need to be considered. The specialized hardware elements often have buffers for their data. The memory-use categories for the main CPU-DSP general shared memory are shown in Table 3.3.

The philosophy should be that modifications to the changeable memory, such as threat parameters, should have no effect on the program residing in memory.

Table 3.3
Memory Categories

Stable	*Changeable*	*Operating*
Program	Threat parameters	Calculated values
Constants	Technique descriptions	Sensor acquired values
	Prioritization	

3.6 SUMMARY

This chapter has reviewed certain specialized knowledge needed for ECM system engineering. A structured system engineering methodology was then described. Criteria for evaluating the quality of straw-man designs were presented. Ways to evaluate and improve microwave, video, and digital elements of the design were presented.

In addition to the structured system design methodology, it was emphasized that great attention should be placed on partitioning, and the partitioning methodology should be based on time scale analysis. Each such partitioned element should have its function defined, and the interfaces should be thoroughly and carefully specified as a key part of the system engineering process.

Is system engineering a science or an art? The answer is that the evaluation of the straw-man candidates is a science, while their generation is an art. When the system design process is started, it is usually impossible to assign parameters to the various components and subsystems. The initiation of the system design can therefore prove frustrating; before certain calculations can be made, certain parameter values are needed, which are in turn dependent on other calculations that need their own driving parameter values, and so on, often back to needing values from the first calculation. To break this interdependence, the straw-man candidate needs to be formulated based on a number of assumptions. The calculations and evaluations will then reveal if the assumptions really made sense. The system designer then uses this experience to revise the component and subsystem parameters, and, if necessary, to restructure the system. The process is then repeated, to iteratively approach the optimum design. Because there are always an infinite number of possible variations, the generation of the candidates must be considered an art.

Chapter 4
EW Receivers and Sensors

4.1 INTRODUCTION

This chapter is organized first, to review EW receiver requirements, then, to describe the simpler or older sensor and receiver design approaches, then, to discuss bandwidth issues and tradeoffs in the context of the more advanced receiver requirements for modern ECM and ESM systems, and finally, to describe more recent EW receiver designs.

Some receivers are tailored for specialized signals or conditions. More commonly, the receiver subsystem must handle a wide range of signals. The general receiver requirements are based on

- EM wave signal characteristics:
  - threat-radar signals,
  - friendly signals,
  - commercial and other signals,
  - signals from the asset being protected, and
  - jamming signals;
- geometric characteristics:
  - transmit antenna patterns,
  - receiving antenna patterns,
  - distances,
  - propagation factors such as atmospheric loss,
  - background clutter,
  - multipath, and
  - horizon and line-of-sight issues;
- output uses:
  - trigger intrapulse or leading edge functions,
  - signal-processing functions, and
  - jamming optimization or set-on functions.

As shown, the EW receiver requirements for ESM and ECM systems can be partly defined by knowledge of the threat-system's radar. This knowledge is sometimes obtained from the actual radar or its operating manual. More commonly, the information is derived from documents that describe intercepted radar signals or from documents based on other sources; or it is inferred from the study of known radars, by documents, measurements and examination, that are purported to be similar; or it is deduced by analyzing the radar's potential characteristics based on the radar's missions and constraints. The ECM system designer may be working for a customer who defines only these signals or else the complete receiver requirements. In actual practice, the threat signal–based EW receiver requirements are defined by a judicious combination of the above. This means that if only selected parameters are available from documents that define the threat system's radar, the analysis determination method would therefore include reverse engineering. This reverse engineering should definitely include such considerations of the threat weapon system that the radar supports as weapon range, portability, *et cetera*. We strongly recommend that the requirements derived from this determination process be based on all the signals that are likely to be intercepted, not just the threat-radar signals. The determination of some parameters of the receiver requirements should also be closely coupled to the determination of the input receiving antenna design and the output data uses. All the following quantitative values derive from generic fundamental considerations.

The EM wave signal and geometric characteristics delineated previously can be combined to characterize the input by defining or describing the following:

- individual intercepted threat signals,
- threat-signal density *versus* frequency, AOA (angle of arrival), *et cetera,*
- self-jamming and self-blanking,
- front-end noise and other interfering signals, and
- external jamming.

Common practice for the EW community has been to pay close attention to the first three considerations; we suggest that all the above need to be carefully considered when defining the receiver's functional requirements.

An ELINT system must accept signals in a general-purpose manner; in fact, the goal is to detect new signals or new characteristics of signals where, however, the burden of real-time processing is relaxed. Electronic warfare support measures (ESM) and ECM receivers have similar input signal requirements but differing output data requirements. An ECM receiver may actually need more sensitivity than an ESM receiver if the operation of the complete ECM system is dependent on tracking the threat signal in the sidelobes. Some ESM systems, however, may also need to detect the sidelobe signals, in order, for example, to highlight a potentially dangerous flight path to a pilot instead of just responding to a present threat.

The part of the requirements determination process based on individual signal characteristics usually starts by classifying signal types, e.g., search, track, *et cetera*. The signals can be classified as coherent or non-coherent, and as having (1) fixed carrier frequency, (2) spread spectrum, or (3) agile carrier frequency. Spread spectrum signals, in turn, are usually classified as chirp or phase coded. The signals can also be classified by duty cycle, which is a key driving factor for EW receiver design. The duty cycles can be classified as CW, high, medium, or low; a low duty cycle is typically on the order of 0.1% (e.g., 1 μs pulses every millisecond, with the millisecond PRI determined partly by horizon limitations). Usually, most of the cost and design expertise is directed toward solving problems with low duty-cycle threat signals, and this chapter describes receiver issues for such signals except where explicitly stated otherwise. The reason for this is that acquisition time and *high probability of intercept receiver* (HPIR) requirements are not as difficult to achieve if the duty cycle is not low. A well managed, tuned, narrow IBW receiver, such as a superheterodyne, can pass on the necessary acquisition, measurement, and update data with sufficient speed and accuracy if the signal duty cycle is adequate. The radar's range capability is determined by the average EM wave power on its intended target; different radar types achieve this average with different duty cycles, but the consequence is a disparate technical and economic burden on EW receivers.

For purposes of this text, the receiver requirements are partitioned based on (1) intercepted signal characteristics and system insertion issues, (2) reception bandwidth and timing issues, and (3) output data needs and output interface compatibility. The output data interfaces will be described in descriptions of digital signal processing in Chapter 6, but some of the functional output data uses will be reviewed later in this chapter. The bandwidth and timing issues will also be described later in this chapter, after first reviewing the simpler or older sensor and receiver design approaches. Selected other requirements are given in Table 4.1.

The electrical parameters in Table 4.1 are interrelated. For example, the minimum pulsewidth for radar detection efficiency is determined by the size of the asset, so that all the reflection points can combine into one net signal, thereby optimizing the radar's sensitivity and, hence, detection and tracking distance, or range. For example, a 50 ft asset would result in about a 100 ns reflected pulse rise time, so by this design constraint the radar pulsewidth should be several hundred nanoseconds. The radar designer actually chooses the pulsewidth to provide adequate pulse energy and average power, adjusted for the requirement for resolution. Tracking radars need higher resolution, so they use shorter pulsewidths than search radars. If there is excess energy available for the application, the radar designer may choose to use a shorter pulse for the sake of resolution, or because he or she needs efficiency only on the smaller targets and can afford to waste energy on large targets.

With regard to the first two items in Table 4.1, the MDS sensitivity and its adjustment, a very sensitive EW receiver may overly burden the DSP. Generally, the most important threat is relatively close, and most radars do not adjust their power levels, so it is appropriate to raise the threshold to unburden the DSP and facilitate rapid acquisition. As stated, a number of the electrical receiver requirements in Table 4.1 hinge on whether or not the optimum ECM operation needs to track the radar signal in its sidelobes. For example, if optimum ECM operation needs

Table 4.1
Selected Receiver Requirements

Parameter	*Typical Value*	*Fundamentally Determined by*	*Other Determination*
MDS Sensitivity	−50 dBm	(1) Distance (2) Cross section of asset	Weakest main-beam signal, or weakest sidelobe signal to maintain track
Sensitivity Adjustment	20 dB	(1) Threat geometry (2) Need to take early actions	Need to thin signal environment
Dynamic Range	25 dB	(1) Distance changes (2) Differences in ERP	Main-beam to sidelobe level
Pulsewidth	0.5 μs	(1) Size of asset (2) Resolution needed	Radar function and type
AOA Resolution	10°	(1) Geometry (2) Jammer ERP needs	(1) Signal separation (2) Other DSP functions (3) Transmitter beamwidth
Output Interface	50 b/pulse	Jamming use	(1) Signal separation (2) Signal ID (3) Jammer set-on values
Cost	10–40% of ECM	(1) Value of asset (2) Degree operated in harm's way	Complexity needed for sensor function
Size and Weight	10–40% of ECM	Nature of asset	Proper balance with other major subsystems

TOA predictions to be made, the TOA predictor tracker in turn may need some time to settle accurately, and if only the main beam is utilized, settling time and stability could be insufficient. Tracking may also be required in the sidelobes to facilitate reliable signal separation, that is, to avoid having constantly to reacquire. Finally, sidelobe tracking may be required so that large-angle-error jamming may be achieved, as shown in Figure 1.2. The requirement for AOA resolution can be based on having a small percentage of the azimuth angular scan to separate the incident signals distinctly, and on allowing a high directivity transmitting antenna to be pointed at the threat; the AOA measurement resolution may therefore be determined by the transmitting directivity. The requirements regarding size and weight are quite strict for fighter aircraft but are generally more relaxed for other aircraft, naval applications, and ground applications, in that order.

The sensors and receivers available for ECM and ESM systems include

- CVR,
- logarithmic-video amplifier (LVA) receiver,

- channelized receiver or SAW or filter bank receiver,
- DIFM or delay-line phase-comparison receiver,
- amplitude comparison AOA sensor,
- phase comparison AOA sensor,
- YIG-tuned receiver,
- Superheterodyne receiver,
- acousto-optic (AO) receiver or Bragg cell receiver, and
- Microscan receiver or compressive receiver.

Each of these will be described in the sections to follow.

4.2 CRYSTAL DETECTOR RECEIVERS

Two types of broadband crystal detector receivers are shown in Figure 4.1: the threshold detect CVR in Figure 4.1a, and the *log video amplifier* (LVA) receiver, in Figure 4.1b. Both receivers rely on the crystal detector, a widely used sensor for EW applications because of its inherent, broad RF bandwidth and fast video output

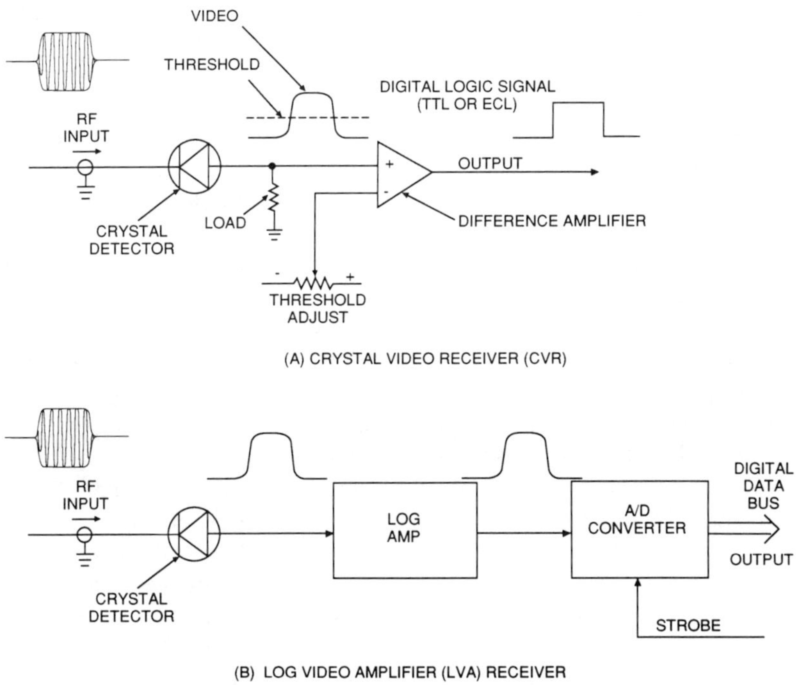

Figure 4.1 Crystal detector receivers.

response. A video signal is an RF signal after envelope or phase detection. The crystal detector is an RF diode which essentially rectifies the incoming RF ac, converting it to a video single-polarity voltage level dependent only on the amplitude, not the frequency or phase, of the RF input signal. It is shown symbolically in the figure as a diode enclosed in a circle. The crystal detector's input is an RF signal, while its output is a video voltage.

The threshold detector CVR has been widely employed because of its relative simplicity and low cost. The function of this receiver is to give a logic signal output indicating the presence of an RF pulse. Sensitivities can approach −50 dBm. Usually the RF input line is filtered to restrict the detector input to a particular range of RF carrier frequencies; this filtering is not shown in the figure. These receivers are intended to trigger real-time digital-logic functions in response to an RF pulse, and to register, or indicate, the TOA of each pulse, although they can also be used to detect CW signals. An example of a real-time hard-wired operation would be the pulsing on of a low duty cycle TWT power RF amplifier. The TOA resolution is quite good; the time value can be registered to a small fraction of the typical radar pulsewidth. This is a common choice for the sensor that feeds the system's digital signal-processor when it performs the functions of PRI-based signal sorting and signal tracking, with or without TOA predictions. Because of their low cost and simplicity, an ECM system may have several of these threshold-detector CVRs, one for each band, and one for each channel; Chapter 1 described the filter multiplexing architecture that steers each band's signals into channels, and Chapter 2 described such filter hardware.

In response to an RF input signal, the crystal generates a video voltage level across the resistive load; this video output voltage level depends on the RF input power. The proper selection of the resistive load is important; if the resistance is large, the CVR will be more sensitive, in that larger voltage levels will result, but the time delay to logic detection will be increased because of a slower rise time on the crystal output; on the other hand, if the resistance is small, the video voltage output level will be attenuated, reducing the sensitivity but shortening the rise time and hence improving the detection response delay. Different types of crystals need different loads, with values generally ranging from several hundred to several thousand ohms. The properly loaded crystal feeds a *difference amplifier*. The difference amplifier is an integrated circuit whose function is to output a logic "1" or logic "0," depending on which of two inputs is more positive than the other. The specific difference amplifier output voltages for these logic states depend on the logic family convention (TTL, ECL, *et cetera*) at the interface. In Figure 4.1a, one of the differential amplifier inputs is from an adjustable reference voltage, and this dc value will be the video threshold on the other differential amplifier input. Hence the MDS sensitivity of the receiver is set by this adjustable reference dc voltage. Difference amplifiers are designed to switch rapidly to the other logic state once the threshold voltage is crossed.

As stated, these receivers are designed to have a fast response. This is achieved by making their video bandwidths and their RF bandwidths wide with respect to the reciprocal of the typical radar input pulsewidth. The real-time operations of the ECM system can then be accomplished near the leading edge of the RF input pulse, independent of the carrier frequency. The key electrical advantages of a CVR receiver, its wide RF bandwidth and its rapid detection output, and these are the cause of the CVR receiver's chief disadvantage: lower sensitivity. It is a common practice to refer to the ECM system's detection sensitivity at the antenna port, because with gain elements and filtering distributed throughout the system, an absolute amplitude level is not meaningful. As explained in Chapter 3, the MDS sensitivity of a system using a CVR, when the RF preamplifier gain is relatively large, is principally determined by five factors: the front-end noise figure, the RF bandwidth, the video bandwidth, the acceptable false alarm rate, and the acceptable probability of detection. The wide bandwidth, therefore, results in lower detection sensitivity than, for example, a relatively narrow-band superheterodyne receiver. This is such a fundamental limitation that improvements in broadband CVR (design) technology will have minor impact on the sensitivity.

The function of the LVA receiver, as illustrated in Figure 4.1b, is to measure the input RF amplitude over a wide dynamic range in a convenient form. Specifically, the desired transfer function of the LVA is to output a video voltage level proportional to the log of the input RF power, independent of the frequency of the carrier. For low duty cycle signals, the complete LVA receiver outputs a digital word for each incident pulse, with the digital code representing the nominal amplitude of the pulse. Some of the reasons for measuring the amplitude of the signal, in an ECM system, are to separate signals based on their amplitude, to respond only to radar threats that are close and that therefore have strong amplitudes, and to measure the scan or lobing pattern and rate. Measurement of the scan or lobing pattern allows the use of ECM techniques such as inverse gain, where the ECM system transmits more power into the threat radar's sidelobe, and less or no power into the threat radar's main beam.

The basic structure of a typical LVA receiver, shown in Figure 4.1b, includes a crystal detector, a video logarithmic amplifier and an ADC. The video output from the crystal detector feeds the logarithmic voltage amplifier. Figure 4.2 shows the transfer function of a typical crystal detector, relating the input RF power to the video voltage output. The crystal output has two basic regions of operation: the square-law region, and the linear region, for which the crystal's video output voltage is proportional to the RF input power and voltage respectively.

The log-video amplifier portion of the LVA receiver has a transfer function that gives a linear output voltage *versus* the log of the input voltage. This accommodates the square law dynamic range response of the crystal. However, larger input signal amplitudes will fall in the linear region of the crystal response, requiring a

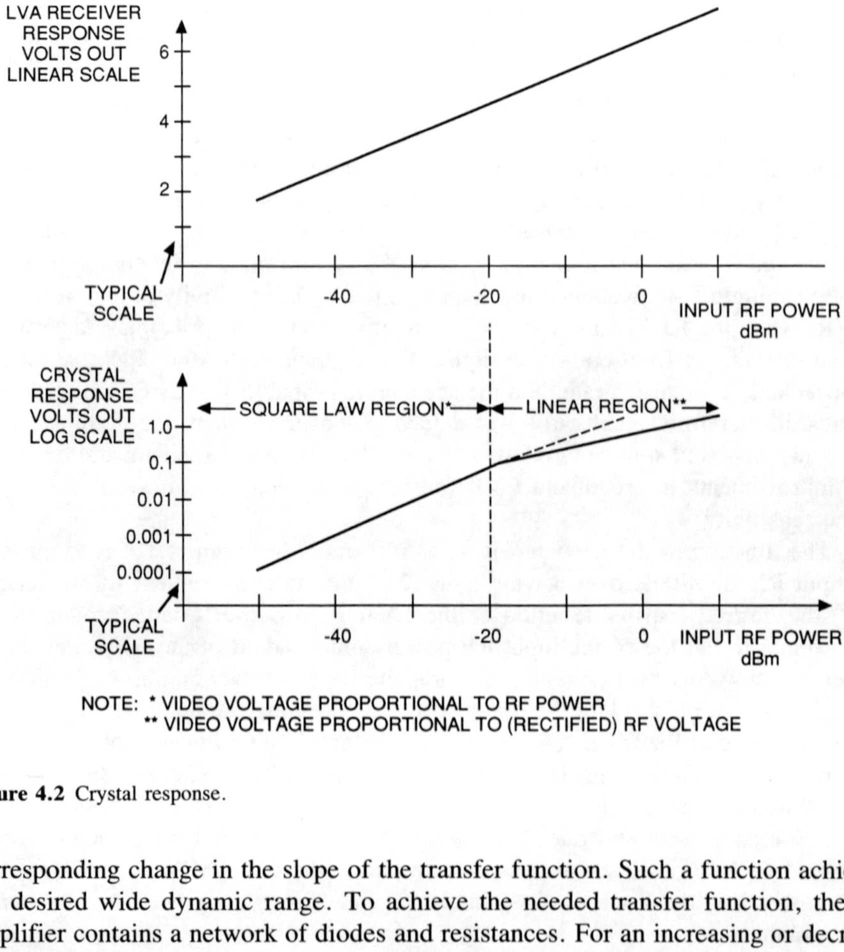

Figure 4.2 Crystal response.

corresponding change in the slope of the transfer function. Such a function achieves the desired wide dynamic range. To achieve the needed transfer function, the log amplifier contains a network of diodes and resistances. For an increasing or decreasing crystal detector video voltage, the transition of each diode from the nonconducting state to the conducting state, or vice versa, changes the impedance and thereby "linearizes" the transfer function of the LVA receiver. Unfortunately, the diode-resistance network can contain considerable stray capacitance and other characteristics that slow the rise time. In other words, the linearizing network tends to reduce the video bandwidth. Therefore, the LVA receiver cannot make as quick a measurement on the leading edge as the CVR can make a logic detection at the leading edge. Fabrication of fast LVAs has been difficult, although good progress has been made. LVAs suitable for measuring the nominal amplitude of pulses, with typical radar pulsewidths, can now operate satisfactorily.

As shown in Figure 4.1, the output of the logarithmic video amplifier feeds an ADC. In modern ECM systems, almost all non-RF functions are performed with digital processing, so the sensor's complete "black box" must include this conversion to digital code words. ADCs require a strobe, unless they convert continually. This strobe signal line is shown as an input to the LVA receiver in Figure 4.1b. The strobe may come from a fast CVR, and is therefore another use for the CVR logic-detect output pulse.

4.3 CHANNELIZED RECEIVERS

The CVR and the LVA receivers exhibit multi-GHz wide RF IBW performance, with sensitivity commensurate with such IBW, and with, at least for the CVR, video bandwidth sufficiently wide to provide an accurate high-resolution TOA trigger. The problem is that wide IBW performance does incur some disadvantages, namely lack of frequency selectivity. In other words, the CVR and LVA receiver outputs can be corrupted by signals that overlap in time. The channelized receiver of Figure 4.3 provides the combined functional benefits of wide IBW, frequency selectivity and coarse frequency measurement. In other words, the function of the channelized receiver is to make coarse frequency measurements even when signals overlap in its reception band. The carrier frequency is one of the most important measured parameters the DSP uses to separate and track signals. The carrier frequency can be used with or without the TOA, and of the two, the carrier frequency is usually preferred because it is a monopulse separation parameter, unless the threat radar exhibits frequency agility.

The channelized receiver is essentially a bank of CVRs. This approach requires that each individual CVR not be too expensive. As shown in Figure 4.3, each "channel" of the channelized receiver contains a crystal detector and difference amplifier, just as in Figure 4.1a. All of the difference amplifier outputs pass together through an OR gate to provide a detector logic output that will give a positive indication that there is a pulse in at least one of the channels. Each difference amplifier also feeds a flip-flop that latches the pulse-presence information for that channel. A *multiplexer* (MUX) interrogates each such flip-flop, resetting the flip-flop latch to effectively make a destructive read. Usually the MUX interrogates each flip-flop in turn, thereby linearly scanning the frequency band, although power management principles applied to the control of the MUX could result in revisiting a given flip-flop more often than some of the others. The output data from the MUX is buffered and placed on the system data bus, often a high speed data bus because of the expected high data flow from this sensor. The strategy for efficiently buffering this information may be simply to list the address, that is the channel number, only of the channels that were latched to a logic "1" since the previous destructive read. The channel number is a coarse measure of the frequency, so a set of frequency digital-code words are effectively passed onto the bus each time the buffer's data is down-loaded to the bus. The buffer needs a queue depth sufficient to hold the expected maximum reasonable

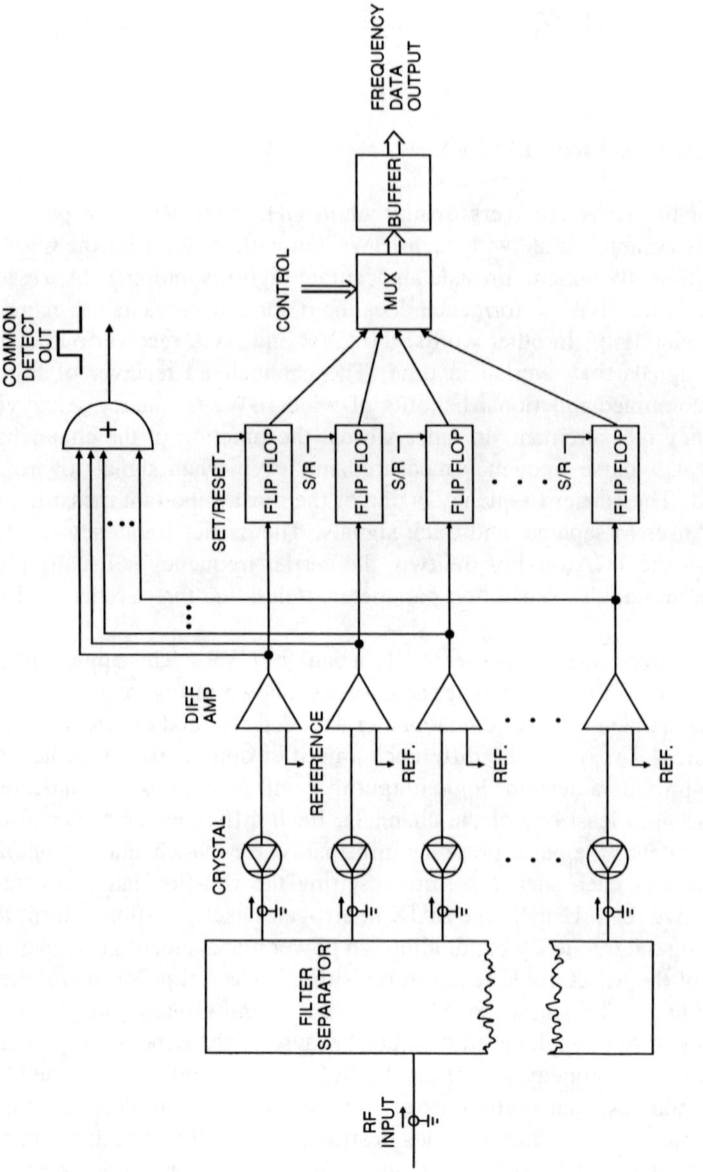

Figure 4.3 Channelized receiver.

number of detections within the time taken to interrogate all the channels. A convenient method is to use a *first in–first out* (FIFO) digital memory in the buffer, with each address (channel having a detection) tagged with the TOA.

The key component in the channelized receiver is the channelizer filter, the function of which is to separate RF signals, entering on a common line, by their carrier frequencies. The channelizer filter frequency-domain response is shown in Figure 4.4. The intent is that the RF power exits from one and only one specific channel filter for a given carrier frequency. However, realizable filters do not have an infinitely steep skirt slope, so, near the crossover point, significant percentages of power will emerge from both channels. It is common to design the channelizer filters for a crossover at the 3 dB points, that is, to match the 3 dB insertion loss points of adjacent channels. Having the crossover point be close to the 3 dB insertion point is more important when the channel separator does double duty: separation for detection purposes, shown in Figure 4.3, and separation to apply distinctive frequency-dependent modulation as a repeater-mode signal to be fed to the transmitter, shown in Figure 1.19. In any case, the important need for an ECM system to have continuous coverage, that is, no gaps in frequency, requires that the crossover point not be too deep down the skirt. The implication of this requirement is the potential for serious interaction between the channels. This is because the input signal power must be shared by the different channels, so the degree to which one channel accepts the signal power influences how much power is available for another channel. There will, therefore, be an interaction between adjacent channels in the shallow portion of the skirt roll-off, as also explained in a previous chapter, so these filters are usually made with an alternating channel structure: one filter separator set has

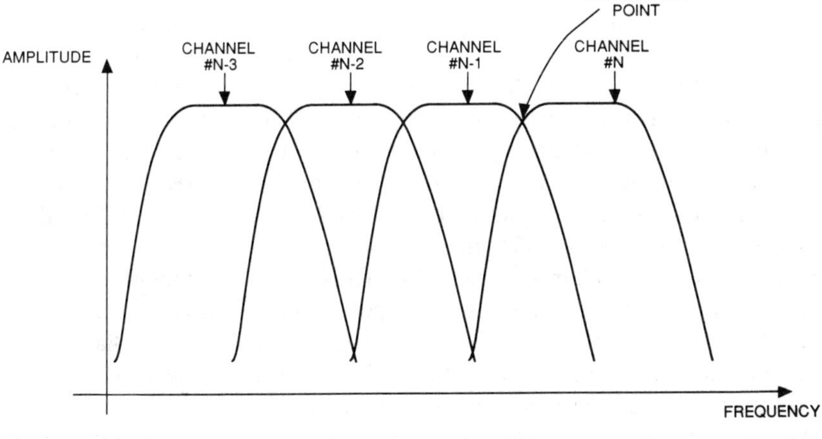

Figure 4.4 Channelizer filtering.

the odd-numbered channels and the other the even-numbered channels. This alternating-channel structure considerably reduces the interaction between nearest-neighbor channels within each set, because the crossover region is then no longer in the shallow portion of each skirt roll-off; that is, the channels are isolated and will not share significant power levels near the crossover.

The technologies considered for the channelization filtering include

- surface acoustic wave,
- magnetostatic wave, and
- lumped element.

The first two of these channel-filtering technologies achieve their filtering function in other media, that is, not as EM waves, and hence require transducers to launch their other-medium waves and to sense this signal. The SAW receiver achieves its filtering by the path length reverberations of an acoustic, or sound, wave, for example. That is, the acoustic power is dissipated unless the doubly, or round trip, reflected acoustic wave is in phase with the original acoustic wave, and such condition only occurs at a certain frequency.

The channel widths of an EW channelized receiver usually are in the range of 5–50 MHz, a choice that is driven by requirements related to

- the typical radar pulsewidth,
- MDS sensitivity,
- frequency selectivity,
- cost, and
- size.

The last three of these are usually the driving factors in making the choice, although the rule that the channel IBW cannot be narrower than the reciprocal of the shortest pulse expected transcends cost and size considerations. The requirement for frequency selectivity derives from estimates of signal density and relative power levels; that is, selectivity derives from the need to make measurements that are not corrupted by signals simultaneously present in the full band. The number of channels can be from twenty to many hundreds. There really is no rule of thumb for the number of channels in such a receiver; the functional objective is to have a large number of channels, but that objective is limited by cost and size constraints. The IBW of the complete channelized receiver is, therefore, many times the receiver's selectivity bandwidth, which is the channel bandwidth. The MDS sensitivity of a channel is set by the channel bandwidth, since the noise power onto each crystal detector is determined by that bandwidth. The common detector line of Figure 4.3 has a sensitivity between the channel sensitivity and the sensitivity of a CVR that would cover the same complete IBW as the whole channelized receiver.

One potential problem with channelized receivers that the designer must face is the problem of transient responses from each channel filter, especially for channelized receivers with relatively narrow channel widths, that is, with channel widths within an order of magnitude of the reciprocal of the input pulsewidths. This transient response, sometimes called the "rabbit ears" response, is illustrated in Figure 4.5. The figure shows the input pulse signal in both the time domain and the frequency domain. The crystal detector video time domain pulse output is shown for filters (1) far removed from the signal carrier frequency, (2) nearby, and (3) centered on the signal carrier frequency. The relationship, in the frequency domain, between the input pulse spectrum position and the channel filter position is also shown. When the input spectrum is centered in the channel filter, the crystal detector video output is seen to be a reasonable pulse, albeit with finite rise and fall times because of the limited bandwidth. As the input spectrum is moved away from the center onto the shallow portion of the filter skirt, the video amplitude in the settled center is reduced, while spikes appear at the leading and trailing edges; hence, the name rabbit ears. In the far skirt, with deep attenuation at the input frequency, only transients at the leading and trailing edges are visible from the crystal detector, and, of course, further out onto the skirt even these transients drop away. A frequency-domain explanation of this phenomenon would point to the energy in the pulse-spectrum sidelobes passing through the main part of the filter, while a time-domain explanation will point to the distortion of the sinusoid carrier that occurs at the leading-edge rise and trailing-edge fall because of the changing amplitude of the envelope.

The rabbit ears phenomenon is a serious problem for channelized receivers, because it means that differential amplifiers associated with channel filters far removed from the true signal carrier frequency will be tripped by the unwanted spike, and often it will be tripped twice by a single sharp-edged pulse. The solution to this problem can involve

- properly shaping the passband and the shallow portion of the skirt,
- pulsewidth integration,
- cross-channel weighting,
- hardwired guard logic,
- subsequent digital processing, and
- wider channels.

Seemingly minor differences in the frequency domain shape of each channel can make significant differences in the time domain response, in terms of resolving the rabbit-ears problem. The pulsewidth integration can be accomplished with relatively simple RC circuitry, that is, with the video voltage passed through a low pass filter. The cross-channel weighting adjusts the threshold by the amplitude in the adjacent channel. The guard logic inhibits all but the center channel when several adjacent channels are activated. Of course, sophisticated algorithms can interpret the data in the digital signal processor subsystem. Wider channels makes the proper channel-detection association easier since it reduces the transient amplitude and width,

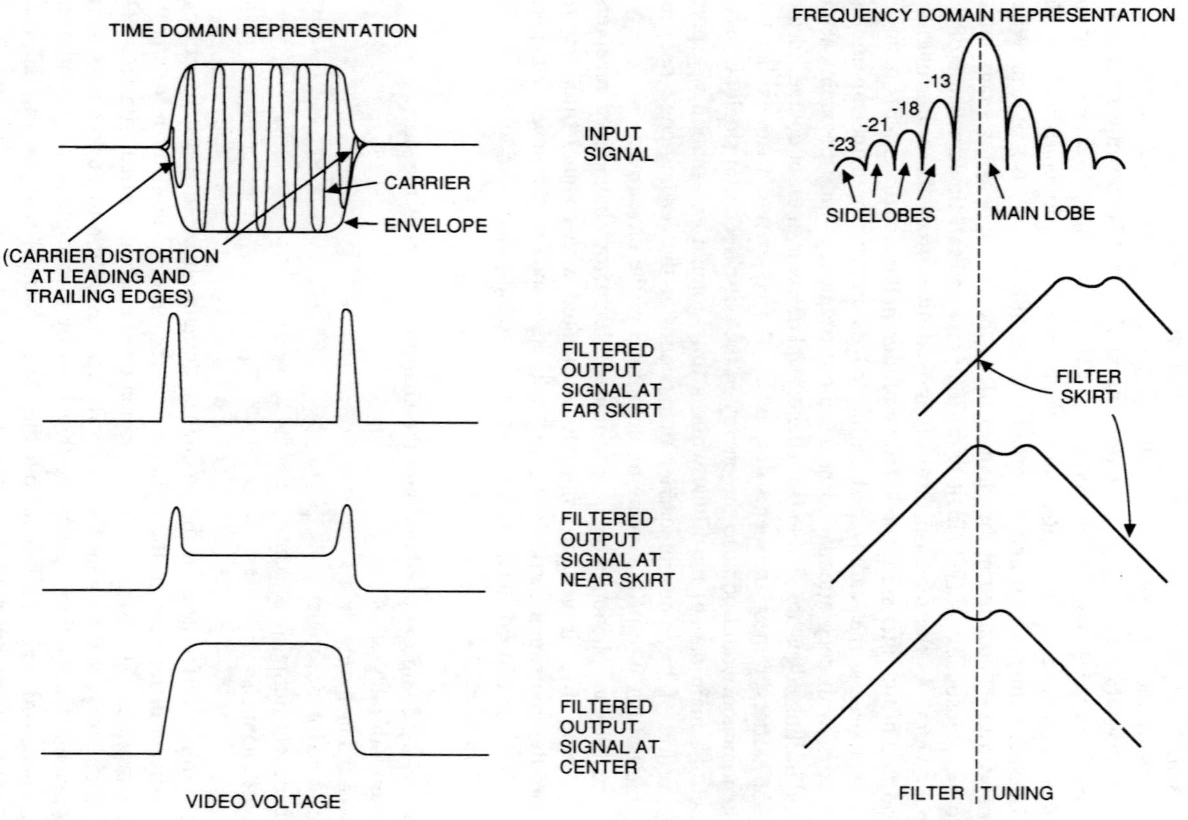

Figure 4.5 The "rabbit ears" phenomenon.

but it also degrades the frequency measurement resolution. Despite all these measures, the requirements for large dynamic range and accurate time-of-arrival measurements result in the rabbit-ears phenomenon causing one of the more difficult design problems.

Figure 4.3 shows a channelized receiver that only outputs a detector logic bit per channel. It is, of course, possible to build a channelized receiver that contains a bank of LVAs instead of just a bank of CVRs, or even contains both. If this is done, the output is far more useful to the DSP, but the cost and size, including the MUX and buffer memory, become more burdensome. It is also possible to use a design that employs various means to make the frequency resolution considerably less than the channel widths.

4.4 FREQUENCY MEASUREMENT RECEIVERS

As stated, the carrier frequency is an excellent parameter to distinguish signals, that is, to separate pulse trains, on a monopulse basis, unless the threat exhibits frequency agility. Therefore, there is an important requirement to measure the carrier frequency of each incoming pulse in the multi-GHz wide radar frequency bands, preferably near the leading edge of each such pulse. Such capability may also allow the measurement of some spread-spectrum signals.

The block diagram of an analog delay line type *instantaneous frequency measurement* (IFM) receiver is shown in Figure 4.6. Although the configuration of Figure 4.6 does not really measure frequency instantaneously, its time response is usually a small fraction of most practical radar pulsewidths, so effectively it is instantaneous. This receiver is fabricated from couplers, a fixed attenuator, a delay line and crystal detectors, all of which have inherently wide bandwidths, and hence this IFM receiver has a wide multi-GHz IBW.

The IFM sensor of Figure 4.6 operates as follows. The input RF signal power is split equally into two paths with a hybrid coupler. One of these paths contains a delay line, while the other path contains a fixed attenuator, a pad, whose attenuation nominally matches the attenuation of the delay line. These two paths are then combined equally with a 3 dB coupler, a hybrid, the two outputs of which feed crystal detectors *X* and *Y*. The response of detector *X*, shown solid in Figure 4.6, and detector *Y*, shown dashed, is graphed as a function of frequency in that figure. At some frequencies, the delay-line delay results in the two signals combining 180° out of phase, and hence providing nil net power to crystal detector *Y*, with all the power instead incident on crystal detector *X*. At other frequencies, the reverse is true, and the pattern repeats across the band. This pattern repetition in the frequency domain repeats at the reciprocal of the delay. In the example in the figure, the delay is 100 ns, so the pattern repeats every 10 MHz. For the output voltage to represent only the input frequency over a given band, (1) the delay-line delay must be less than

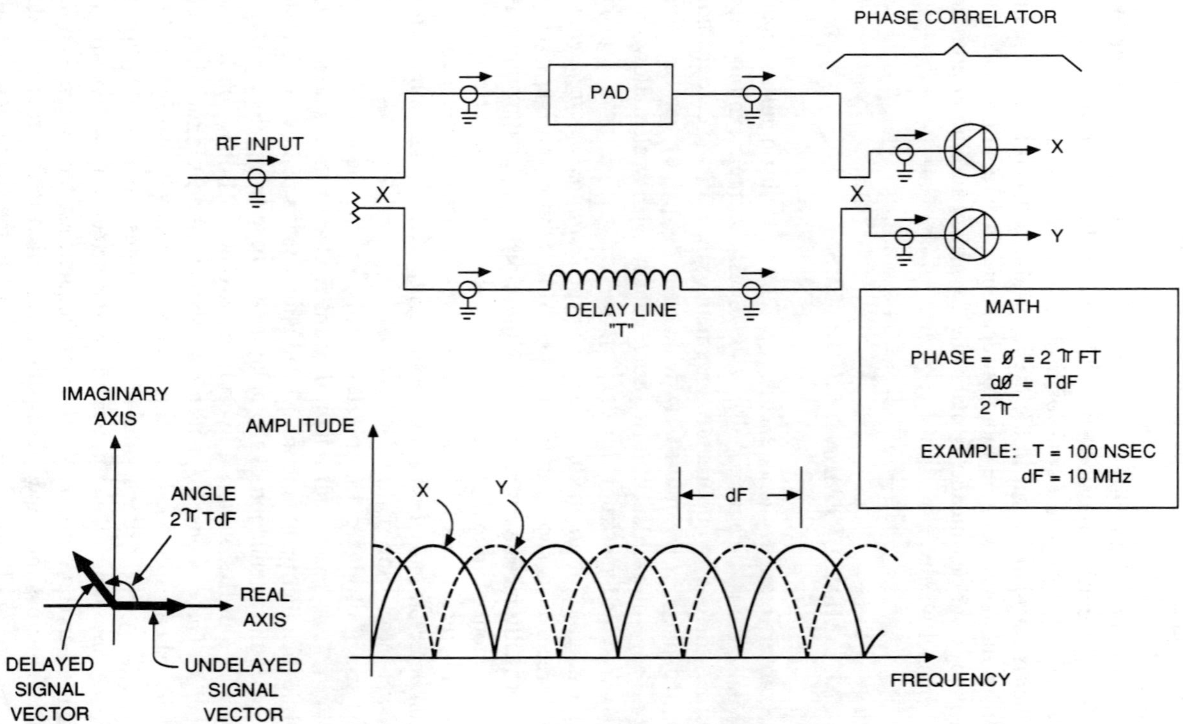

Figure 4.6 Frequency measurement.

half the reciprocal of the desired unambiguous bandwidth, (2) the phasing must be such that the video voltage is single-valued, and (3) the input amplitude must not vary with frequency. The third restriction is unacceptable for EW applications, because of the wide DR of signals experienced by ECM and ESM receivers. A means needs to be included to normalize the voltage.

Figure 4.7 shows the block diagram of a more practical analog IFM, with automatic DR normalization. Again, the IFM sensor consists of a delay line, a matching pad, couplers and crystal detectors. The reciprocal of the delay line length determines the bandwidth that gives unique voltage values for each input frequency and amplitude. There are twice as many crystal detectors as in the unnormalized configuration, the outputs of which are summed in pairs to give *X* and *Y* output video voltages. Each pair of matched crystal detectors typically feeds alternate inputs of an *operational amplifier* (op amp) that then sums them as opposite polarity signals, so that the *X* and *Y* output voltages can be positive or negative. The arrangement of four couplers, four detectors, and the two summing op amps, not shown in the figure, together make what is known as an RF phase correlator. In this case, this sub-unit correlates the phase between the delayed and the undelayed paths. Phase correlators have a number of applications, in addition to being a sub-unit of an IFM, and are themselves sold by vendors as subsystem units. Another application of phase correlators, for example, is to measure the phase difference between two antenna elements, thereby sensing the AOA.

The video output of the phase correlator of Figure 4.7, fed by a delayed and undelayed path, is shown in Figure 4.8, using the conventional polar display. The axes are the *X* and *Y* video voltages. This can be displayed quite easily on most laboratory oscilloscopes by shutting off the scope sweep, connecting the IFM *X* output voltage to the horizontal drive input of the scope, connecting the *Y* output to the

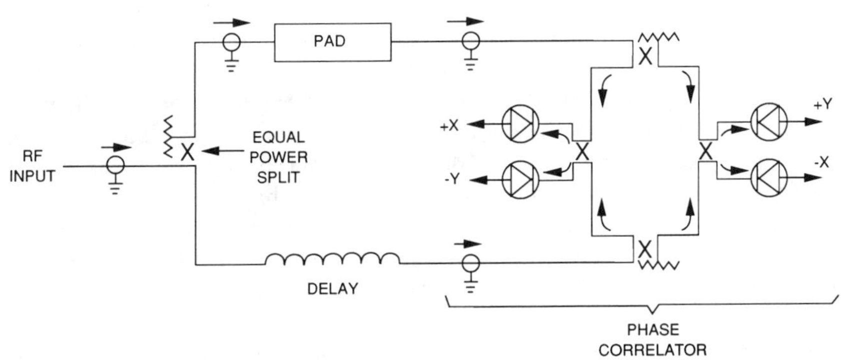

Figure 4.7 Normalized frequency measurement.

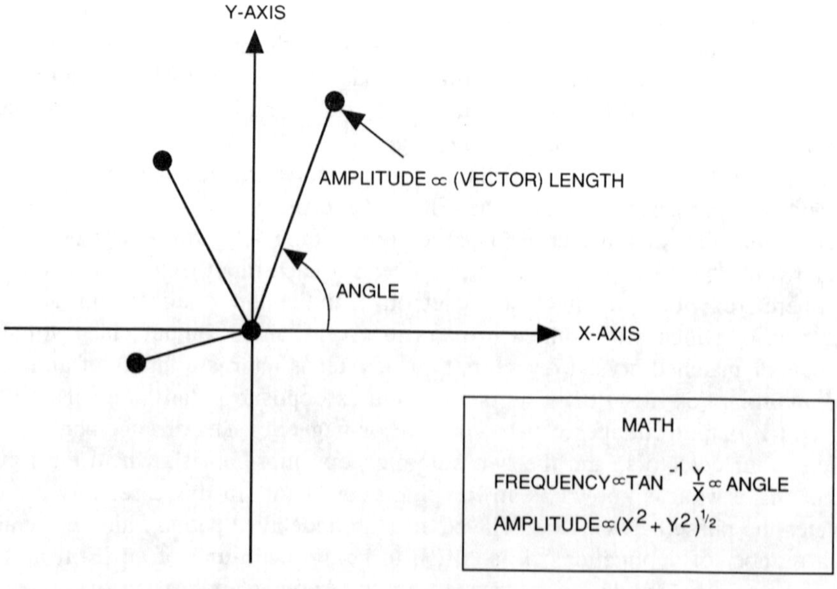

Figure 4.8 IFM polar display.

normal scope probe, adjusting the deviation-per-volt sensitivity for both to be the same, and centering the display. With no RF input, there will be a bright dot at the center of the display. As shown in the example illustrated in Figure 4.8, with three relatively low-duty cycle (low pulse overlap probability) RF pulse trains present, three additional dots will be present on the screen, each with an intensity determined by that pulse-train duty cycle. The distance from the center represents the RF input amplitude, while the angle represents the carrier frequency. The fact that this angle is independent of amplitude is the reason that this correlator approach can be considered to have automatic normalization. Such an IFM is relatively simple to implement, and the polar display instrumentation can be employed by a human operator to distinguish easily the individual signals even when many interleaved pulse train signals are present, provided each duty cycle is not too high, whereas the output of an LVA monitored with the same scope will show hopelessly jumbled signals. When pulse signals overlap, the resultant video voltage is the vector summation of the voltages of the separate signals; however, if the frequencies are close (e.g., <10 MHz) a polar beat may result, similar to the amplitude beat of an LVA.

Although the human brain can interpret the polar display readily, it has proven somewhat difficult to translate the sensed voltage electronically to a digital format suitable for use by a DSP. Figure 4.9 shows the general approach to create a digital output representing the RF carrier frequency. This configuration is the digital in-

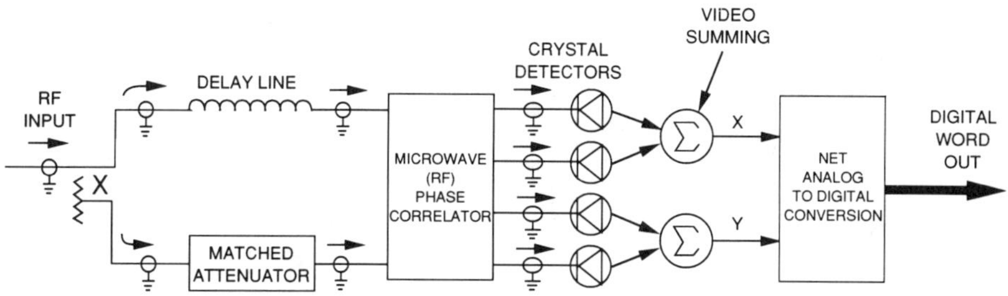

Figure 4.9 DIFM receiver.

stantaneous frequency measurement (DIFM) receiver. It includes the delayed and undelayed path, the phase correlator, the crystal detectors, and the opposite polarity summing network to provide the X and Y voltages, which are then digitized. If high resolution is needed, then multi-bit ADCs are employed. The conversion from the two ADC rectangular coordinate code words to a polar angle code word representing linear frequency can be accomplished with a one-to-one correspondence digital look-up table within the DIFM, or with the arctangent of the Y/X division computations made subsequently within the DSP. In either case, the data words are placed onto the data bus for use by the DSP. The DIFM receiver of Figure 4.9 can be thought of as the digitized version of the analog IFM receiver of Figure 4.7.

The conversion speeds for ADCs as well as various calibration considerations have generally resulted in the configuration of Figure 4.9 not being utilized for high-resolution frequency measurement. Some ECM system applications require that the DIFM receiver generate a digital word near the pulse leading edge, or at least within a small fraction of a typical radar pulse. This capability is used to generate an appropriate response within the pulse time interval. Other ESM and ECM applications require the rapid output of the digital word to facilitate the handling of the data throughput, including pulse bunching. In either case, the configuration of Figure 4.10 is a practical DIFM receiver that meets this need. The Figure 4.10 DIFM can be considered the high-speed and, usually, the high-frequency-resolution version of the Figure 4.9 DIFM. This receiver configuration contains several channels, each of which includes the power splitter, the delayed and undelayed paths, the correlator, and the digitizer. In addition, there is the RF power-splitting network that feeds identical signals to all the channels. A modest frequency-resolution DIFM would have three channels, whereas a high (i.e., fine) frequency-resolution DIFM would have five channels. The output resolution varies from six to twelve or more bits, depending on the number of channels. The high speed frequency measurement is achieved by using difference amplifiers simply to tell the polarity of the X and Y voltages, as illustrated in Figure 4.11. That is, the differential amplifiers indicate

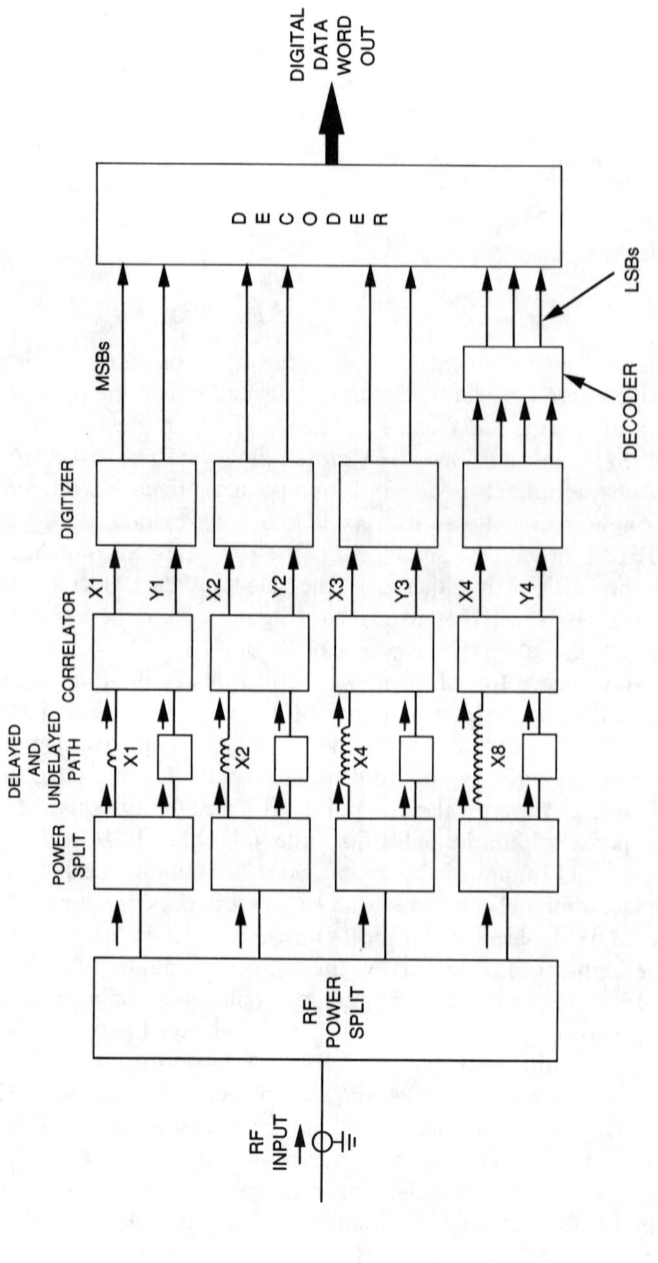

Figure 4.10 Multi-channel DIFM.

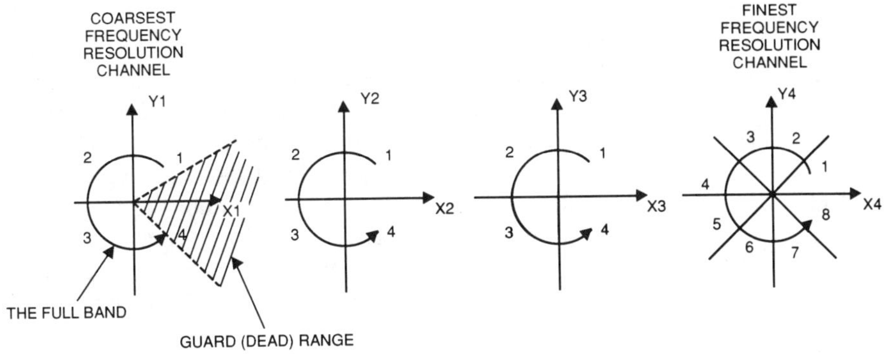

Figure 4.11 DIFM "gas meter" representation.

the quadrant, and no further data conversion is necessary. For the example shown in the figure, the coarsest channel, which has the shortest delay line, gives two bits of data for the full band frequency range; there is usually at least 45° of dead range to avoid ambiguities, that is, to ensure a single-valued response so that the upper frequency end is not confused with the lower frequency end of the band. The subsequent channels, typically with successively double the delay for a conservative design, also provide two bits of data, identifying the quadrant, but span the full 360°, without a dead range, several times across the band. The final channel, with the longest delay line, has the finest resolution; this channel's output is often divided more finely than into just quadrant slices because there is no next channel to be correlated with. One way to perform this division is to have a differential-amplifier compare the X and Y voltages, to give a 45° cut, for example. If such finer cuts are utilized, then a digital look-up table is needed to convert this data to a linear function. If the ratio of the channel delays does not match a multiple of the individual channel code resolutions (e.g., $4X = 2$ bits), then a master decoder is employed to generate a suitable frequency word.

Fabricating a DIFM with multiple channels makes it practical to achieve both excellent frequency resolution and speed of measurement. Were it not for the need to determine the frequency word rapidly, one or a few IFM channels could be employed with high resolution ADCs, the outputs of which would be digitally processed to compute the frequency. Because of the requirement for speed, however, more expensive parallel RF IFM channels are employed, the outputs of which have faster differential amplifiers or lower-resolution flash converters. For most subsystems within ECM systems, the design rule of thumb is to minimize the expensive RF circuitry and use video or digital circuitry instead; because of the need to output data words rapidly, that rule of thumb has proven inappropriate for mass-produced DIFMs.

Because the components used to fabricate DIFMs—RF couplers, delay lines, pads and crystal detectors—have wide RF bandwidths, the DIFM has multi-GHz bandwidth capability. Likewise, the crystal detector exhibits sufficiently wide video bandwidths and the differential amplifiers provide fast response. The DIFM ADC and data transfer to the system digital bus is triggered by either (1) a CVR, (2) a differential amplifier detection of the polar magnitude of the video voltage, or (3) some external trigger. The external trigger can come from either another ESM receiver or a TOA prediction from a pulse-train tracker in the system's DSP. Employing a CVR for triggering is common, with the bandwidth limiting front-end filtering in front of both the DIFM and CVR, and the overall MDS sensitivity determined by the CVR.

The DIFM receiver has been widely used in ECM systems because of its high performance, that is, multi-GHz wide RF bandwidth and submicrosecond delay, with reasonable cost and size.

4.5 ANGLE OF ARRIVAL SENSORS

An important requirement is that the ECM system's DSP be able to separate incoming pulse trains. Amplitude is often too erratic to reliably differentiate signals. The carrier frequency has been the most commonly used of the single-pulse measurement parameters for differentiation, because the radars are spread over a wide range of values, and to extract that parameter is reasonably cost-effective extracting. Pulsewidth is another option, but the range of pulsewidths does not provide suitable differentiation between signals. The carrier frequency and pulsewidth parameters are under the control of the threat radar, albeit with constraints. One parameter not under the radar control is the AOA, sometimes called the *direction of arrival* (DOA). To exploit this parameter for DSP purposes is therefore considered useful. Furthermore, the AOA can be used for other purposes. Signal separation is important because it is a necessary first step in assessing the criticality of the threat associated with the signal, measuring its parameters, generating the optimum response, and predicting the next pulse arrival for multiplexing purposes. In addition to being utilized for such signal separation purposes, AOA often can also be utilized by the asset being protected to take evasive maneuvers or to direct fire. Unfortunately, AOA is usually a time-varying parameter.

Amplitude comparison or phase comparison monopulse antenna systems are used to fabricate AOA sensors. A simple amplitude comparison approach is shown in Figure 4.12. Two antennas are aimed at somewhat different angles, with their main beams overlapping a significant amount. The signals from these two antennas are fed to crystal detectors, the outputs of which are summed with opposite polarities by an operational amplifier *integrated circuit* (IC). This video output is then con-

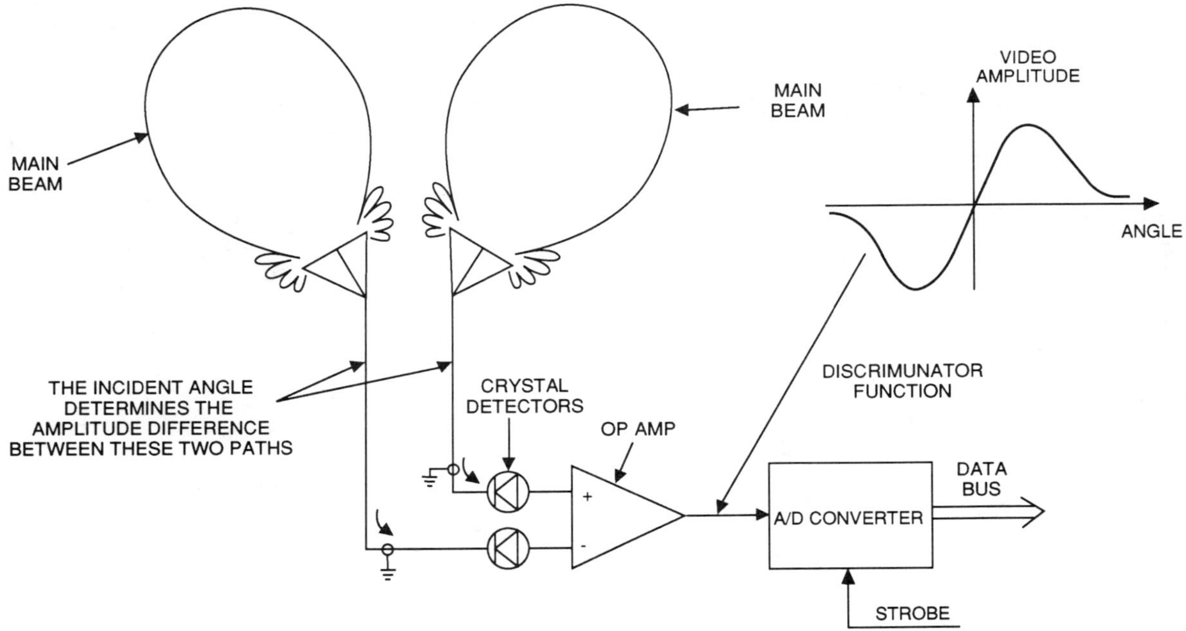

Figure 4.12 Amplitude comparison monopulse AOA.

verted to digital format and passed on to the data bus. A CVR (not shown) can be used to trigger the ADC.

As shown in the illustration, a classic discriminator curve is available from this sensor. This sensor output can be used to feed a servo system, with, for example, negative voltages driving it to the left in azimuth, and positive voltages to the right in azimuth. An EM wave signal source can therefore be tracked in angle. The simple arrangement in the figure does not include amplitude normalization, so the signal's input amplitude will alter the slope of the discriminator curve near the crossover point. To achieve stable servo operation, use of amplitude normalization is essential. The normalizing voltage can be obtained from the same polarity sum of the two crystal detector voltages.

Amplitude-comparison and phase-comparison monopulse antenna systems both have their advocates; both approaches have advantages and disadvantages. A simple phase comparison monopulse AOA system is shown in Figure 4.13. Two or more spatially separated antennas are aimed at the same angle. The signals from these two antennas are fed to a phase correlator. The illustration gives an example of a quite simple phase correlator, consisting of a hybrid coupler, two crystal detectors and an

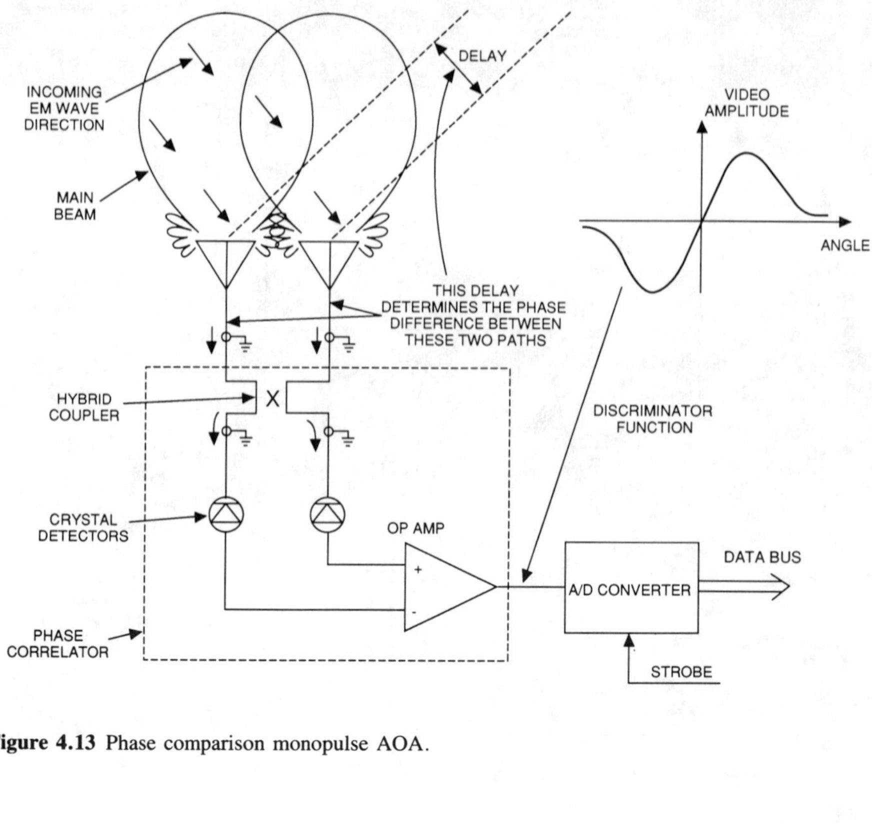

Figure 4.13 Phase comparison monopulse AOA.

operational amplifier to sum the video voltages with opposite polarities. This simple correlator does not include automatic amplitude normalization, which is needed for most applications.

If the EM wave approaches such a phase-comparison monopulse AOA antenna from boresight (i.e., head-on), the phases of the RF signals on the two lines entering the phase correlator will be equal. If the EM wave approaches at an angle, as illustrated in Figure 4.13, the phases will be unequal, because the RF EM wave takes longer to reach the further antenna. Hence, the phase correlator is comparing the phase of a delayed and undelayed path, and the operation is quite similar to the operation of an IFM as illustrated by Figure 4.6. The incident angle determines the delay, and given that delay, the response of the phase correlators of Figures 4.6 and 4.13, as a function of frequency, is identical. Likewise, to include an automatic amplitude normalization capability to an AOA system, the phase correlator employed for the IFM block diagram of Figure 4.7 can be used as a perfect guide.

The phase comparison AOA system also can provide a discriminator response. The discriminator response is used for servo tracking purposes or just as a single-valued transfer function.

A monopulse AOA system is essentially a miniature phased array. The spacing of the elements, that is, the individual antennas, is critical. The length of the delay line in a DIFM determines the unambiguous bandwidth that the DIFM will operate over. Similarly, the separation of the two antennas determines the angular range that the sensor can unambiguously cover. DIFM receivers can use multiple channels, each with its own distinctive delay line and identical phase correlator, to achieve wide-frequency coverage, fine-frequency resolution and fast response. Similarly, phase-comparison monopulse systems can use a number of antennas, with each pair having distinctive spacing and feeding identical phase correlators, to achieve wide angular coverage, fine angular resolution and fast response. In addition to these antenna element spacing considerations, more antenna elements can be added to improve the total aperture, and hence the sensitivity, directivity and gain of the system; ultimately, with a large number of such antenna elements, this becomes a true phased array, or active aperture, system.

Whether or not an amplitude-comparison or phase-comparison monopulse system is employed, requirements often dictate the need to measure AOA in both azimuth and elevation. This can be accomplished by adding antennas, or elements, in both measurement planes, and often at least one of the antennas is used in both planes.

4.6 TUNABLE RECEIVERS

The EW receivers and sensors described above have multi-GHz IBWs useful for detecting and measuring RF signals anywhere in these wide microwave bands. A tunable, or swept, receiver has a narrow IBW in relation to its operating band. The combination of tunability, narrow IBW and wide operating range is quite useful for ensuring adaptive selectivity in signal measurements if it can be properly managed. The tunable receiver is needed to make optimized measurements that are not corrupted when signals overlap in the wide operating band. To ensure such accurate AOA sensing, tunable receivers can be substituted for the broadband crystal detectors in Figures 4.12 and 4.13. The tunable receiver is most commonly used to supplement the data from CVR, LVA, channelized, and DIFM receivers. The tunable receiver will supplement a channelized receiver if the channel widths and amplitude measurement capabilities were determined by economic and size constraints rather than functional SNR, DR, and selectivity needs.

Two types of tunable receivers are illustrated in Figure 4.14. The YIG-tuned receiver is simply a broadband CVR or LVA receiver, or both, preceded by a YIG-tuned filter. The signal incident on the crystal detector is filtered at direct RF, rather than at an intermediate frequency; this feature facilitates a high amplitude DR without spurious signals. One of the significant advantages of the YIG-tuned receiver is that the frequency tuning is very linear and predictable, the frequency being proportional to the current driving the electromagnet that creates the magnetic field for the YIG sphere. The tuning transfer function however, is characterized by hysteresis, as is common in many magnetically driven ferrite applications; this hysteresis phenomenon is illustrated in Figure 4.15. When tuning up the band, a given current driving the electromagnet will cause a certain carrier frequency to be received; when tuning down the band however, that same current will cause a higher carrier frequency to be received. In most applications, knowledge of the absolute frequency is critical, so the hysteresis offset needs to be eliminated. This can be accomplished by driving the YIG filter to well past one end of the band, with either a spike impulse or step function of current, before tuning to the new desired current. This ensures that the historical information in the hysteresis phenomena is lost, or, more precisely, that the hysteresis offset is always the same; hence, the frequency tuning will be accurate.

The YIG-tuned receiver is characterized by very wide (multioctave) RF tuning ranges with excellent linearity, the hysteresis phenomena aside. The instantaneous bandwidth stays approximately constant, becoming slightly wider at the higher bands, with 30 MHz being a typical value. Because an electromagnet is needed to tune the receiver, it is relatively slow and therefore is not compatible with many sophisticated power-management functions, which typically dwell at a carrier frequency for about twenty times the typical radar pulsewidth, that is, about 10 μs. The YIG-tuned receiver's speed is one to two orders of magnitude slower than this power-management requirement, so it has limited applicability despite its excellent linearity and wide tuning range, except when used to judiciously supplement the data from other receivers.

The superheterodyne receiver (also known as superhet) is illustrated in Figure 4.14b. The superheterodyne is one of the oldest receiver designs there is. When the

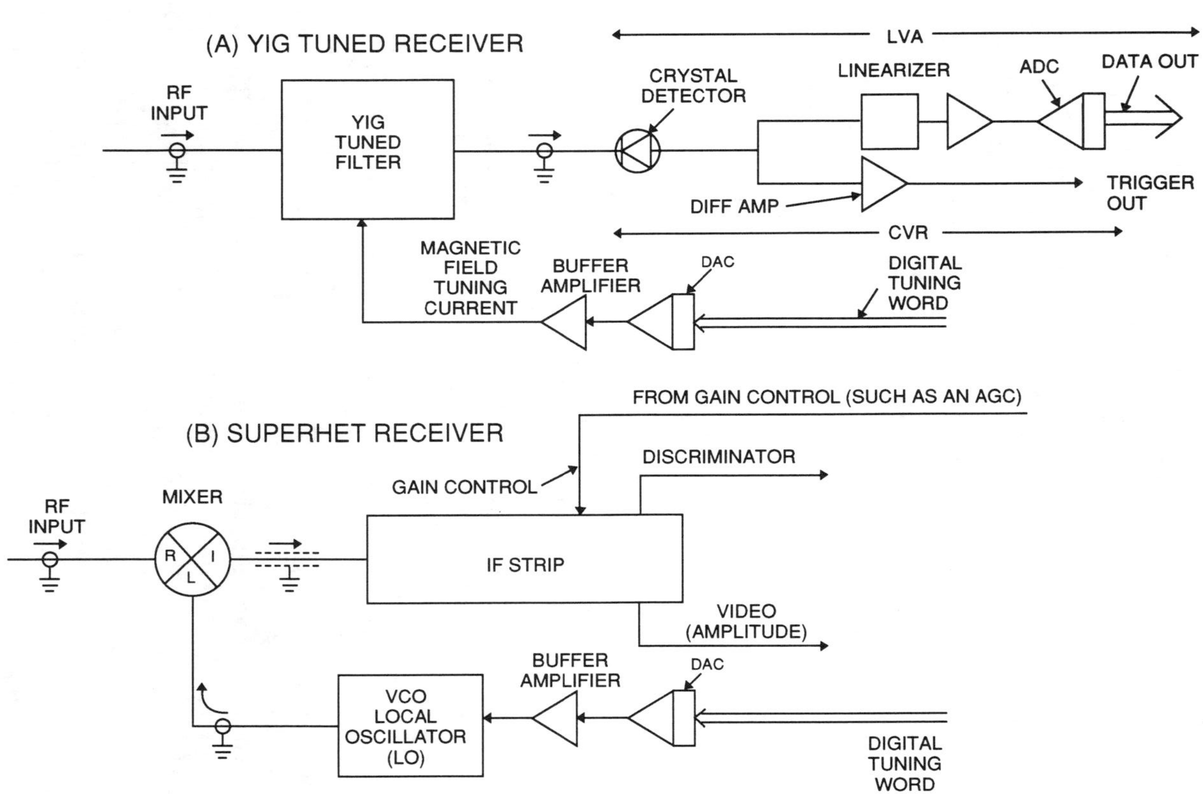

Figure 4.14 Tunable receivers.

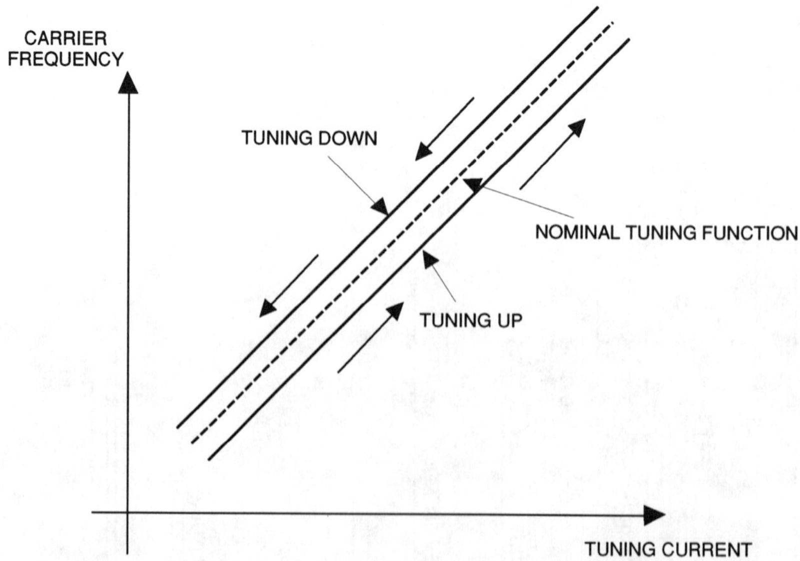

Figure 4.15 The hysteresis phenomenon.

superheterodyne approach was developed, RF gain was much more expensive, so this approach relies on amplification at an intermediate frequency. The RF input signal is beat or mixed to the IF with a frequency converter or mixer. This IF signal is then amplified in the IF amplifier known as an IF strip which supplied two outputs: the discriminator voltage and the video voltage. These two outputs are illustrated in Figure 4.16. The video output indicates the signal strength within the receiver's IBW, and is used as an error signal for an AGC servo. The discriminator output indicates the relative frequency within the receiver's IBW, and may be used as an error signal for an automatic frequency control servo. The discriminator voltage is proportional to frequency within the bandpass for a constant signal amplitude, with the null voltage at the center of the IBW, as shown.

The frequency converter or mixer is fed by both the relatively weak input signal and the saturating LO signal. In a superheterodyne the LO signal is usually generated by an RF VCO. In the Figure 4.14b illustration, the VCO is tuned by a DAC which translates input digital words from the digital-data bus. The VCOs available from component vendors can be tuned, unlike the YIG filter, at rapid rates that are quite compatible with power management needs. Unfortunately, the linearity of these VCOs, and hence of the receiver itself, is poor, unlike the YIG filter, which has excellent tuning linearity. Highly linear YIG-tuned oscillators are available, and can be used for superheterodyne LO purposes; however, they have the slower tuning speeds of the YIG filter.

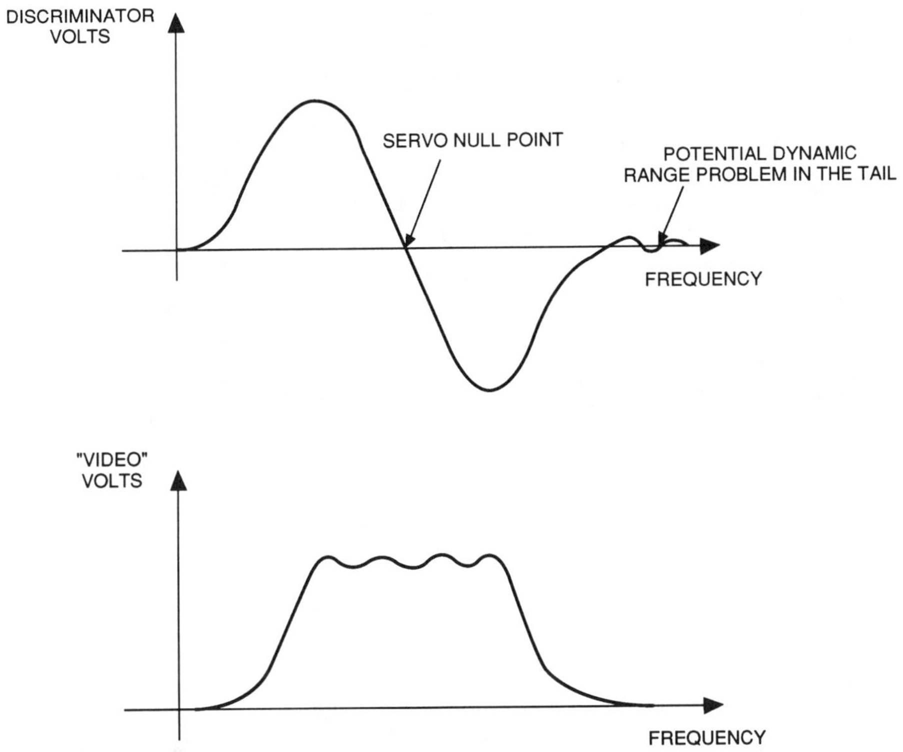

Figure 4.16 IF strip outputs.

The IBW of the superheterodyne is centered in frequency at the RF carrier frequency that is offset from the LO by the IF nominal center. Typical IF center values are 30, 60, or 120 MHz. Because these IF center values are relatively low in relation to the very wide operating bands needed to be covered for ECM, ESM, or general EW purposes, there often is a problem with the image. The image is the other RF carrier frequency also offset from the LO by the IF nominal center value; that is, the image is on the opposite side of the LO. For example, if a 60 MHz IF strip with 10 MHz IBW were used, then when the LO passed through 3000 MHz the receiver's IBW would be from 3055 MHz to 3065 MHz and from 2935 MHz to 2945 MHz. This image can be troublesome for some applications, and in such a case a SSB mixer (described in Chapter 2) is substituted for the simpler DSB mixer to suppress the unwanted image.

The IBW of a superheterodyne is completely independent of its tuning, since the IF strip determines the IBW. Unlike the YIG-tuned receiver, which does its filtering at direct RF, the filtered IF is always at the same frequency (e.g., 60 MHz). This means that the filter response is identical everywhere in the band.

The YIG-tuned receiver can be designed to have both a discriminator and video output, just like a superheterodyne, by using a pair of offset YIG filters. In either case, the receiver typically is swept through the RF band until the video output indicates that a signal is present, at which time the sweep is turned off and the servo takes over to track the signal in frequency. The discriminator output gives positive or negative voltages as an error signal for this servo. The servo is typically an operational amplifier with a resistor and capacitor circuit that operates as a voltage integrator in such a way that a fixed error voltage input causes a voltage ramp output. The servo drives in the direction to null the discriminator error voltage. The servo will settle at the null point indicated in Figure 4.16.

There is a problem with this simple mechanization, however, especially if amplitude normalization or AGC is employed. Note that Figure 4.16 shows that the polarity of the signal is not reliable in the tail, or far from the center, of the discriminator curve. Quite often the discriminator voltage polarity has one or several sign reversals. The reason the polarity is not reliable is that the operational amplifier bias point may differ from the expected null value, or the analog summation of the two offset filter responses may cause a slight error, or the transient response in the filter skirts may be unpredictable. As the receiver sweeps toward a signal frequency, the sweep could be turned off prematurely in the tail of the video response because the AGC servo is commanding maximum gain because no strong signal is present. Because the error voltage does not necessarily have the correct polarity, it therefore will not drive toward the true servo null point, near which the error responses are proper. The wrong error polarity is much more serious than the wrong servo loop gain. The AGC worsens the problem by increasing the gain in the filter skirt, which is the discriminator tail. This is a problem that designers of swept receivers must address.

4.7 EW RECEIVER REQUIREMENTS

The above descriptions of the older EW receiver design approaches will serve as background for a further description of EW receiver requirements, with emphasis on the bandwidth and timing issues which have driven the development of newer EW receiver design approaches such as the AO and microscan receivers.

This chapter has described CVRs and LVAs that have multi-GHz RF instantaneous bandwidths, with noise levels and hence MDS sensitivities commensurate with such bandwidths as described in Chapter 3, and with good video bandwidths, that is, video bandwidths somewhat wider than the reciprocal of the shortest expected input pulsewidth. Although these receivers have the ability to detect and measure pulses that occur anywhere in the radar band, as defined in Chapter 1. There are important disadvantages associated with this important feature. The problem is that more than one signal may be incident on the system at one time, thereby corrupting either the sensor-receiver or the subsequent digital signal processor outputs. The most troublesome case is, of course, attempting to sense weak signals in the presence of high-amplitude high duty cycle signals, especially CW signals. Fortunately, the average transmitted power is the true limit for radar systems, just as it is for ECM systems, so CW and other high duty cycle signals generally have less amplitude than low duty cycle pulsed signals. Nevertheless, because of (1) distance disparities, and (2) the strong probability of pulse overlap in dense signal environments—the condition for which proper ECM operation is the most important—receivers less vulnerable to being blinded are needed. For a CVR, that means that logic pulses need to be output to the data bus for each new pulse incident on the system even when there is a strong CW or other overlapping signal present. This can be accomplished, to a large degree, with proper AC coupling within the CVR. The LVA, which is supposed to measure the amplitude of the incoming signal, also can, in principle, be ac-coupled so the presence of a CW signal does not reduce its sensitivity to pulsed signals. In principle, a CW signal should not disable an ac-coupled CVR at all, but in practice there are two practical problems: (1) the noise on CW signals firing the leading edge detector, and (2) the finite ac-coupling recovery time at pulse edges causing mutual interference. The practical problems for the LVA are even more difficult to overcome, since the amplitude prior to the ac-coupling will be the sum of all the signals instantaneously present. The problems include noise on CW signals, pulse edge recovery time and the possibility that the crystal will not be in the square law region. Likewise, the IFM receiver will report an erroneous frequency based on the vector sum of multiple signals present. Therefore the property of frequency selectivity is needed, for large DR operation, in addition to the property of wide instantaneous bandwidth. We emphasize, again however, that ECM systems indeed cover very wide bandwidths, and receivers able to detect a pulse anywhere in that frequency range provide critically important real-time data.

A tuned receiver, such as a YIG-tuned receiver or a superheterodyne, also can be used to monitor wide bands by sweeping the relatively narrow IBW across that band. Indeed, most laboratory test equipment spectrum analyzers operate that way. The relatively narrow IBW of such receivers gives it the property of good (i.e., fine) frequency selectivity, so that the entire band can be analyzed, even if multiple signals, pulsed or CW, are present. Laboratory spectrum analyzers generally have automatically adjusted IBW so that the ratio of the total bandwidth to be displayed will not be too large; otherwise, the sweep, or scan, will take too much time. The problem of sweep time also limits the usefulness of tuned narrow-IBW receivers for EW reception. The problem is illustrated in Figure 4.17. In the example shown, the receiver cannot be swept much faster than 5 MHz/ms to be sure that a pulsed RF carrier with a PRI of 1 ms, a typical value a radar designer would select, would be detected. If the sweep were faster, then it would be likely that the receiver would

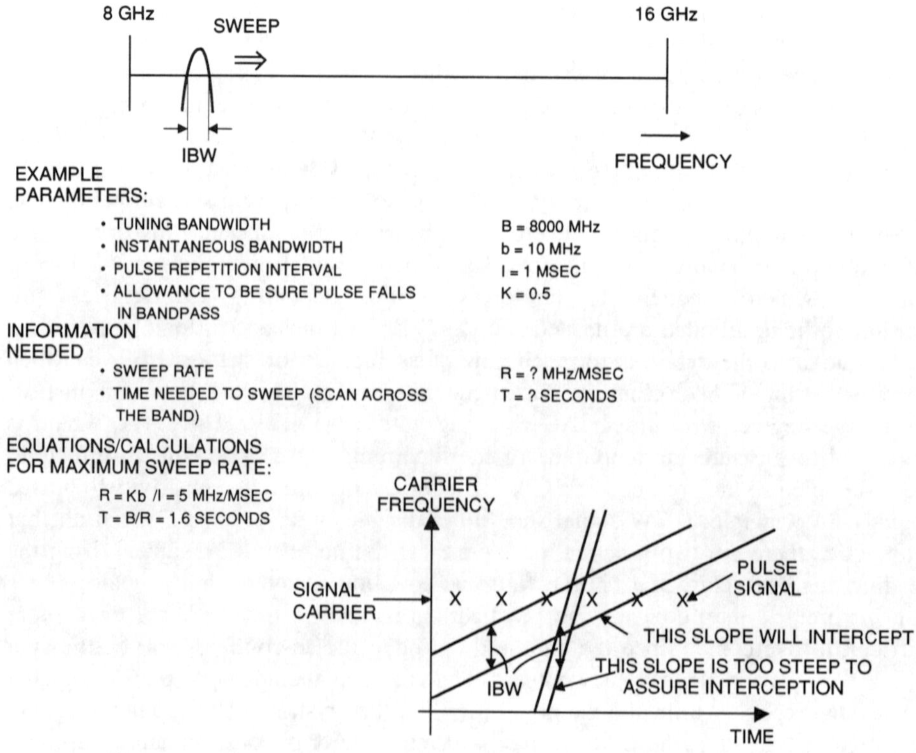

Figure 4.17 A tuned receiver sweep rate example.

sweep through that carrier frequency in between such pulses, so that there would be no assurance that the receiver would detect strong signals within one sweep. Such detection assurance is considered necessary so that both the average and maximum acquisition time can be predicted. The Figure 4.17 example shows that for 8000 MHz of bandwidth coverage the acquisition time would be 1.6 s. This is unacceptably long. The ECM receiver acquisition time should not be much longer than a radar dwell so that the active-ECM jamming can commence before the radar gets a "free look"; the radar's angular scan typically paints the target with about ten pulses in the main beam target dwell. Furthermore, if the signal environment is dense and many CW and pulse train signals are present, the DSP-CPU will have to stop and hold the tunable narrow band ECM receiver at each threat carrier frequency for some time (at least 3–5 PRI, and probably longer) to perform the initial assessment of each signal.

By way of contrast, it should be noted that a tuned receiver searching for a CW signal with a sweep function can theoretically sweep at a rate equal to the square of its IBW, albeit a factor of two slower would be prudent to allow some tolerance. Therefore, for example, a 1 MHz IBW swept receiver could cover 8000 MHz of bandwidth in $((8000 \times 10^6)/(1 \times 10^6)^2)/2 = 4$ ms. This is why the CW signal acquisition time, and the general data throughput, are not difficult problems for tuned narrow-IBW receivers.

With these bandwidth and timing considerations in mind, it is now appropriate to describe the uses of the receiver output data. Table 4.2 shows a summary of this data output, organized into the three functional groups: trigger, intra-pulse measurement and inter-pulse measurement. The trigger receiver function is used to alter the state of the ECM system at the leading edge of the pulse, to initiate ADC operation of other sensors (e.g., LVAs, AOA monopulse, *et cètera*), and to provide a TOA measurement strobe for the DSP. The reason a simple ECM receiver can trigger on the leading edge, while the radar needs and is designed for an entire pulse, is that the EM wave only travels one way to the ECM receiver, but must travel two ways to the radar receiver; therefore, the ECM receiver signal input is much stronger,

Table 4.2
Receiver Output Data Usage (for sensed RF pulses)

Sense Function	*Use or Parameter*	*Time Scale*
Trigger	Pulse tube Gain switching Path switching Modulation switching TOA measurements ADC initiation Miscellaneous	Tens of nanoseconds
Intrapulse measurement	Frequency Amplitude Pulsewidth AOA Miscellaneous	Hundreds of nanoseconds
Interpulse measurement*	Lobing-Scanning Rate Angle Track Condition and Stability PRI, PRI Switching, and PRI Modulation Range	Tens of microseconds to hundreds of milliseconds

*Indirectly obtained from either the trigger or intrapulse measurements.

and the leading edge can be reliably detected with simple mechanizations. It is not unreasonable to expect the ECM system to use the trigger to operate high speed microwave switches and the like, to restructure automatically its RF functions at the leading edge of an input RF pulse. The TOA measurement derived from the trigger, on the other hand, is not used "instantaneously." The TOA measurement is used by the DSP to analyze the PRI and, when appropriate, to predict the arrival of the next pulse for radars selected as being associated with threatening systems.

The next function listed in Table 4.2 is the intrapulse measurement. The DSP needs to use the receiver's sensed data to analyze the signal environment and to track the important, pulsed signals. To accomplish these goals, the DSP needs to be able first to distinguish the signals. Quite often this can be accomplished with sophisticated algorithms that analyze the PRI patterns derived from the TOA data obtained from the receiver's trigger output. Such algorithms can be remarkably powerful in accomplishing their goal. Nevertheless, it is necessary to support the signal separation function with the receiver's intrapulse measurements. That is, whereas the DSP's PRI sorter used the relationship between the TOA triggers to separate the individual pulses into separate distinct pulse trains, the intrapulse measurement data is used to separate the signals based on the character of the individual pulses, that is, the intrapulse data is used for monopulse sorting. A variation on that process is to characterize incoming pulses based on selected intrapulse parameters, and then check for matches to separate the signals. The most common parameter used for this purpose has been the pulse's RF carrier frequency value. This is because the carrier frequency is the one parameter that generally can most easily be utilized to distinguish between the different radar signals. The nominal carrier frequency distinguishes the threat radar because of the wide range of carrier frequency values, unlike the rather indistinguishable pulsewidths, for example. If the threat radars congregate near one frequency, then that would indeed make the task of distinguishing the signals more difficult; it would also however, facilitate effective jamming, with barrage noise, for example. The AOA, although it can sharply distinguish between individual radar signals, is often a time varying function. In most situations, either the asset defended by the ECM is moving and changing its orientation, or the threat radar is moving; any of these eventualities will cause the AOA-sensed digital word to vary with time. However, if the threat radar employs frequency agility, the technical problems associated with the AOA measurement, and its time-varying nature, may be preferable to using frequency for the principal monopulse parameter to separate signals. The AOA is also a good choice because the threat radar has no control over it. These intrapulse measurements are derived from combinations of sensors and ADCs, the DIFM receiver being a notable exception. The ADC function is initiated at the leading edge, and flash ADCs will convert within the pulse, higher resolution ADCs may take hundreds of nanoseconds to complete the conversion and ADCs employed with sample-and-hold functions take even longer than the pulsewidth.

Finally, a number of interpulse measurements are made, such as the PRI just mentioned. These measurements are indirectly obtained, generated by the DSP based on either the trigger or the intrapulse measurements, or both. That is, these measurements are not direct outputs from the receiver, they are outputs from the receiver and DSP combination.

The intrapulse and interpulse measurement data values are employed by the DSP to separate pulse train signals. Separating the signals is necessary so that they can be individually examined for identification purposes, and generally to assess the need to direct active ECM jamming against each radar generating such signals. If a jamming response is deemed appropriate, the data are also employed to optimize the jamming technique. Separating the signals is also needed to facilitate tracking and making TOA predictions, so that the ECM system can efficiently multiplex the jamming between the threats present in a power management fashion. The triggers, once separated into individual impulse trains, are utilized for such tracking and TOA prediction functions.

Given these sensor functional requirements, the parameters of interest for EW receivers can be examined. These parameters are tabulated in Table 4.3 for bandwidth and timing, including five different bandwidths, and in Table 4.1 for other parameters. The bandwidth parameters that receivers are generally characterized by are the operating bandwidth, the instantaneous bandwidth and the video bandwidth. However, to assess the ability of an EW receiver in an ECM or ESM system to meet the functional requirements described above, the selectivity bandwidth and resolution bandwidth also must be considered. These two bandwidths are not necessarily the same as some of the other bandwidths, especially for the newer EW receiver designs.

For purposes of this description, the operating bandwidth is the frequency range over which the receiver can make measurements. The instantaneous bandwidth is the frequency range over which the receiver can make measurements without tuning or other control, just prior to reception. The selectivity bandwidth is the bandwidth that determines the noise power, that is, the SNR, and hence the MDS sensitivity; it also defines the capability to distinguish between simultaneous signals. The resolution bandwidth is the frequency separation needed to obtain distinct digital frequency-measurement output words from the receiver. The video bandwidth, usually less than or equal to the other bandwidths, determines the minimum pulsewidth and the measurement response time.

The absolute frequency accuracy is shown combined with the resolution bandwidth in Table 4.3. Good, or fine, resolution does not necessarily imply good absolute accuracy of frequency measurement. Nevertheless, if a receiver has both good frequency resolution and good frequency linearity, or at least predictable nonlinearity, then various real-time calibration schemes are possible so that the CPU can predict the absolute frequency, if that is necessary.

Some EW receivers use RF delay lines to delay input pulses long enough to make a coarse frequency measurement, and to steer resources to that portion of the

Table 4.3
Bandwidth and Timing

Parameter	*Typical Value*	*Determined by*
	BANDWIDTH	
Operating BW	16,000 MHz (tracking radars)	(1) Antenna size (2) Propagation factors (3) Transmitter technology
IBW	3000 MHz (DIFM)	(1) Threat frequency distribution (2) Threat density (3) Acquisition time (4) Throughput (5) Component tolerances (6) Cost and size
Selectivity BW	20 MHz (channelized)	(1) Threat density in frequency (2) Sensitivity (3) Threat pulsewidth (4) Cost and size
Frequency Accuracy and Resolution BW	1 MHz (DIFM)	(1) Threat density (in frequency) (2) Radar BW (for set-on jamming) (3) Cost and size
Video BW	20 MHz (CVR)	Threat pulsewidth
	TIMING	
Tuning Response	1 μs	(1) Power management control (2) Average PRI (3) Threat density
Recovery Time (from strong signals)	0.5 μs	(1) Average PRI (2) Threat density
Throughput Data Rate	10–200 kpulses/s	(1) Average PRI (2) Threat density (3) Scenario (4) Acquisition time

band for fine frequency measurement. Since no prior tuning is necessary, this design approach provides the functional equivalent of fine frequency measurement over wide instantaneous bandwidths. The purpose is to achieve the functional capability in a cost-effective manner. In general, however, the above definitions, though highly useful to differentiate receiver capabilities, should be used with care when applied to steering or switching mechanizations.

The EW receiver requirement for operating bandwidth, shown in Table 4.3, is derived from the need to match the band of operation of the threat radars. If both the threat tracking radars and the search radars are to be handled, the bandwidth must be extended to lower (under 2 GHz) carrier frequency values. On the other hand, if the ECM is to operate in space, the threat radars will exploit the increased antenna directivity from a given antenna size, by operating in higher bands, without suffering atmospheric propagation losses.

In our judgment, the signal density drives the receiver bandwidth, timing, and other requirements more than the parameters of any one individual threat. Indeed, in certain parts of the world, the threshold will need to be deliberately raised well above the theoretically best MDS level in order to thin the signal environment, and thereby unburden the DSP.

4.8 RECEIVER TRADE-OFFS

The above text has described the functional properties of an ECM receiver. The generic ECM receiver parameter requirements derivation sources are tabulated in Tables 4.1 and 4.3, and the receiver output parameter measurement-resolution requirements for the DSP input will be described in Chapter 6. However, there are just four properties that are the key technology drivers, and these are tabulated in the illustration of Figure 4.18. The first property is the capability for instantaneous wideband reception, so that when a pulse enters, at any time, with any carrier in that band, the receiver will correctly pass the needed digital-word descriptors over the data bus for the DSP's use. This capability will keep the threat radar from having a free look. The next capability is the simultaneous signal reception capability, so that this sensing of the signal, especially its carrier value, will not be corrupted or blinded by other signals even if they are stronger. The third property is the compatibility with digital data bus passage of the data, including the simultaneous signal case; this last case requires a data queue and buffer, although multiple output buses, one for each signal overlapped, may be a practical if not an idealized solution because the number of such simultaneous signals will be limited. The challenge is to obtain all three properties from one receiver; generally these three requirements have been viewed as conflicting with one another. The final property in Figure 4.18 is the cost effectiveness of the receiver. For avionics applications in particular, the cost effectiveness must consider size as well as performance, since each cubic inch of

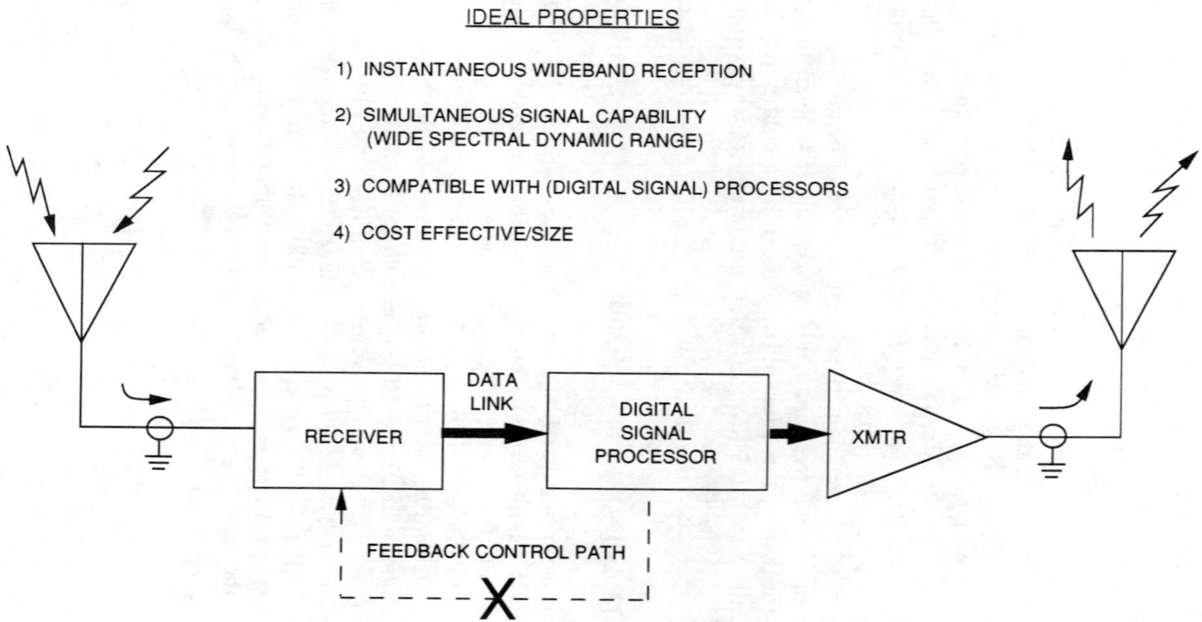

Figure 4.18 Properties of an ideal ECM receiver.

volume in the avionics suite is costly. For many other ECM applications the cost effectiveness is not as dependent on size.

Figure 4.18 symbolically shows the compatible data link from the receiver to the DSP. A dashed line is also shown for feedback control. The properties of an ideal receiver imply, although they do not quite necessitate, that there be no required feedback to the receiver. A tunable receiver is a classic case of a receiver that requires feedback control. The problem with feedback is that it makes the measured data dependent on the DSP's assumptions about the input environment in the first place. These assumptions may not always be correct, and hence a distorted view of the signal environment may result. An example of this would be the DSP-CPU control of a superheterodyne receiver in a power management fashion; for each pulse that is expected, the superheterodyne would be commanded to the expected carrier frequency a few microseconds before the pulse was due, and then the exact TOA and carrier values would be sensed and passed on across the data bus. In this example, however, if another pulse occurred synchronously and simultaneously at a different carrier frequency, the DSP would not be aware of that information. Therefore, dependence on feedback control implies inferior performance, since the measured data would be dependent on the expected results. Furthermore, any receiver that truly had the combination of the first three properties in Figure 4.18 would not need feedback.

Table 4.4 grades the different receiver technologies with regard to their bandwidths. The laboratory test equipment spectrum analyzer generally has good properties, except that it has narrow IBW. The superhet receiver is rated identically with the spectrum analyzer, since the only difference is that a superhet, in a power-managed ECM system, is jumped around the band in a sophisticated fashion, in anticipation and in search of key data, whereas the spectrum analyzer's tuned narrow-IBW receiver blindly sweeps across the band. The spectrum analyzer therefore cannot cope with complex signal environments, unless the signal environment does not change

Table 4.4
Bandwidths vs Technology

Receiver Bandwidths	*Spectrum Analyzer*	*Superheterodyne Receiver*	*Crystal Video*	*Delay-line DIFM*	*Transform Receiver*
Operating BW	Good	Good	Good	Good	Good
Instantaneous BW	Poor	Poor	Good	Good	Good
Frequency Resolution	Good	Good	Poor	Good	Good
Selectivity BW	Good	Good	Poor	Poor	Good
Video BW	Good	Good	Good	Good	Good

during the sweep time. In other words, a spectrum analyzer provides information slowly. The superhet receiver, although it can be rapidly jumped around under power management control, cannot sense simultaneous signals. The superhet receiver's use also cannot guarantee rapid acquisition in wide operating bandwidths even for light signal densities, as was illustrated in Figure 4.17.

As shown in Table 4.4, the CVR has good instantaneous bandwidth, identical to its operating bandwidth. However, it does not have any frequency resolution or selectivity, except that if a detection is made, the signal must have been in-band. The difference in ratings in Table 4.4 between the delay line type of DIFM and the CVR is that the DIFM has excellent frequency resolution despite its wide IBW. However, the DIFM selectivity is similar to the CVR, and it also can be blinded by strong simultaneous signals.

All the receivers shown in Table 4.4 are rated as having good video bandwidths, and hence the video bandwidth is generally not a trade-off issue. A video bandwidth is good if it is somewhat wider than the reciprocal of the shortest expected pulsewidth.

The EW receiver design options generally provide less than satisfactory performance in meeting the needs of an ECM system, at least on a cost-effectiveness basis. However, the use of two or more receiver types can synergistically provide reasonable cost effective performance. For example, a superheterodyne receiver and a DIFM can be used together, with the DIFM providing the wide IBW and the superhet providing the selectivity and sensitivity. In such an arrangement, the DIFM will provide most of the measurement data, and the superheterodyne will provide selected confirmation or high quality alternate data. The DIFM usually makes the initial detection because of its wide IBW. The superheterodyne can be used to continue the track on the signal in the radar-antenna sidelobes, where the signal strength is reduced. The superheterodyne can be used to make measurements when the DIFM data is erratic; it is not usual for the DIFM data to be both incorrect and unchanging.

The combined use of different receiver types, therefore, tends to provide the advantages of each, and minimize the weakness of each, generally to achieve reasonably satisfactory performance. There is still a strong incentive, however, to develop a receiver that has the ideal EW receiver properties described above, such as completely divorcing the output data quality from the proper management of the receiver. Most of the remainder of this chapter addresses receivers that attempt to achieve that goal.

This text will define a class of receivers as *transform receivers* having a certain set of bandwidth properties. The purpose is to clarify the functional equivalence of certain receiver approaches, which widely differing technical approaches tend otherwise to obscure. Proper classification is important for technical design approaches for the same reasons that it is important to classify biological species. The term "transform" is a mathematical term used to describe the conversion of a function in one domain to its equivalent in another domain. The most common transformation used by the electronics engineering community is the transformation between time and frequency. Many mathematical calculations and analyses are more easily accomplished in one of these domains, but would be difficult or cumbersome in the other domain. Likewise, certain signal measurements by hardware sensors are more easily accomplished in the frequency domain. One of the key purposes of an EW receiver is to facilitate the separation of pulse train RF signals based on their carrier frequencies, a requirement specified in the frequency domain. A discriminator converts a single sinusoidal waveform to a voltage proportional to frequency. When multiple signals are simultaneously present, the time-domain function does not lend itself easily to distinguishing the individual signals, whereas the frequency-domain function clearly does so, as shown for the three CW signals in Figure 4.19. Therefore, we consider appropriate the choice of transform receiver as a classification for frequency measurement EW receivers. One note of caution, however, is that a simple frequency domain amplitude *versus* frequency representation, such as displayed on spectrum analyzer test equipment, for example, does not include pulse TOA information, although pulsewidth is revealed, given sufficient resolution. A phase *versus* frequency response is also needed to describe completely the function, and that is difficult for hardware sensors to extract.

For purposes of this discussion, a transform receiver is defined as a receiver that is rated good for all the bandwidth measures listed in Table 4.4, and also Table 4.3. Specifically, a transform receiver is defined to have these properties:

- wide instantaneous bandwidth,
- selectivity BW << instantaneous BW,

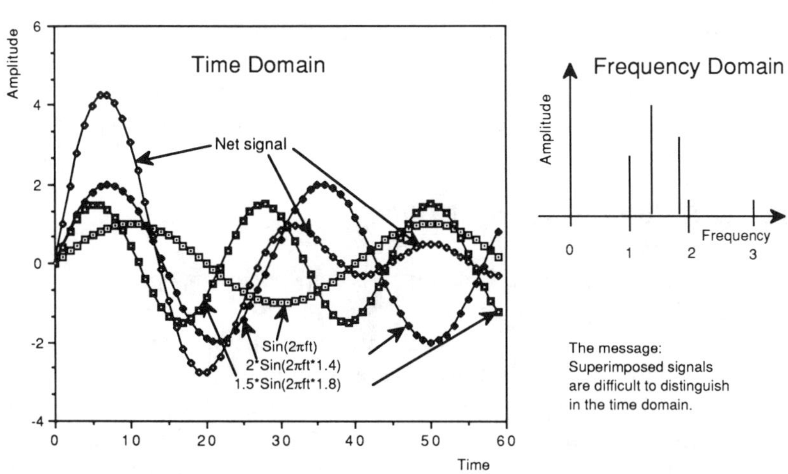

Figure 4.19 Distinguishing simultaneous signals.

- time of arrival approximated, and
- an effective digital data output interface.

In other words, the transform receiver, as so defined, has a set of key properties that matches the above concept of the properties that an ideal receiver would have. The simultaneous signal capability is obtained by having both wide IBW and narrow selectivity bandwidth, and appropriate means to efficiently buffer the data. The TOA resolution must be within an order of magnitude of the reciprocal of the lessor of the video or selectivity bandwidths. Certain false range or pulse target techniques and certain other ECM techniques require highly accurate TOA predictions, which in turn require that the TOA resolution be less than the reciprocal of the lessor of the video or selectivity bandwidths; this further restricts the reasonable-technology alternatives.

The selectivity bandwidth of a transform receiver is the bandwidth that

- sets the frequency measurement resolution,
- sets the MDS sensitivity,
- sets the minimum input pulsewidth, and
- does not set TOA uncertainty.

The requirement to have an effective digital data interface in order for a receiver to be classified as a transform receiver is necessary to accommodate simultaneous signals. For example, pulse *A* may have its leading edge after the leading edge of pulse *B*, but have its trailing edge before the trailing edge of pulse *B*. Circuitry needs to be included in the receiver unit to extract the appropriate information in a logical format. That is, a buffer needs continually to organize this data into a queue and pass it onto the data bus.

One technical approach that can have the properties of a transform receiver has already been described: the channelized receiver. Two other technologies that hold the promise of exhibiting these transform receiver properties are the AO receiver and the microscan receiver. These will be discussed in the following sections.

4.9 THE ACOUSTO-OPTIC RECEIVER

The acousto-optic, or Bragg cell, receiver is dependent on optical, acoustic, RF (microwave), video, and digital technology for its construction. That is, quite a varied set of electrical engineering specialists are required for its design. A symbolic block diagram is shown in Figure 4.20. A laser is used as a source of coherent light. A collimating lens forms this light into a beam, and a transform lens refocusses the light onto a point on an image plane. Between the collimating and transform lens the beam passes through a Bragg cell, where the laser light interacts with the signal. The RF input signal is applied to an acoustic transducer that converts the incident voltage into a sound amplitude or intensity. The result is that the input RF EM signal

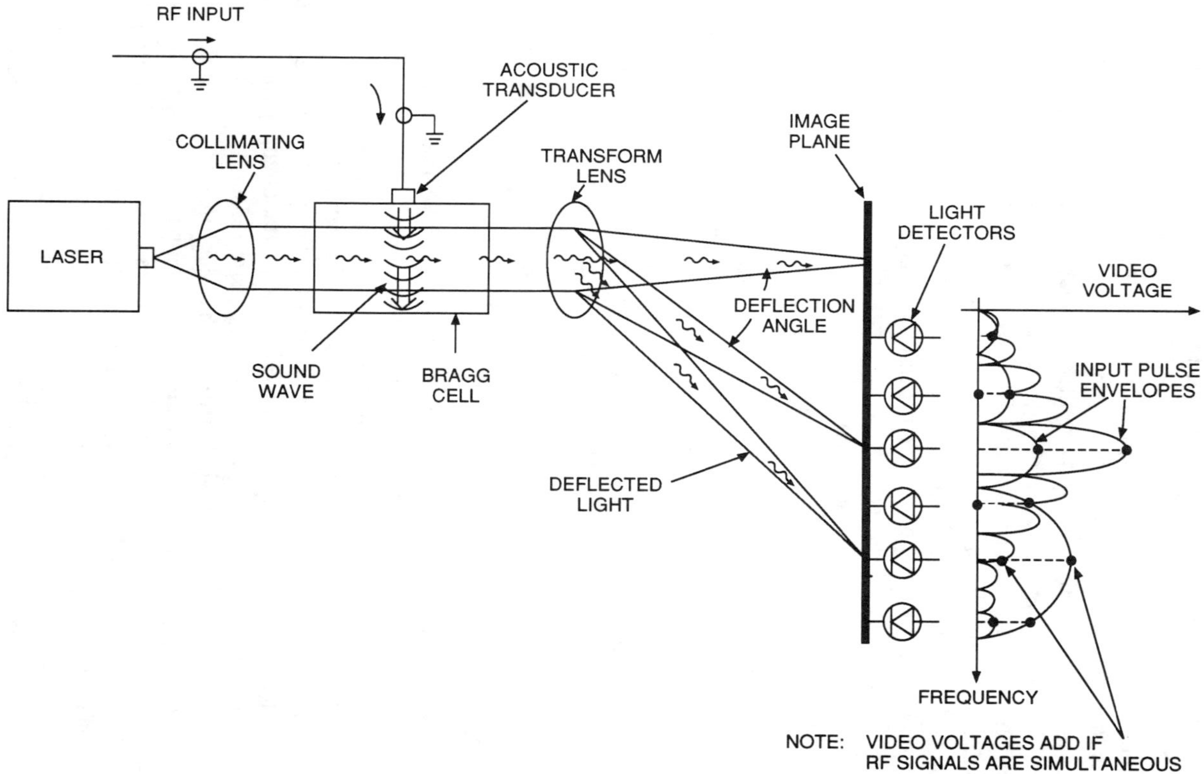

Figure 4.20 The acousto-optic receiver.

is converted into a sound wave at the identical frequency and at a proportional amplitude. A typical center frequency for this RF input is 1000 MHz, so frequency converters are needed to beat the signal down from the higher bands to this IF. In any event, the transducer has converted the RF EM signal into a sound wave that propagates through the Bragg cell perpendicular to the coherent laser beam. The Bragg cell is fabricated in such a way that this sound wave locally distorts the crystal in a manner dependent on the acoustic amplitude, which, in turn, distorts the phase front of the laser beam. These phase changes in the beam result in a deflection of a portion of the laser light beam when it reaches the transform lens. The theory explaining this deflection is similar to the angle steering of an RF phased array, although the wavelengths are radically different. The intensity of this deflected beam of light is proportional to the sound-wave amplitude, which, in turn, is proportional to the RF signal amplitude. Over most of the input signal's DR, only a small fraction of the laser beam intensity is so deflected. Generally most of the laser beam is incident on the null spot on the image plane. The deflection angle is linearly dependent on the sound wavelength, and hence linearly dependent on the RF carrier frequency. Light detectors are spaced along the image plane, usually continuously, to sense the deflected but still focused light wave. Just as an RF crystal detector senses RF carrier signals, these detectors sense the sum of all the signals simultaneously causing light rays to be incident on the light detectors. The resulting integrated voltage from the light detector is read by subsequent video circuits, and the detector usually must be deliberately discharged to make the next reading. The net result is a response similar to an RF channelized receiver, with its bank of RF crystal detectors. One caution regarding Figure 4.20, however, is that it is not to scale; indeed, the AO receiver typically uses a number of mirrors to deflect the light, usually at 90°, several times so that the optics will not require a long housing.

The block diagram in Figure 4.21 symbolically shows some of the AO receiver's light-detector interrogation circuits and output data buffering. The typical partitioning between the RF, optics, and acoustics on the one hand, and the video and digital on the other hand, is also shown. Just as the channelized receiver employs a MUX, a MUX is used in the AO receiver to scan the light-detector array, interrogating each detector in turn. More complex video and digital circuitry may be used to control the interrogation sequence in a more sophisticated fashion, as indicated in the figure. There is a strong parallel between the interrogation of the detectors and the tuning control of a superheterodyne receiver, although the signal-sensing capture of the light intensity energy by the light detectors has no corresponding parallel in the superheterodyne. Often, the amplitude magnitude information is desired, so an ADC is used to digitize the selected light detector's voltage. Just as for a channelized receiver, it is possible to multiplex the ADC or to have an ADC dedicated to each light detector; the second option raises the complexity and cost but facilitates a higher data throughput rate.

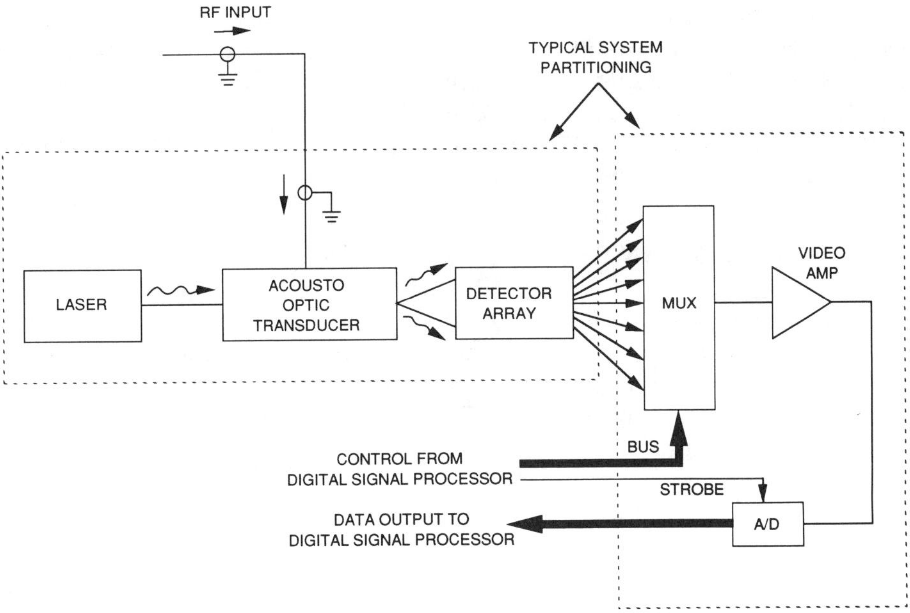

Figure 4.21 AO receiver outputs.

Because AO receivers have not yet been widely used, there are no typical parameters for the IBW and number of cells, or light detectors, but there have been IBW values from under 400 MHz to over 2000 MHz and numbers of cells from under 50 to over 3000.

The AO receiver design would be for input pulsewidths with spectra somewhat narrower than is implied in Figure 4.20, which is intended to illustrate the spectral overlap phenomenon. The TOA data word can be obtained in several ways: (1) the TOA resolution can be simply the time to interrogate all the cells, or (2) data latches associated with each cell can latch the TOA word (from the system clock counter) when the threshold is crossed for each cell, or (3) a control line can steer to a high priority cell, which then acts as a tuned filter receiver.

In summary, the AO receiver is considered a HPIR. This is because the Bragg cell operates in a linear fashion, with an overlapped, simultaneous signal condition causing no adverse impact to the detection process, albeit requiring complex channel interrogation circuitry to organize, queue, and pass on the information with sufficient throughput. The DR is set by the integrated RF noise in the channel width at the low end of the DR, and by either the detector or the laser total light intensity at the high end of the DR. The spacing of the detectors along the image plane is compatible with ECM channelized receiver requirements, that is, the spacing sets the channel IBWs, which are typically 5–25 MHz. AO receiver designers are challenged to fuse the diverse technologies into a small cost-effective receiver.

4.10 THE MICROSCAN RECEIVER

The block diagram and typical waveforms for the microscan receiver, or compressive receiver, is shown in Figure 4.22. A high speed sweep circuit generates a sawtooth waveform, often with a period well under a microsecond. This sweep voltage is applied to the VCO, as shown. This VCO signal is beat in a mixer with the RF input

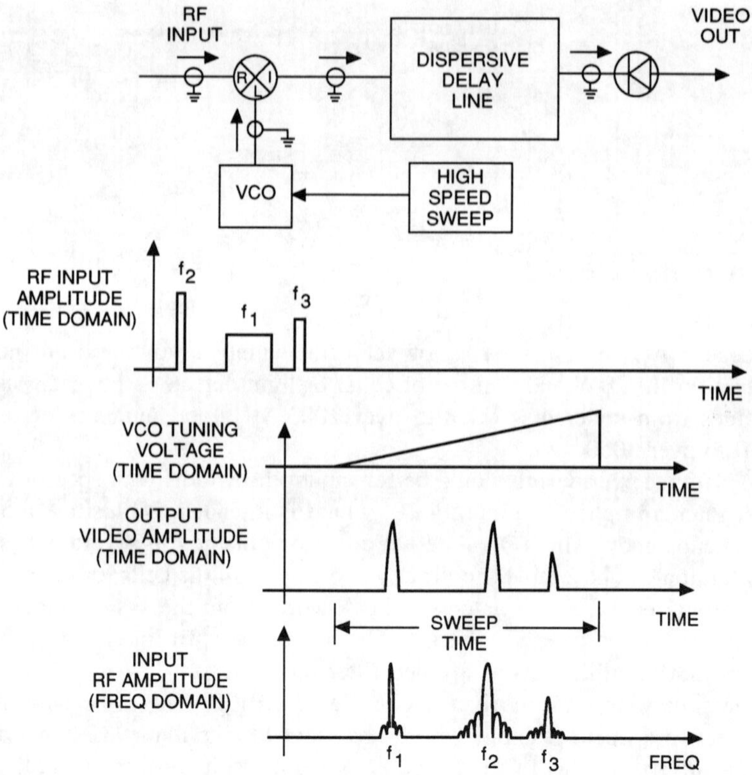

Figure 4.22 The microscan receiver.

signal. The result, for a fixed RF carrier input, is a swept intermediate frequency. This swept IF signal—constant amplitude for a constant amplitude RF input—passes through a dispersive delay line. The dispersion is intended to match the VCO sweep inversely as illustrated in Figure 4.23, which shows the delay line transfer function. The net transfer function from RF input to crystal detector time delay is inversely proportional to time for a fixed original RF carrier.

The dispersive delay line is a critical component. Dispersion means that the delay is dependent on frequency. The delay line used in the microscan receiver is designed to exhibit substantial dispersion. The dispersive delay line is usually fabricated with one of the "exotic" technologies, such as *surface acoustic wave* (SAW) or MSW technology. The FM imposed on the IF carrier, when it interacts with the delay line's dispersion, will continually reduce the delay through the dispersive delay line. As such, the portion of the signal entering later in the sweep will be delayed less because of its new IF value. Therefore, if the VCO sweep and dispersive delay line are properly matched, the IF signal exits the dispersive delay line at a time independent of when the signal entered the receiver, provided it entered within that sweep. The time at which the signal exits the dispersive delay line is only dependent on the original RF carrier frequency. As shown in the block diagram of Figure 4.22, the IF signal exiting from the dispersive delay line goes to a crystal detector which in turn generates a video voltage. Figure 4.22 also shows an example input, with three pulses having distinctive amplitudes, pulsewidths, and carriers. The input signals are shown in both the time and frequency domains. It can be seen that the crystal detector's output video voltage time-domain pattern aligns with the input RF spectrum pattern when the abscissas of both graphs are appropriately scaled and aligned,

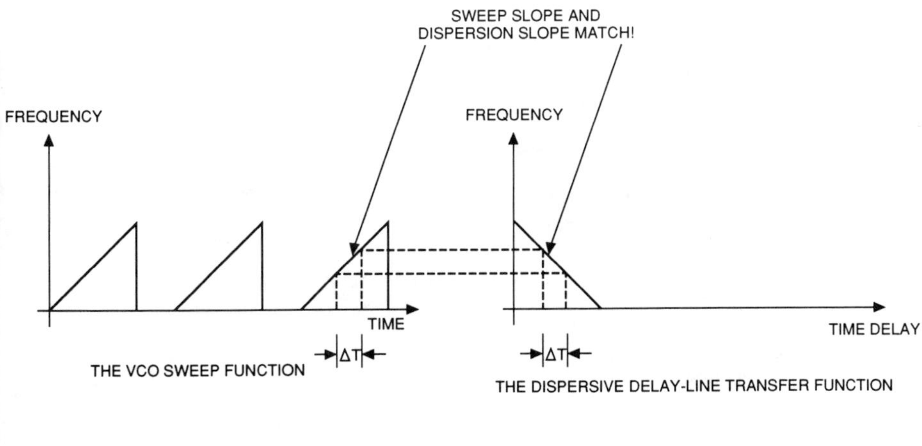

Figure 4.23 Utilizing dispersion.

and similarly their amplitude patterns align for appropriate ordinate scaling and alignment. The crystal detector's output video voltage, displayed as a function of time, therefore represents the spectrum of the input that occurred during the sweep time. In other words, the crystal detector output voltage can reveal the RF input spectrum on an oscilloscope just as a commercial laboratory spectrum analyzer reveals an RF input spectrum, except that this microscan output responds many orders of magnitude more quickly to input signal changes.

Figure 4.24 further illustrates the transfer function of (1) the time delay from input to the RF crystal detector *versus* (2) the input time of arrival within the sweep period, and *versus* (3) the input original RF, not IF, carrier frequency. Since three parameter variables are involved, the graph is shown as a three-dimensional drawing. At time *T*1, with input signal carrier F, the delay to the RF crystal detector is *D*1. At time *T*2, this signal experiences delay *D*2, *et cetera*. The microscan receiver is designed so that *D*1 − *D*2 = *T*2 − *T*1, that is, so that the signal will exit the dispersive delay line independent of when it entered, provided it entered within the sweep. The detected output voltage amplitude is therefore dependent on two things: (1) the input signal amplitude, and (2) the percentage of time the input signal was

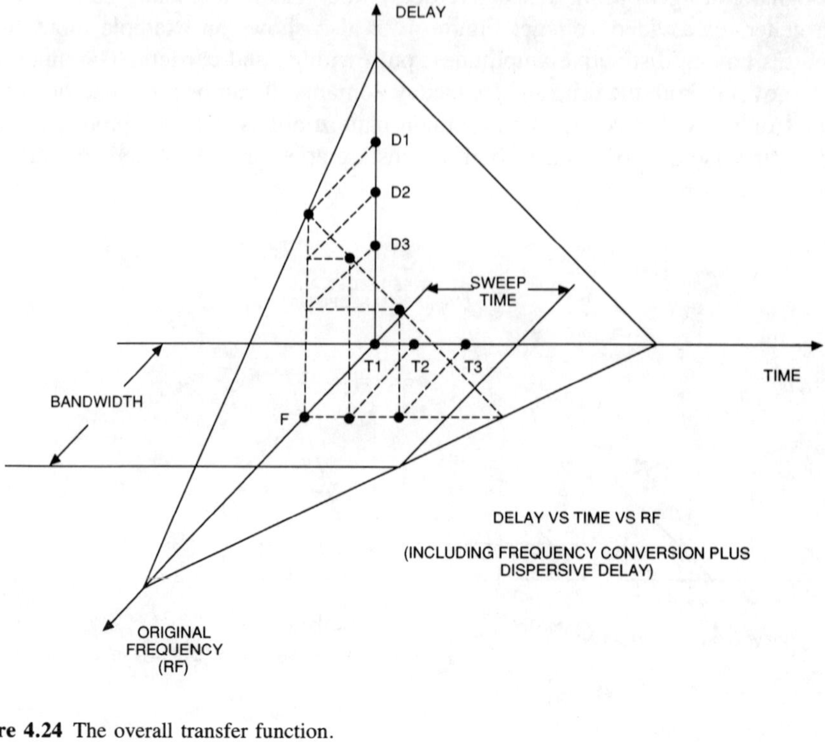

Figure 4.24 The overall transfer function.

present during the sweep. As shown in Figure 4.23, there is typically a dead time between sweeps to allow the dispersive delay line to empty and refill properly. As such, to ensure high probability of intercept operation, a dual channel unit is needed. There are two dispersive delay lines, two mixers, and two VCOs, with sweeps displaced by one-half period.

Fabricating a linearly dispersive delay line, with wide bandwidth, that is properly matched to a short VCO sweep, is a challenge. The shorter the sweep time, the less the TOA uncertainty, so it is important to minimize that sweep time. The other challenge is to digitize the signal in a form compatible with being passed to the DSP across a data bus. The time period of each individual count can be considered a cell, functionally equivalent to the bandwidth of the frequency channels of a channelized receiver. The complete-count period of all the cells is then functionally equivalent to the full bandwidth of all the channels of a channelized receiver. The count for each detection is latched, in a manner that is functionally equivalent to the flip-flop, or latches, shown in Figure 4.3 for the channelized receiver. One way this is accomplished is to use a high-speed counter as the source word to be latched for each detection, indicating the presence of a signal within that cell. Multiple difference amplifiers can be utilized to latch different amplitude thresholds; that is, a flash converter can be used to digitize the amplitude. Latching and buffering the data at these speeds is not simple. Many of these high speed counter latching problems and solutions have parallels to channelized receiver problems of latching a single channel for pulsed inputs; weighting circuits and guard logic can be used to inhibit latching a detection that is really centered on a nearby cell.

The achievement of wide IBW is limited by the capabilities of the dispersive delay line. The achievement of moderate size and cost effectiveness, compared to channelized receivers, appears realizable.The microscan receiver is also a HPIR, since the mixer and dispersive delay line actions are linear. The sensitivity is set by each count cell's IBW, just like the channelized receiver's sensitivity is determined by each channel's IBW. In other words, the description in Chapter 3 of how to calculate the MDS sensitivity of a CVR, and hence of each channel of a channelized receiver, can, to a reasonably good approximation, be also used to calculate the idealized MDS sensitivity for each channel of an AO receiver and each cell of a microscan receiver, despite the radically different technologies involved.

4.11 EW RECEIVER SUMMARY

This chapter has briefly described EW requirements and the implementation of several EW receiver technologies. It has also attempted to show the linkage between several receivers with widely differing technologies that nevertheless have functional similarities, by defining the transform receiver. The transforming receiver, unlike a spectrum analyzer, has wide IBW but narrow selectivity bandwidth. The definition explicitly included the data latching and buffering needed to place the data on a data bus in an orderly fashion, with one frequency word at a time, with tagged TOA words, despite the signals entering the receiver simultaneously. Therefore, according to this concept, a SAW receiver, an MSW channelized receiver, a lumped circuit filter bank, an AO receiver, and a microscan receiver can all be evaluated as transform receivers with the same set of tests; given the same set of black box requirement parameters, they can be considered functionally equivalent. The SAW, MSW, AO, and lumped filter circuit internally all use multiple detectors, one per channel, with a following interrogation MUX network, whereas the microscan receiver multiplexes one detector across many channels. In all cases, and despite the differing technologies, the sensed rise time is approximately the reciprocal of the channel IBW, and likewise the SNR is also dependent on the channel's IBW as shown in the MDS description in Chapter 3. The acoustic reverberations that create the channel filtering in a SAW receiver, the optic frequency-dependent deflection in the AO receiver and even the time-dependent amplitude swings in the microscan receiver are all functionally creating the same selectivity filtering despite the widely differing technologies.

The idealized EW receiver has wide IBW, narrow selectivity bandwidth and HPIR characteristics without external control. Barring such a receiver asset, the ECM system can employ a mix of receiver types, and properly manage them.

The preceding text has listed the parameters that should be defined properly to specify an EW receiver. This chapter will conclude with a summary of issues to watch out for when purchasing or specifying EW receivers. These issues include

- settling time of tunable receivers,
- past history phenomena of tunable receivers,
- potential problems in the tail of discriminator curves due to a wide DR,
- simultaneous signal capability,
- IM products or 3d order intercept,
- the response to combinations of strong and weak signals
- recovery time, especially after strong pulses,
- a proper definition for sensitivity,
- long and short pulse responses, an AC coupling consideration, and
- frequency resolution and minimum separation (these are not the same specification).

The entire class of transform receivers can and should be specified and tested independent of technical approach. The issue of properly assigning a detection to a single channel, over the full DR, and not being blinded by strong signals in other channels, should be carefully assessed. Output data rates and signal overload are also an important issue. Finally, for any receiver, and especially for the transform receivers, parameters of operation (e.g., number of channels, DR, *et cetera*) should definitely not be compared unless the sensed signal information is efficiently placed on a digital data bus; otherwise misleading and unfair comparisons are likely.

Chapter 5
Microwave Memory Loops

5.1 APPLICATIONS

This chapter is concerned with the theory and applications of the special phenomenon of feedback loops operating within a regime where the loop delay is much longer than the carrier period. Although feedback theory is well documented in the technical literature, such documentation is generally intended for a regime where the feedback delay is considerably shorter than the carrier period; this limitation is usually not stated explicitly. Although much of the theory in the literature applies to this regime, such operation in practice exhibits unique attributes. This chapter, therefore, will describe appropriate theory, calculation methodology, and attributes.

Three ECM-related applications within this regime of feedback systems will be covered: (1) the unusual results of operating a repeater system with finite antenna isolation, as illustrated in Figure 5.1a, (2) VSWR phenomena, as illustrated in Figure 5.1b, and (3) RML transponders as illustrated in Figure 5.2. The first two applications of the theory are for linear phenomena, while the third is for a nonlinear subsystem, which includes electronic control to initiate, sustain, and end the operations.

Although not covered in this text, the theory and calculation methodology also can be applied to many other phenomena and applications, including

- filter responses (e.g., SAW receiver filtering),
- laser operation, and
- financial stock valuations.

5.2 CALCULATION METHODS

The classic linear feedback circuit is shown in Figure 5.3, including the definition of symbols. This text will consider both signals and noise, so coherent and non-coherent analyses will be applied to these respectively. Assuming coherency, the transfer function of the circuit of Figure 5.3 is derived, as shown in Figure 5.4, by

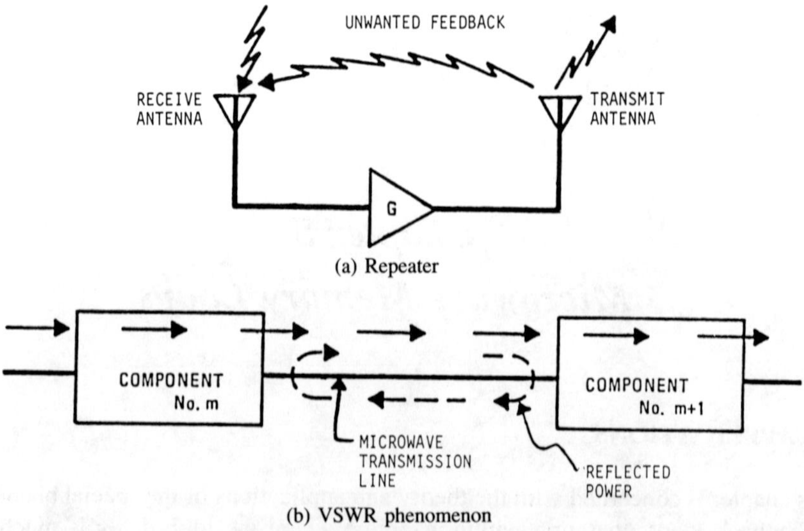

Figure 5.1 Short wavelength feedback applications.

describing the output voltage as if the feedback voltage were a second input, and then appropriately substituting and grouping terms. The mathematical result is that the normalized output is inversely proportional to the reciprocal of one minus the loop gain. The loop gain can be represented as a complex number, with magnitude and phase. Three numerical examples, using this derived transfer function, all with loop gains of −6 dB, but with different loop phase angles, are illustrated in Figure 5.5. Notice that the output phase can be pulled (i.e., shifted), and the output amplitude can be either larger or smaller than the straight-through output, that is, larger or smaller than the output with the feedback path cut.

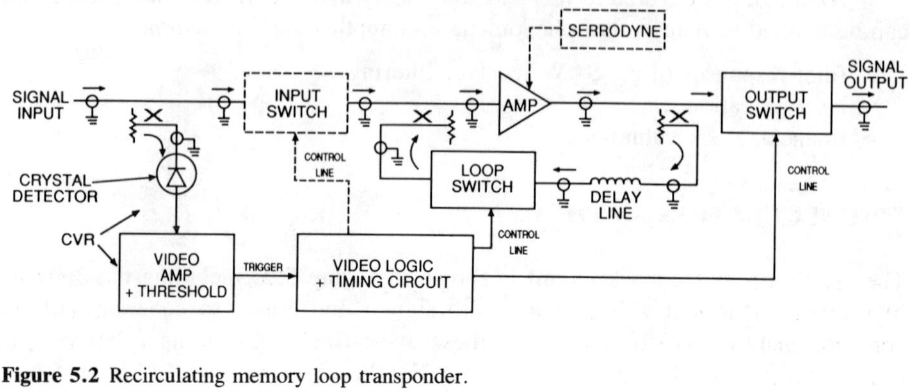

Figure 5.2 Recirculating memory loop transponder.

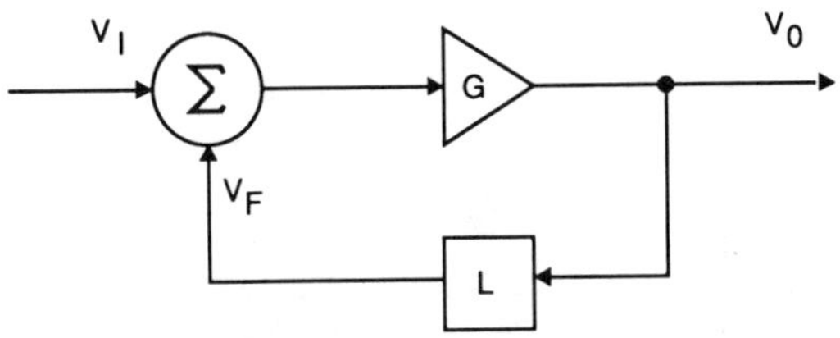

DEFINITIONS

G = FORWARD VOLTAGE GAIN

V_I = INPUT VOLTAGE

V_O = OUTPUT VOLTAGE

V_F = FEEDBACK VOLTAGE

L = FEEDBACK VOLTAGE GAIN ($L < 1$ FOR GIVEN APPLICATIONS)

A = NORMALIZED (TO OPEN LOOP) OUTPUT VOLTAGE AMPLITUDE

X = LOOP VOLTAGE GAIN (= GL)

Figure 5.3 The linear feedback circuit.

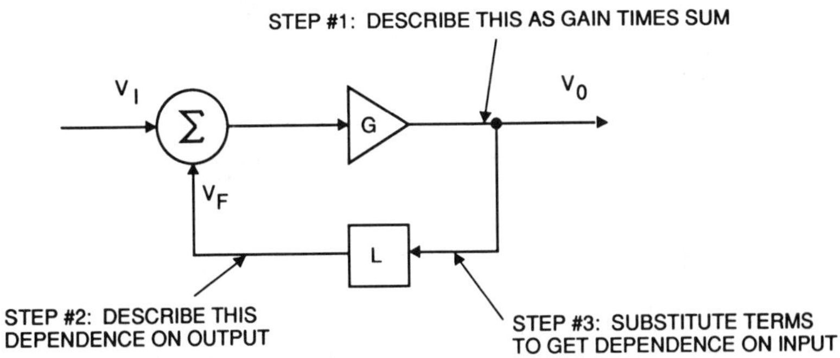

STEP	COMMENT	EQUATION
1	OUTPUT:	$V_O = G(V_I + V_F) = GV_I + GV_F$
2	BUT:	$V_F = L\ V_O$
3	SUBSTITUTING:	$V_O = GV_I + GLV_O$
4	GROUPING:	$V_O - GLV_O = V_O(1\text{-}GL) = GV_I$
5	DIVIDING:	$\frac{V_O}{V_I} = \frac{G}{1\text{-}GL} = G\frac{1}{1\text{-}GL}$ (TRANSFER FUNCTION)
6	NORMALIZED AMPLITUDE:	$A = \frac{1}{1\text{-}X}$

Figure 5.4 The coherent transfer function derivation.

EXAMPLE #	FEEDBACK	FEEDBACK IN dB	OUTPUT AMPLITUDE CALCULATION	OUTPUT AMPLITUDE IN dB	AMPLITUDE BAR GRAPH
1.	$X = GL = 1/2$	-6	$A = \frac{1}{1 - 1/2} = 2$	+6.02	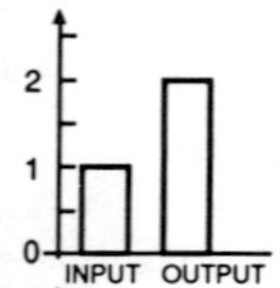
2.	$X = 1/2 \angle 180^{\circ} = -1/2$	-6	$A = \frac{1}{1 - (-1/2)} = 0.66$	-3.52	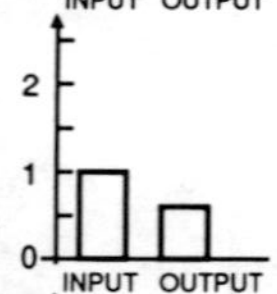
3.	$X = 1/2 \angle 90^{\circ}$	-6	$A = \frac{1}{1 - 1/2 \angle 90^{\circ}} = 0.89 \angle 26.6^{\circ}$	-0.97	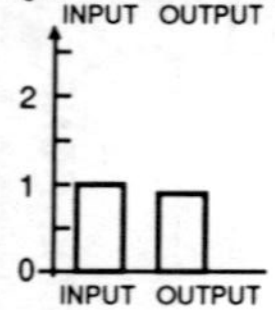

Figure 5.5 Coherent feedback numerical examples.

Some further comment regarding loop gain is in order. Small signal loop gain is defined by the following: if, somewhere within the loop, a transmission line was cut, and a generator's RF signal inserted into the cut transmission line, a power meter was connected to the other end of the cut transmission line, and a phase meter connected between these points, then the loop gain would be the measured ratio, or difference in dB, between the power meter power reading and the generator output power level, and the loop phase would be the measured phase difference.

One way to solve these feedback problems quantitatively is to use the transfer-function equation derived in Figure 5.4, and simply substitute as shown in the examples in Figure 5.5. There are, however, other ways to calculate the output of such a system, some of which are appropriate to transient solutions. One method is to use the result of long division as illustrated in Figure 5.6. This long division gives an infinite series for the $A = 1/(1 - X)$ normalized amplitude of Figure 5.4. It is quite reasonable to solve such series by using programmable calculators or a computer to sum a sufficient number of terms for the accuracy desired. Each term in these series can be perceived as the contribution of another pass around the loop by the original signal. Each time the signal goes around the loop, it experiences, or gets

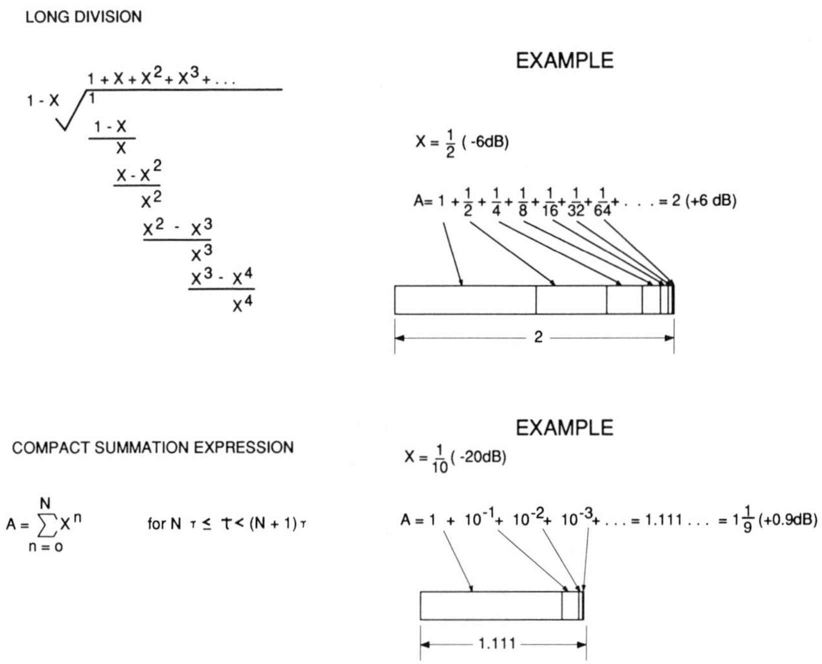

Figure 5.6 Long division.

multiplied by, the loop gain X. Figure 5.6 also shows two numerical examples of using such a series to evaluate specific problems. One example is for a feedback loop with zero degrees loop phase angle and −6 dB loop gain; the problem and answer are the same as shown in Figure 5.5, Example 1. Figure 5.6 also shows a more compact notation, using the summation symbol.

A third way to calculate the output is to use vector analysis, as illustrated in Figures 5.7 and 5.8. These two figures show the vector representation of the signal at one point in the loop, usually the output, normalized to unity and zero degrees phase angle for the zeroth circulation, or loop delay interval. The zeroth circulation is the time from the leading edge of the signal till the time of the feedback signal leading edge. In these vector representations, a constant frequency signal is repre-

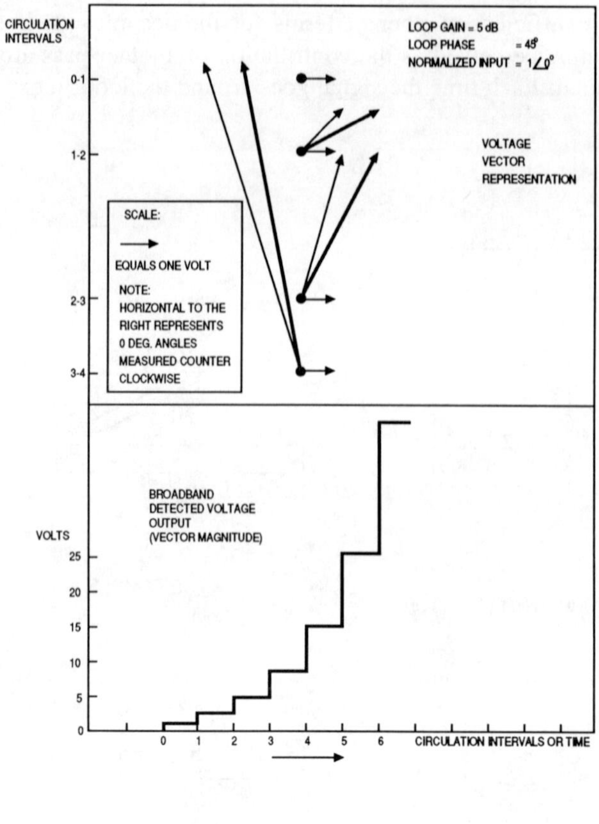

Figure 5.7 Vector analysis, 45° example.

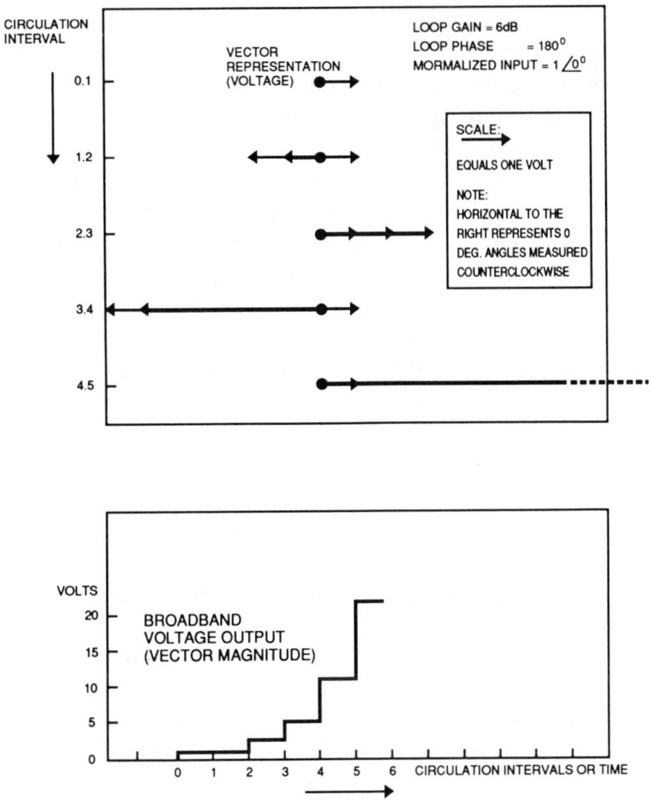

Figure 5.8 Vector analysis, 180° example.

sented as a vector with constant phase angle, so the vector diagram can be normalized to only one frequency.

The vector analysis example in Figure 5.7 is for a loop with 5 dB loop gain and 45° loop phase angle. A situation that would correspond to this normalized vector example would be, for instance, an RF loop delay of 100 ns and with an input signal having a carrier of 5001.25 MHz, or 5011.25 MHz, or 5021.25 MHz, *et cetera*, all of which have slightly in excess of 500 cycles of the signal in between the discontinuity points. The discontinuity points are the points in time where the amplitude and phase simultaneously jump to new values, which are held until the next discontinuity, a loop-delay, or circulation.

As shown in the example in Figure 5.7, the first circulation has a net output vector, shown as the heavy vector, which is the sum of the original signal plus the feedback signal, which is 5 dB larger in magnitude and shifted 45°. The parallelogram rule can be used to sum these two vectors. Note that the new net phase angle is less than 45°. For the second circulation interval, the unit vector at zero degrees, representing the original input signal, is summed with a feedback vector that is 5 dB larger than, and shifted 45° from, the net signal during the first circulation. The results for the third circulation are also shown. It can be seen that the vector magnitude is rapidly becoming large and the phase angle is being pulled from zero. At the bottom of Figure 5.7, the vector magnitude, that is, the broadband detected output voltage, is graphed as a function of time, out to the sixth circulation. Note that the amplitude stays constant between the discontinuity points, and the result looks like a staircase with ever larger steps.

Figure 5.8 shows the interesting case of an RF loop with 6 dB loop gain and 180° loop phase angle. This vector analysis example would correspond to, for instance, a loop with 100 ns loop delay and an input signal carrier of 5005 MHz, or 5015 MHz, or 5025 MHz, *et cetera*, again with over 500 cycles of the signal between phase discontinuities. Each circulation interval vector picture is for these five hundred-plus cycles of the signal, and characterizes the RF magnitude and RF phase for that interval. The vector magnitude is also plotted through six circulations in the lower part of the figure. Notice that the magnitude of the signal in the first circulation remains the same as it was for the zeroth circulation, even though the phase has changed radically. The ever-larger step staircase starts after these first two circulations.

For both examples of Figure 5.7 and 5.8, the signal grows without bound.

We expect that the reader, after examining these sets of vector diagrams, would be able to calculate the net vector for any case of loop gain, loop delay or circulation interval and frequency at any point in time. The reader should calculate the loop phase angle based on the loop delay and frequency, then calculate the new net output signal as the summation of the straight-through and feedback signals. This net output remaining constant, albeit as a carrier sinusoid, for the loop delay interval, then abruptly jumping to a new magnitude and phase, which remain constant until the end of that circulation interval, and so on.

In summary, this section has discussed the theory of coherent feedback for a regime in which the input frequency is much higher than the reciprocal of the loop delay, and has described three ways to calculate the output

- derived equation,
- infinite series, and
- vector analysis.

The derived equation is only valid for steady-state solutions, when the loop gain is less than zero decibels, and is not appropriate for calculating either the transient phenomenon or the runaway responses. If it is known, however, that the steady state solution is appropriate, the derived equation is the simplest to employ. The vector analysis approach is most appropriate for achieving insight into why the loop responds the way it does, although it can be a relatively cumbersome calculation method. Making these calculations may not be necessary because the outputs for the most common and important conditions are shown in graphical form in this chapter.

5.3 LINEAR COHERENT DYNAMIC RESPONSE

Figures 5.9 through 5.13 present the coherent signal feedback transient and runaway cases, that is, the dynamic response, in three-dimensional graphic form for a wide range of parameters, and hence should be a useful graphical solution reference so that actual calculations are not needed, or are only needed for final precision. Each figure graphs the results for loop gains from −5 dB through +5 dB, in steps of 1 dB, *versus* time from zero through twenty circulations, and *versus* amplitude from −10 dB through +40 dB. This sequence of graphs then shows the results for loop phase angles from zero through 180°, in steps of 45°.

In Figure 5.9, with the feedback perfectly in phase, for loop gains less than zero decibels, the response quickly settles to a steady-state amplitude; with loop gains

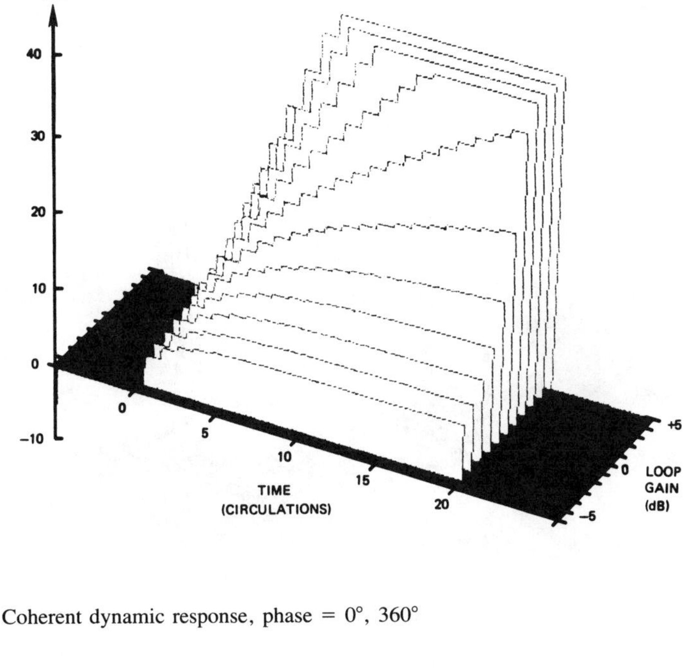

Figure 5.9 Coherent dynamic response, phase = 0°, 360°

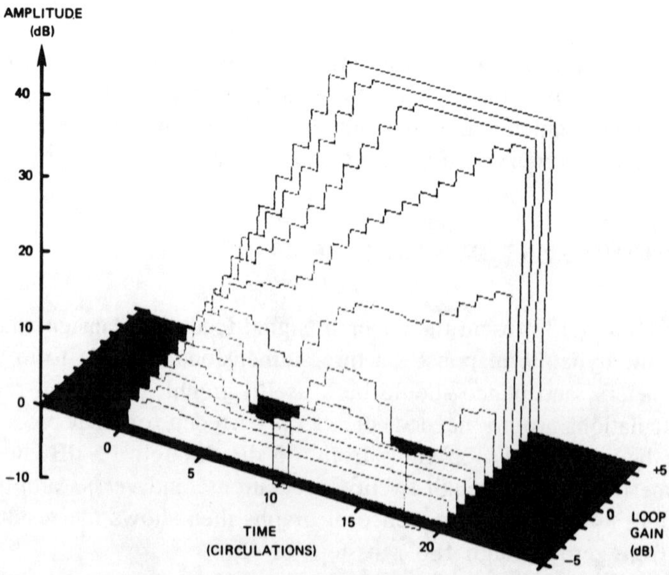

Figure 5.10 Coherent dynamic response, phase = 45°, 315°

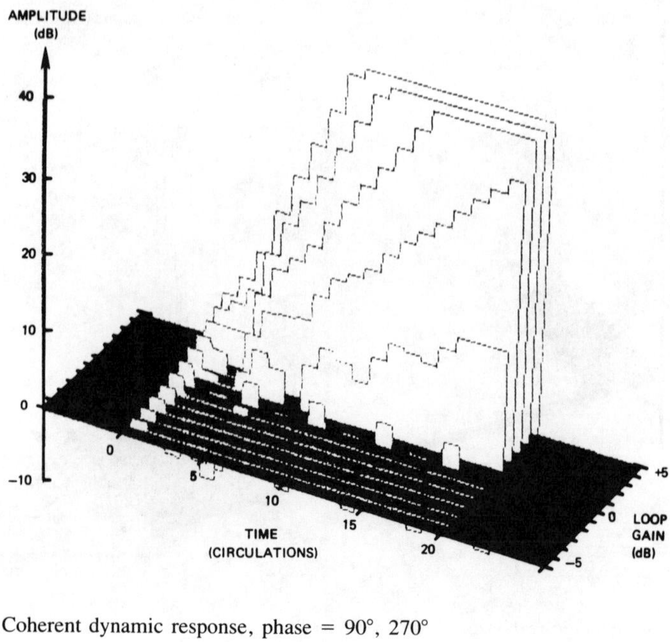

Figure 5.11 Coherent dynamic response, phase = 90°, 270°

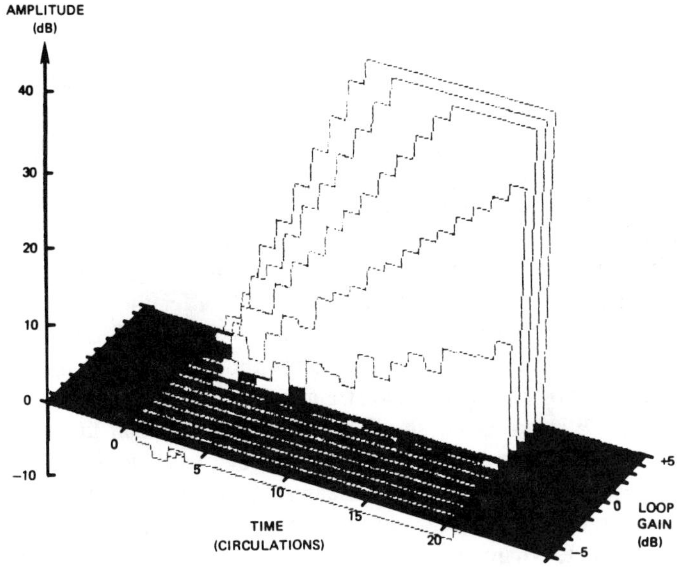

Figure 5.12 Coherent dynamic response, phase = 135°, 225°

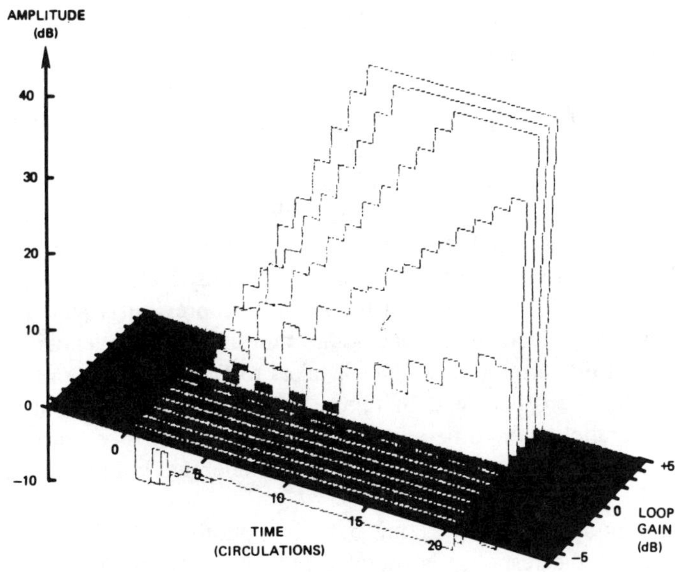

Figure 5.13 Coherent dynamic response, phase = 180°

greater than zero decibels the signal climbs to infinity, that is, it is a runaway condition; at zero decibels loop gain the signal does not quite reach a steady state solution.

In comparing the graphs in Figures 5.9 through 5.13, it can be seen that two general rules apply: 1) for loop gains less than zero decibels, a steady-state amplitude is always the result, and 2) for loop gains greater than zero decibels, the signal always eventually climbs toward infinite amplitude, that is, it is in a runaway condition. The in-phase case, with zero degrees loop phase, always has an amplitude larger than the straight-through reference amplitude after the zeroth circulation; in general, however, for all phase angles, the steady-state amplitude can be larger or smaller than the straight-through reference amplitude. The closer the loop-feedback phase is to being out of phase with the straight-through reference signal, the greater the tendency for the steady-state amplitude to be smaller than the reference amplitude. We can also see that for the larger, or out of phase, loop phase angles there is a tendency, especially for loop gains near zero decibels for the amplitude to oscillate between larger and smaller values; however, following the rule just stated, if the loop gain is greater than zero decibels, the amplitude will eventually break free and climb to infinity. This oscillation or alternating between larger and smaller values is the result of the feedback cancelling the straight-through signal, and hence cancelling the subsequent feedback on the next circulation, so that the straight-through signal will not be cancelled on that circulation, and so on.

Figures 5.9 through 5.13 are accurate graphical solutions for the case of feedback delays much longer than the signal carrier period. Because three-dimensional graphs are somewhat awkward to use, however, and because the graphical solution is needed over an even further range of parameters for the two extreme cases of 0° loop phase and 180° loop phase, these cases are shown in two-dimensional format in Figure 5.14 and Figure 5.15 respectively. In these figures, loop gain is scaled along the abscissa from below −12 dB to above +14 dB; time, in normalized circulation intervals, is logarithmically scaled along the ordinate up to 1000 circulations, and the different amplitude-level curves, from −5 dB to +60 dB are appropriately labeled. In these figures, therefore, a steady-state condition is a straight vertical line. Figure 5.14 clearly illustrates the rule previously stated, that for loop gains less than zero decibels a steady-state amplitude will eventually result. This figure is useful to determine how long it takes to stabilize the amplitude. For loop gains greater than zero decibels, in Figure 5.14, a line projected straight up from the abscissa, parallel to the time axis, will successively cross each of the amplitude curves; this means that as time passes the amplitude gets larger and larger, without limit.

The amplitude growth in Figure 5.14 is monotonic. In Figure 5.15, for 180° loop phase angle, the response is quite complicated; the amplitude response is not a monotonic single valued function of time. Some curves are for even circulations and some are for odd. For problems with loop gains greater than zero decibels, where the loop phase angle is unknown, Figure 5.15 is appropriate to determine the longest

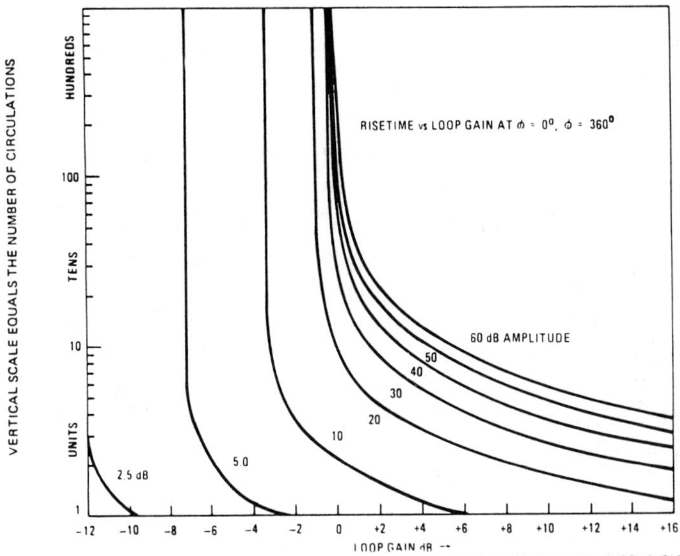

THIS GRAPH REPRESENTS THE TIME, IN CIRCULATIONS, NEEDED TO REACH VARIOUS AMPLITUDE LEVELS, AS A FUNCTION OF LOOP GAIN.

THIS GRAPH IS VALID FOR: (1) A COHERENT CW SIGNAL WITH LOOP PHASE OF ZERO DEGREES, OR (2) A NOISE COMB LINE.

Figure 5.14 Coherent dynamic response, 2D, phase = 0, 360°

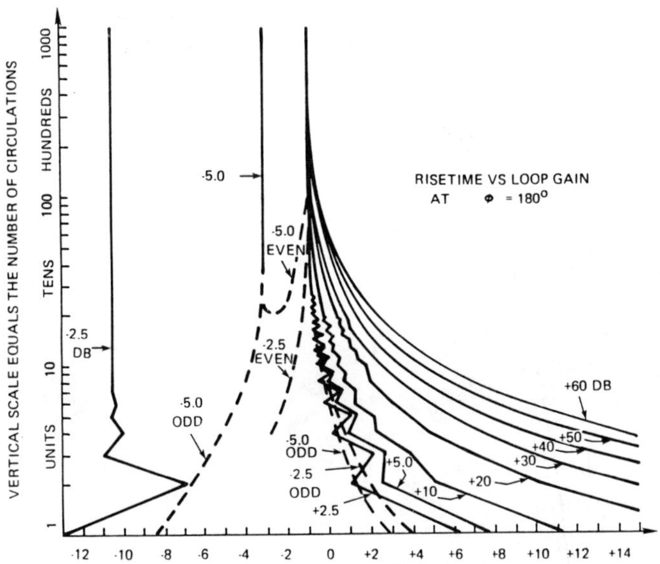

This Graph Represents The Time, In Circulations, To Reach Various Amplitude Levels, As a Function Of Loop Gain.

This Graph Is Valid For: (1) A Coherent CW Signal With Loop Phase Of 180 Degrees, or (2) Noise Amplitude Midway Between Two Comb Lines.

Note: The Dashed Lines Are Only Valid For An Odd or Even Number Of Circulations, As Indicated.

Figure 5.15 Coherent dynamic response, 2D, phase = 180°

time it will take for the amplitude to reach a given amplitude, while Figure 5.14 is appropriate to determine the shortest time.

5.4 LINEAR STEADY-STATE RESPONSE AND RIPPLE

The above text has described three methods to calculate the coherent response of a linear feedback loop with loop delay much longer than the signal's carrier period, and has also given numerous graphical solutions, the use of which is really a fourth way to quantify the operation of such a loop. These graphical solutions were intended to exhibit the character of the dynamic response, although they also reveal steady-state solutions. However, problems related to steady-state conditions often demand even more precision than the previous graphical solutions provide. Figure 5.16 gives the general graphical solution, with sufficient resolution, for the coherent steady-state response of a loop with feedback delay much longer than the signal's carrier period. We consider this graph probably the most useful graph in the entire text,

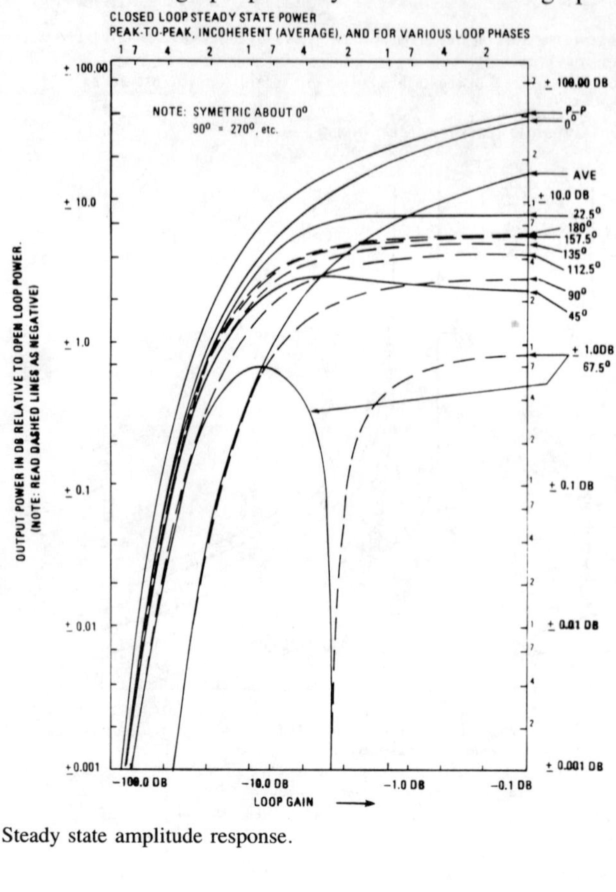

Figure 5.16 Steady state amplitude response.

and it is appropriate for use on a wide variety of problems. The loop gain is scaled logarithmically in decibels on the abscissa to obtain a wide range for this parameter, while the amplitude is scaled logarithmically in decibels on the ordinate to obtain a similarly wide range, but necessitating that the dashed curves be read as negative; different curves are plotted for the different loop phase angles, every 22.5°, and are appropriately labeled. There are two additional curves on this graph which are very important: the average (AVE) curve and the peak-to-peak (P-P) curve. The AVE curve gives the average amplitude as a function of loop gain, where the average is taken over all loop phase angles. The P-P curve gives the difference in decibels between the extreme cases of 0° loop phase, which gives the maximum amplitude, and 180° loop phase, which gives the minimum, or most negative on a log scale, amplitude.

The P-P, 0°, AVE, and 180° curves are particularly useful for voltage standing wave ratio problems. For such VSWR phenomena, the recirculation is up and down the transmission line between components, as shown in Figure 5.1b. The reader may find the use of the graph in Figure 5.16 easier and more appropriate than the traditional Smith Chart for solving such VSWR problems. The Figure 5.16 P-P curve represents the peak-to-peak ripple, or the difference in decibels between the maximum and minimum gain or loss across the band, caused by the VSWR phenomena. Similarly, the zero degree curve gives the maximum peak amplitude, and the 180° curve gives the minimum amplitude possible because of VSWR phenomena. Likewise, the AVE curve represents the average amplitude as a function of frequency resulting from the VSWR-induced ripple. The P-P 0° peak, 180° minimum, AVE, and all the other curves are with respect to what the transmission would be if there were an ideal lossless isolator in the transmission line.

To use the graph of Figure 5.16 to solve VSWR problems, the loop gain is needed. Figure 5.17 shows how to obtain the loop gain given the connecting transmission line loss and the interface impedances of the components specified by either VSWR or reflection coefficients. Knowledge of the components' reflection coefficients is the most convenient, but it is traditional to specify the component impedance in terms of VSWR. The VSWR value for the generator is actually a misnomer. The reference impedance is 50 Ω, unless stated explicitly otherwise. The reflection coefficient can be obtained from the component VSWR with the equation given in Figure 5.17, or from the graph in Figure 3.6: the loop gain, X, is the loss through the transmission line, plus the load reflection loss, plus the loss through the transmission line in the opposite direction, plus the reflection loss of the generator, so that the net loss over the complete closed path has been summed. The equivalent equation for the magnitude of X, the loop gain, is shown in linear form, as opposed to logarithmic form, in Figure 5.17. The equation for the normalized amplitude A is also shown in Figure 5.17, as a function of X; it is the same equation as in Figure 5.4. The normalized amplitude A is with respect to what the power delivered to the load would be if an ideal (loss-less, infinite isolation) isolator were in the transmission

EQUATIONS

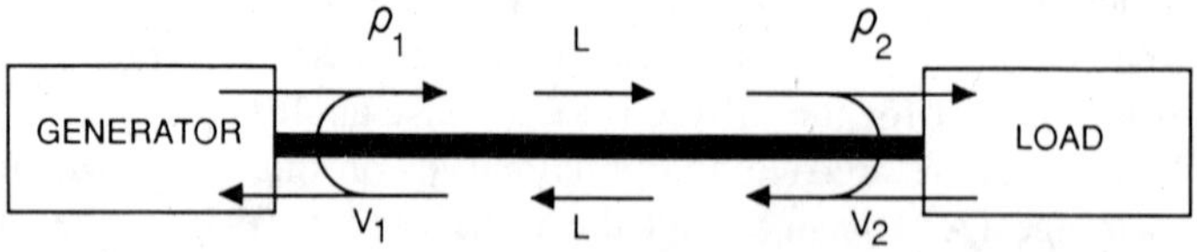

$$V = \frac{1+\rho}{1-\rho} \qquad \rho = \frac{V-1}{V+1}$$

$$|X| = \rho_1 \rho_2 L^2 \qquad A\angle\theta = \frac{1}{1 - X\angle\text{Ø}}$$

DEFINITIONS

V = VSWR = VOLTAGE STANDING WAVE RATIO
ρ = REFLECTION COEFFICIENT ($\rho < 1$)
L = TRANSMISSION LINE ONE-WAY GAIN ($L < 1$)
X = LOOP GAIN
A = NORMALIZED (TO OPEN LOOP) AMPLITUDE
Ø = LOOP PHASE ANGLE
θ = RESULTANT SIGNAL PHASE
L' = TRANSMISSION LINE ONE-WAY GAIN IN DECIBLELS ($L' < 0$)
L' = $20\ \mathrm{LOG}_{10}(L)$

Figure 5.17 VSWR equations and definitions.

line. The loop gain *X*, which must be less than unity, that is, less than zero decibels, for a physically realizable system, is a complex number whose phase angle is the sum of both reflection coefficient phase angles plus the phase value given in Figure 5.18 resulting from the round trip delay up and down the transmission line. The figure is generic, because it illustrates the phase angle resulting from any delay. The Δf value gives the frequency span after which the phase angle, and hence the steady-state amplitude, will repeat. For example, a 7 ft, 10 ns transmission line, which takes 20 ns round trip for VSWR phenomena, will have its VSWR-induced loop phase, and hence amplitude ripple, repeat every 50 MHz.

Given the value of *X*, the loop gain, and the phase phi, for any loop, such as the repeater case of Figure 5.1a, or the VSWR case of Figure 5.1b, the graphical solution for the amplitude is given by Figure 5.16. For the repeater use of Figure 5.1a, the normalized amplitude is the amplitude as if the antenna isolation were infinite.

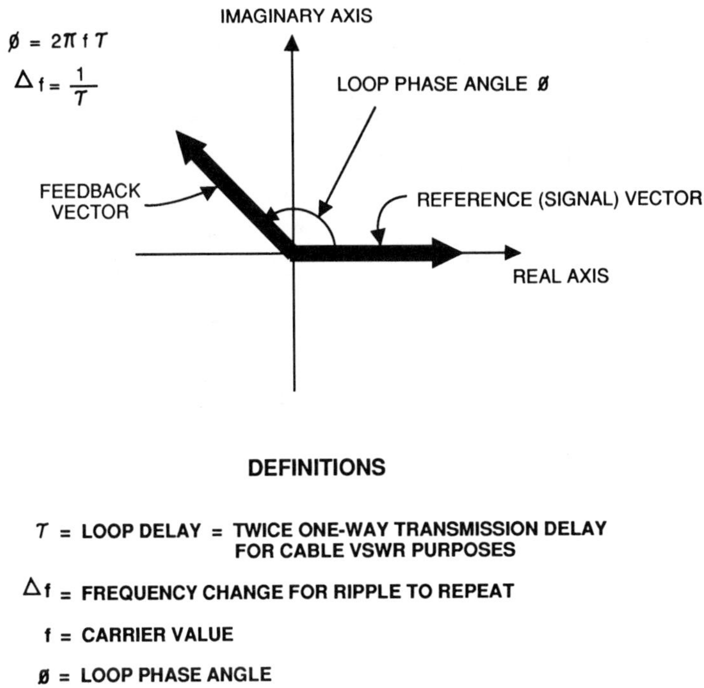

Figure 5.18 Loop phase angle.

Two examples of this VSWR ripple phenomena are given in Figure 5.19, for two sets of component VSWR values and transmission line losses, with symbols corresponding to Figure 5.17. Such a calculation is made by first converting both the VSWR values to reflection losses in decibels, then adding these to twice the transmission line loss in decibels to get the loop gain. Usually the reflection phase angles are not known, but, as shown in Figure 5.18, the ripple pattern repeats in frequency at a rate of twice the reciprocal of the cable one-way delay. Figure 5.16, or one of the calculation methods described, then gives the desired ripple value. The graph for Example 1 (Figure 5.19), of VSWR induced ripple *versus* phase, is identical to a graph for, say, a repeater corresponding to Figure 5.1a, with straight through repeater gain of 60 dB and antenna isolation of 63 dB, of isolation induced ripple *versus* phase. The graph for Example 2, for VSWR-induced ripple, is identical to a graph for a repeater with straight-through repeater gain of 60 dB and antenna isolation of 70 dB, for isolation-induced ripple. The calculation methodology for both is identical, once the loop gain and loop phase are determined. Furthermore, the calculation of the ripple period, in frequency, which is dependent only on the loop

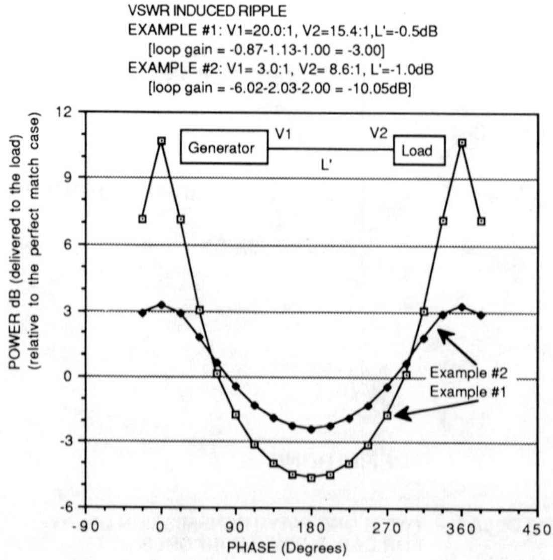

Figure 5.19 Ripple examples.

delay, is also identical, given such delay, for both the finite isolation repeater and VSWR applications. In other words, the abscissa of Figure 5.19 can be scaled in frequency instead of the normalized phase that is shown.

5.5 LINEAR NONCOHERENT STEADY-STATE RESPONSE

The classic linear feedback circuit is shown in Figure 5.20, including the definition of symbols suitable for noncoherent signals, especially noise. Because noise is noncoherent, voltages cannot be added vectorially with fixed phase angles. Therefore, noncoherent signals, such as noise, when brought together at the summing point shown in Figure 5.20, must be added on a scaler power basis. Given that constraint, the transfer function is derived as shown in Figure 5.21. Figures 5.20 and 5.21 for the noncoherent noise case should be compared to Figures 5.3 and 5.4 for the coherent signal case. The equations for both the transfer function and the normalized average output power as a function of loop power gain are shown. An example calculation of the average power is shown in Figure 5.22a; for a loop gain of -3 dB, the normalized output power will be $+3$ dB. Figure 5.22b shows the mathematical relationships between the coherent and noncoherent cases. The interesting thing about these relationships is that coherent calculations can be utilized to predict noncoherent output power! As shown in Figure 5.22b, the noncoherent output power is the integral of the square of the coherent amplitude response when integrated over

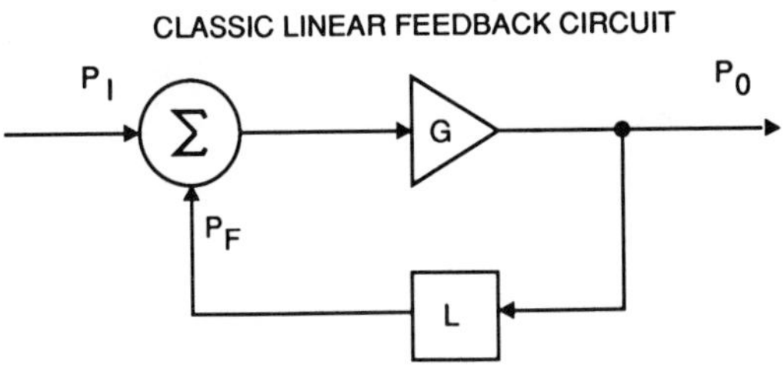

DEFINITIONS

G^2 = FORWARD POWER GAIN

P_I = INPUT POWER

P_O = OUTPUT POWER

P_F = FEEDBACK POWER

L^2 = FEEDBACK POWER GAIN ($L^2 < 1$)

P_{AV} = AVERAGE NORMALIZED (TO OPEN LOOP) OUTPUT POWER

X^2 = LOOP POWER GAIN

Figure 5.20 Noncoherent feedback.

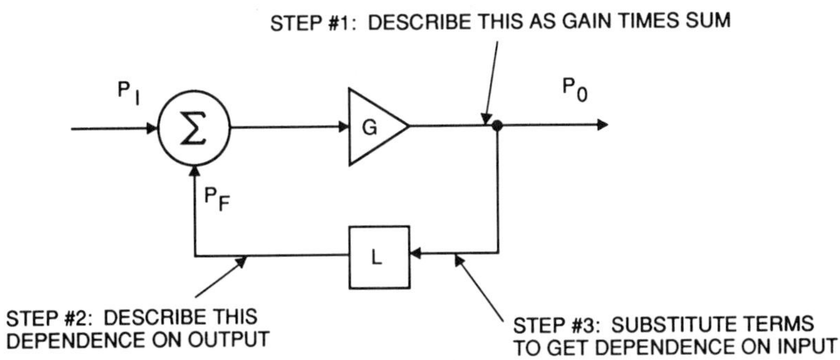

1 OUTPUT DESCRIPTION: $P_O = G^2(P_I + P_F) = G^2 P_I + G^2 P_F$

2 BUT: $P_F = L^2 P_O$

3 SUBSTITUTING: $P_O = G^2 P_I + G^2 L^2 P_O$

4 GROUPING: $P_O - G^2 L^2 P_O = P_O(1 - G^2 L^2) = G^2 P_I$

5 DIVIDING: $P_O/P_I = \dfrac{G^2}{1-G^2L^2} = G^2 \dfrac{1}{1-G^2L^2}$ (TRANSFER FUNCTION)

6 NORMALIZED AVERAGE POWER: $P_{AV} = \dfrac{1}{1-|X|^2}$

Figure 5.21 The noncoherent transfer function derivation.

FIGURE A: NONCOHERENT EXAMPLE

GAIN $X^2 = 1/2;\ X = 0.707$ (-3 dB)

AVERAGE OUTPUT POWER $P_{AV} = \dfrac{1}{1-(0.707)^2} = 2$ (+3 dB)

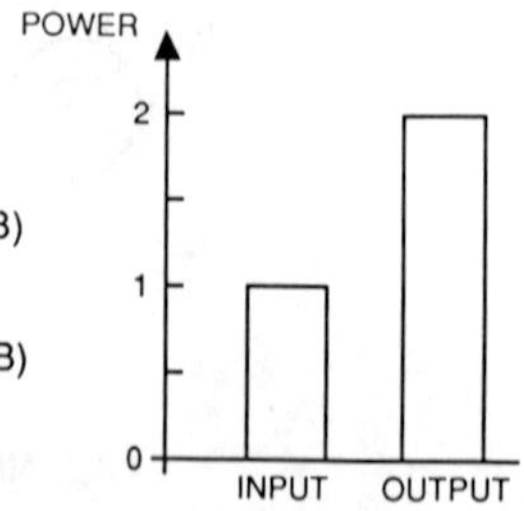

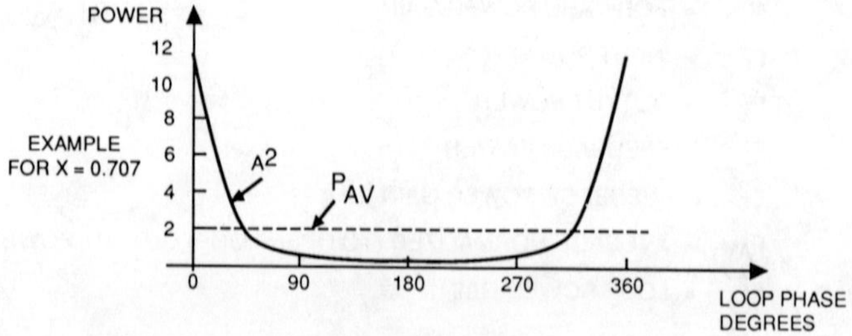

FIGURE B: COMPARISON OF EQUATIONS

$$A\angle\theta = \frac{1}{1 - X\angle\phi} \qquad P_{AV} = \frac{1}{1 - |X|^2}$$

$$P_{AV} = \frac{1}{2\pi}\int_0^{2\pi} |A|^2\, d\phi = \frac{1}{2\pi}\int_0^{2\pi} \left|\frac{1}{1 - X\angle\phi}\right|^2 d\phi$$

Figure 5.22 Coherent *versus* noncoherent response.

360°. Therefore the AVE curve of Figure 5.16 gives the noncoherent power output! In other words, if a coherent signal is swept in frequency, the average output power across the band will be equal to the power for a noncoherent noise signal with the same input power. For this statement to be accurate, the frequency sweep must correspond to multiples of 360°, with the frequency-to-phase relationship determined according to Figure 5.18.

It is also interesting that noncoherent noise will appear to have coherent qualities, and it is possible for thermal noise to exhibit a ripple effect similar to the coherent signal ripple examples shown in Figure 5.19. When measuring noise, the measuring instrument's bandwidth is important, whereas that is not usually an issue for coherent signals. To measure ripple in the noise floor, a narrowband swept receiver must be employed with bandwidth much less than the ripple period. That is, using a broadband thermal noise input to produce such a ripple curve that closely tracks, albeit with an amplitude offset, the ripple curve produced by a swept coherent signal tone, the bandwidth of the receiver measuring the noise must be much less than the reciprocal of the loop delay. As this narrowband receiver is swept across the band, the measured noise power will agree with the values given by the coherent transfer function of Figure 5.4. The normalized amplitude is simply the *kTbFG* value. If, on the other hand, the measuring receiver bandwidth *b* were a multiple of the reciprocal of the loop delay, then the measured noise power would not vary across the band, and would be equal to the power predicted by the noncoherent transfer function of Figure 5.21.

How can thermally generated broadband truly random noise exhibit a ripple that matches the ripple of a coherent signal swept across the band? An example should help clarify the phenomena. Suppose that a loop with $G = 23$ dB forward gain, 26 dB feedback loss, and 100 ns loop delay were being measured for noise output levels. In this case, $X = 23 - 26 = -3$ dB loop gain. If the measuring receiver had a bandwidth *b* of 10 MHz or 20 MHz or 30 MHz, *et cetera,* then the measured noise power would not vary with frequency and would be 3 dB greater than given by the *kTbFG* equation, according to AVE curve of Figure 5.16 or as shown in the example of Figure 5.22, based on the normalized average power equation in Figure 5.21. If, however, the receiver had, for example, a bandwidth of $b = 1$ kHz, then there would be a measured ripple that repeated every 10 MHz. A 1 kHz receiver bandwidth is sufficiently narrow with respect to this ripple period of 10 MHz for the curves of Figure 5.16 to apply. The peak of this measured ripple would be 10.7 dB, the minimum would be −4.7 dB, and the average would be +3.0 dB, all with respect to the *kTbFG* noise level, according to the 0°, 180° and AVE curves of Figure 5.16 or the equations of Figure 5.4 for the peak and minimum voltage, and according to the equation of Figure 5.21 for the average power. With the receiver bandwidth *b* placed at any one frequency point in band *B*, to monitor the noise's random amplitude and phase, the filter's output signal amplitude or phase would never change very much in less than a millisecond; that is, a 1 kHz receiver takes about a millisecond to change its output fully, regardless of how fast the input changes. This receiver is thus only sensitive to 1 kHz bandwidth components of the noise. These components will have truly random amplitude and phase over time intervals much longer than one millisecond. However, because the loop feedback takes just 100 ns, that is, 0.0001 milliseconds, that feedback will be coherent, and will reinforce or cancel. Although the long-term amplitude and phase is random, if the amplitude and phase are known at one instant, the corresponding value 100 ns later cannot be very different, as measured by this 1 kHz bandwidth receiver, even for noise. Hence, 100 ns feedback will coherently interact with the straight-through noise. A 10 MHz bandwidth receiver sees no such coherent response, just the scalar addition of feedback power predicted by the equation in Figure 5.21.

5.6 LINEAR NONCOHERENT DYNAMIC RESPONSE

Figure 5.23 presents the linear noncoherent noise signal feedback transient and runaway cases, that is, the dynamic response in graphic form for cases where the operating band has carrier periods much shorter than the loop delay. This figure is similar to the coherent dynamic response of Figures 5.9 through 5.13 for the various loop phase angles, except, of course, that there is no phase angle associated with Figure 5.23, because the feedback is noncoherent. Just as for Figures 5.9 through 5.13, the curves are graphed for each integer loop gain value from −5 dB through +5 dB, for time from 0–20 circulations, and for amplitude levels from −10 dB through +40 dB, normalized to the amplitude as if the feedback path were cut. According to the discussion of the previous section, Figure 5.23 is not appropriate to predict noise power levels as measured by receivers with instantaneous bandwidths that are narrow with respect to the reciprocal of the loop delay. Indeed, for Figure 5.23 to be accurate for predicting the loop's power amplitude output of true thermal noise, the receiver bandwidth must be precisely an integer multiple of the reciprocal of the loop delay. For practical purposes, however, a large multiplier need not be precisely an integer; for example, the error caused by a receiver that was 100.3 times the reciprocal of the loop delay instead of precisely 100.0 would be quite slight.

An examination of Figure 5.23 reveals that the same two rules also apply for the noncoherent feedback case: (1) for loop gains less than zero decibels, a steady-

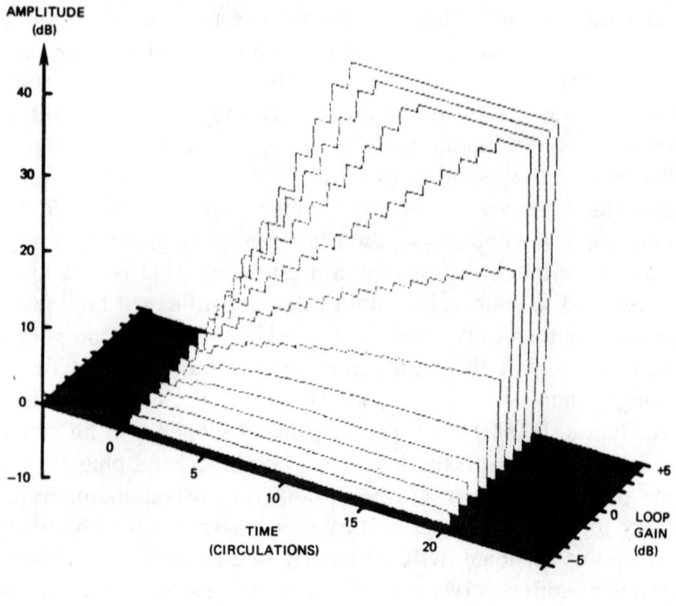

Figure 5.23 Noncoherent dynamic response.

state amplitude is always the result, and (2) for loop gains greater than zero decibels, the signal always climbs toward infinite amplitude; that is, it is in a runaway condition. Unlike the general coherent cases, two further rules apply for the noncoherent case: (3) after the zeroth circulation, the amplitude is always larger than zero decibels, and (4) the amplitude increases monotonically with time, in a step-like fashion, each circulation. Because the graph in Figure 5.23 is intended to predict noise amplitude *versus* time *versus* loop gain, it must be remembered that the prediction is for the nominal noise amplitude; the input used in the transfer function is a randomly time-varying signal, and such variations will occur at the output, depending on the measuring bandwidth, during each circulation, even though the nominal amplitude, graphed in Figure 5.23, is constant for the duration of each circulation.

A graphical solution for the noncoherent dynamic response over an even wider range of parameters, is presented in the two-dimensional graph of Figure 5.24. This graph is the noncoherent equivalent of the graphs of Figures 5.14 and 5.15, except that phase is not meaningful for Figure 5.24. It can be seen that all amplitudes are greater than zero decibels. The curves become perfectly vertical, indicating that the response has reached its steady-state non-time-varying condition, only for loop gains less than zero decibels. Just as in Figures 5.14 and 5.15, loop gain is scaled along

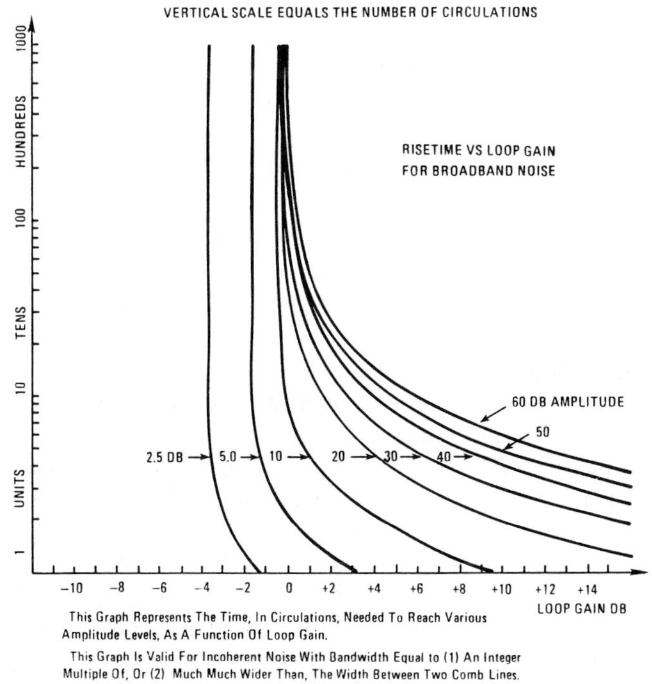

Figure 5.24 Noncoherent dynamic response, 2D.

the abscissa from below −12 dB to above +14 dB; time, in circulation intervals, is logarithmically scaled along the ordinate up to 1000 circulations; and the different amplitude level curves are appropriately labeled from +2.5 dB to +60 dB. This figure is useful to determine how long it takes the noise amplitude to stabilize, once the feedback path is closed, or once an amplifier feeding the loop is turned on and the amplifier's noise enters the feedback loop. There must always be some noise input to every system, even if it is the thermally generated RF power on the input cable. For loop gains greater than zero decibels, a line projected vertically from the abscissa of the graph of Figure 5.24, parallel to the time axis, will successively cross each of the amplitude curves; as time passes the amplitude grows monotonically without bound.

5.7 SUMMARY OF LINEAR FEEDBACK THEORY

The theory of linear feedback systems with loop delay much longer than the carrier period, as presented above, is summarized in Figure 5.25. In cases where significant dynamic response is evident, the spectrum of the output will be significantly altered with respect to the input. For such systems, dynamic amplitude changes are generally accompanied by dynamic phase changes, both of which impact the spectrum. Once steady state conditions are achieved, the spectrum, in response to a coherent CW input signal, will be pure. The output noise level of the system, however, is always degraded by feedback, as viewed by wideband receivers. As shown in Figure 5.16, the output noise level of a system, as measured by a receiver with bandwidth much less than the reciprocal of the loop delay, may actually be improved by feedback.

5.8 RECIRCULATING MEMORY LOOP TRANSPONDERS

The typical block diagram of a *recirculating memory loop* (RML) transponder is shown in Figure 5.26. The key waveforms are shown in Figure 5.27. The input signal enters at the left, in Figure 5.26, and passes through a coupler and optional input switch. The signal then passes through another coupler (the feedback summing point, as indicated in the preceding discussions), passes through the loop amplifier, passes through the feedback coupler and is then incident on the output switch. The initial state of this switch depends on the jamming technique waveform. The waveforms of Figure 5.27 are consistent with a RGPO waveform at a point in time when the delayed-transponder pulse has been delayed a little more than a pulsewidth. For such a response the output switch will initially be off, that is, in a high attenuation state, when the signal is first incident on it. The other path, from the coupler fed by the loop amplifier, has the signal incident on the RF delay line. The length of this delay is a critical issue for the RML designers, and the considerations for selecting that delay will be discussed below; typically the loop delay is 10–25% of

CONDITION: THE CARRIER PERIOD ≪ LOOP DELAY

- APPLICATIONS
  - REPEATER WITH FINITE ANTENNA ISOLATION
  - VSWR PHENOMENA
  - NONSATURATED AMPLITUDE RESPONSE OF AN RML TRANSPONDER
  - FILTERS
- ANALYTICAL TREATMENT
  - CLOSED FORM EQUATION
  - INFINITE SERIES
  - VECTOR ANALYSIS

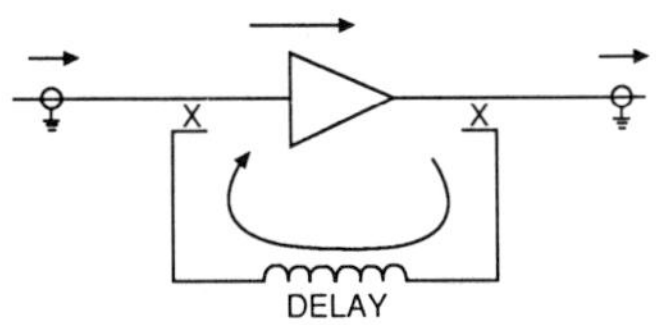

- AMPLITUDE RESPONSE
  - LOOP GAIN DEPENDENCE:
    - CONTINUAL GROWTH WHEN $|X| > 0$ dB
    - REACHES STEADY STATE WHEN $|X| < 0$ dB
  - LOOP PHASE DEPENDENCE
    - INFLUENCES RATE OF GROWTH WHEN $|X| > 0$ dB
    - RESULTS IN STEADY STATE CUSP SHAPE WHEN $|X| < 0$ dB
  - BROADBAND NOISE VS. A COHERENT SIGNAL CENTERED IN b
    - POWER VS. PHASE /FREQUENCY IDENTICAL WHEN $b \ll \frac{1}{\tau}$
    - POWER VS. PHASE/FREQUENCY IS FLAT FOR NOISE WHEN $b \gg \frac{1}{\tau}$
- SPECTRAL RESPONSE
  - PULSE SPECTRUM SHAPE ALTERED WHEN $T \approx \tau$
  - A CW INPUT RESULTS IN A PURE SINGLE SPECTRAL LINE
  - NOISE LEVEL ALTERED
  - CW (STEADY STATE) SIGNAL GAIN ALTERED
  - NOISE FIGURE ALTERED UNLESS $b \ll 1/\tau$

NOTES: b = MEASUREMENT BANDWIDTH; τ = LOOP DELAY;
T = PULSE WIDTH; X = NET LOOP GAIN

Figure 5.25 Linear feedback summary.

the input pulsewidth. The loop delay is considerably shorter than the input pulsewidth, but orders of magnitude longer than the input-carrier period. When the leading edge of the input signal emerges from the delay line, it will be incident on the loop switch.

The RF signal input initially passes through an RF coupler, as stated above, and the other leg feeds a crystal detector as part of a CVR that outputs a digital pulse in response to the presence of an RF pulse somewhere within the wide band of operation. The IBW times the loop delay typically has a value of several hundred. All the RF components shown in the block diagram of Figure 5.26, including the crystal detector, have this wide IBW. The digital logic detector output triggers the

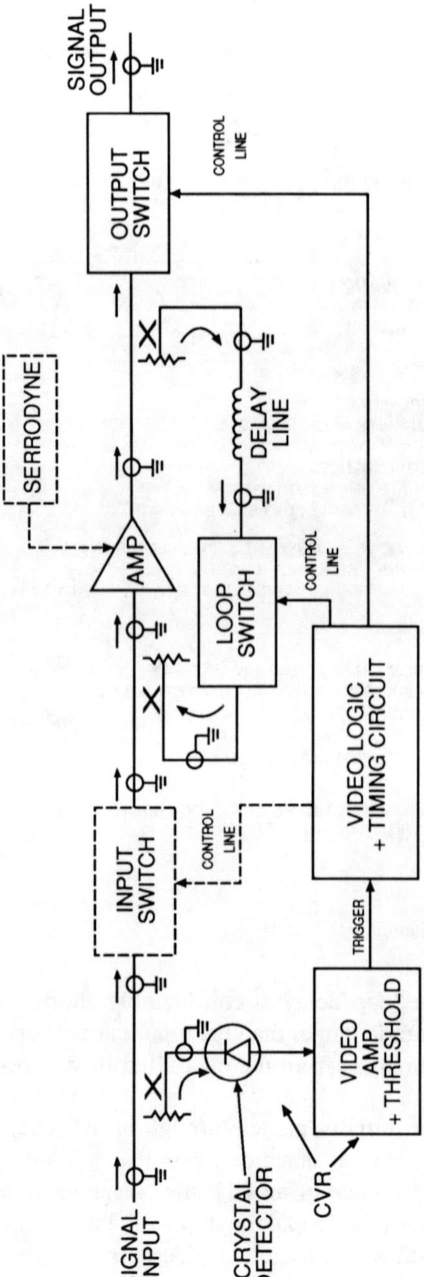

Figure 5.26 Recirculating memory loop transponder.

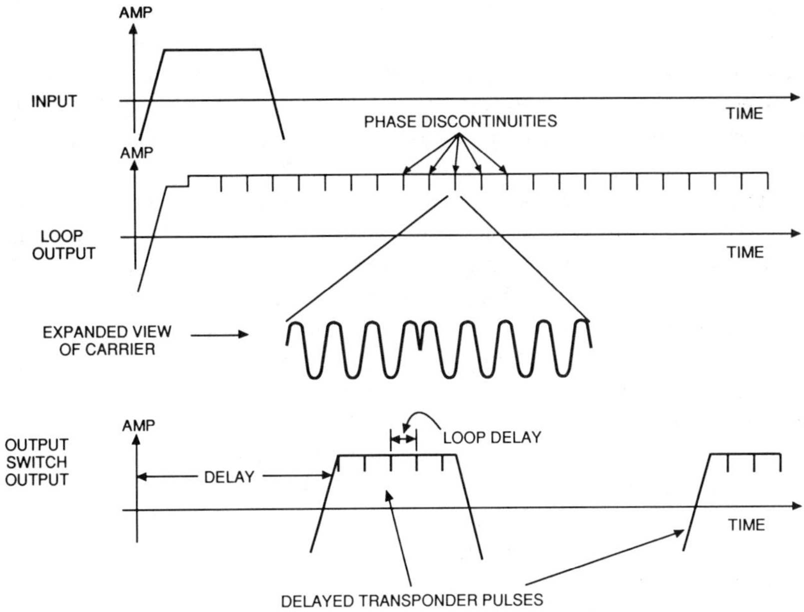

Figure 5.27 RML transponder waveforms.

video logic and timing circuit, which puts out a command to turn on the loop switch, usually immediately. The loop switch has usually just turned on as the leading edge of the signal emerges from the RF delay line. Note that the loop delay is not long by video design standards, so little if any deliberate delay needs to be added to the video path controlling the loop switch. However, the *output switch* remains in its initial off state for deliberate range delay, as in Figure 5.27. To continue the RF signal flow description, the RF signal, from the delay line, passes through the loop switch and into the coupler, the coherent summing point. The output of this coupler is on the main path described just above. The signal then continues to recirculate around the loop, maintaining the input carrier even well past the trailing edge of the input pulse. The envelope of this recirculating signal is shown in Figure 5.27, labeled "loop output," as the stretched signal that is incident on the output switch. The output switch, under the control of the video logic and timing circuit, then gates this stretched pulse. As shown in Figure 5.27, the output switch is gated to create output pulses, such as a moving delayed transponder pulse and a hook-delayed transponder pulse. After the video logic and timing circuit has gated out the last false pulse, the output switch is left off, ready for the next cycle, and the loop switch is turned off, thereby extinguishing the recirculating signal in preparation for the next cycle, that is, in preparation for the next input pulse.

Figure 5.27 shows the typical timing relationships between the loop delay, the input pulsewidth, and the delays generated by the timing circuits. The result gives the appearance that the RF pulse, with its particular carrier value, has been delayed. The RGPO waveform is shown in the "time *versus* time" format in Figure 1.15, with corresponding walkoff pulse and hook pulse shown in Figure 1.15 and Figure 5.27.

The optional input switch of Figure 5.26, if employed, is commanded off just as the loop switch is commanded on. This ensures that the feedback and straight-through signals do not really sum, and hence the output amplitude *versus* time function is not dependent on the loop phase angle. Therefore, the use of the input switch would give predictable amplitude rise times independent of the input carrier value; however, as the linear feedback theory graphical solutions indicate, the typical rise time would be improved if the input were not cut off.

The loop amplifier is shown with a serrodyne drive video control signal, to control the loop phase angle. When RMLs were first designed and deployed in large quantities, the loop amplifier was a TWT. TWTs could have their electrical phase length varied with a video input voltage; this input voltage was usually a sawtoothed waveform known as a serrodyne signal, as shown in Figure 2.14. Modern RML systems would certainly use an SSA; such amplifiers have no convenient phase control. To implement loop phase control with an SSA, a digital phase shifter, or digilator, could be put within the loop; the sawtoothed-equivalent waveform is shown in Figure 2.16.

To summarize the operation of the RML transponder of Figure 5.26, the input signal is recirculated around a loop with delay significantly less than the input pulsewidth, converting it to a continuous full power signal incident on the RF output switch with approximately the original carrier frequency. The output switch is used as a gate, the on time of which is the output pulsewidth. Various pulse configurations can be generated by properly controlling the output switch.

The loop is designed in such a way that with the loop switch off while the RML is awaiting a new pulse, the loop gain is well below zero decibels, and with the loop switch on when the input carrier has just been detected, the loop gain is initially well above zero decibels. Since the gain is greater than zero decibels during initial storage, linear feedback theory predicts unlimited growth. Sustained growth cannot really be maintained in practice; the loop amplifier is the device in the RML transponder that saturates, limiting the growth. The saturation of the loop amplifier drops its gain sufficiently to bring the net loop gain to precisely zero decibels, provided the loop is stable in this condition. Therefore, if the initial small signal loop gain is very high, the loop amplifier will drive hard into saturation. This means that the loop output, when the output switch is gated on, will be a constant power output, and will be independent of the power level of the input. This is advantageous because the ECM system will transmit at maximum power over a wide dynamic range.

5.9 RML STORAGE TIME

Figure 5.28 illustrates the output of an RML transponder with four photographs, as viewed by a broadband crystal detector, preceded by a filter, for four different input power levels separated by 10 dB. The RML illustrated did not include an output switch to gate the signal into false pulses. We can see that the output amplitude does not vary significantly over this 30 dB dynamic range. By way of contrast, the input signal is barely visible when it is 10 dB down from the first case. The trailing edge of the RF memory is determined by the loop switch. When the input is 20 dB down, at −30 dBm, the amplitude can be seen to droop slightly toward the tail end of the RF memory storage. When the input is 30 dB down, at −40 dBm, the signal memory ends prematurely, not when the loop switch is commanded to the off state.

How long will an RML store the input carrier? What happens when the RML ceases to store the input carrier? As linear feedback theory predicts, once the loop switch is closed, the recirculating signal and the noise in the loop drive toward saturation. Once saturation occurs, linear feedback theory is no longer sufficient. Suffice it to say, however, that as long as the loop's amplitude remains stable, the power output will remain at full power until the loop switch is opened. During this period, the signal and the noise are struggling for control of the loop. "Having control of the loop" means being the primary source of RF power for the saturating component within the loop—normally, the loop amplifier. The noise almost always wins in the end.

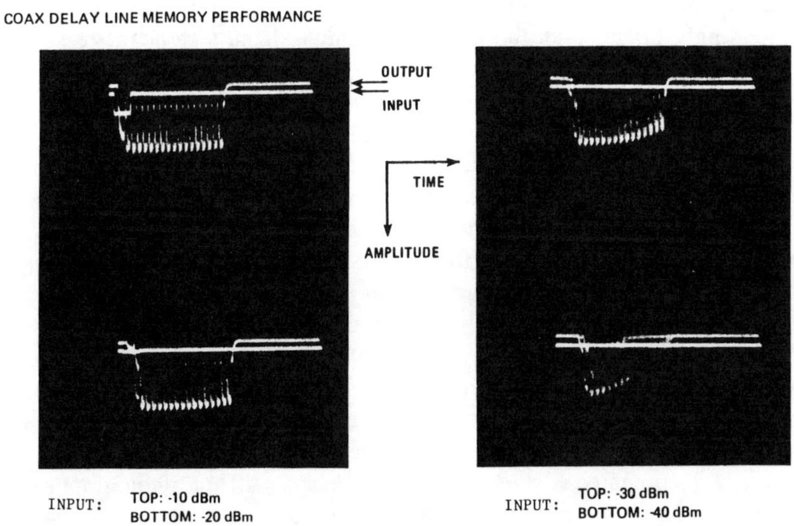

Figure 5.28 Output amplitude *versus* input amplitude.

Initially, when the loop switch is closed, both the signal and the KTbFG noise race for the saturation point. Because the signal input is at a higher level than the noise, it has an advantage and saturates the loop first, nominally dropping the loop gain to zero decibels, which nominally stops the growth of the noise. The amplitude rise of the signal is illustrated by the linear feedback theory coherent dynamic response graphs of Figures 5.9 through 5.13. If the signal carrier is such that the loop phase, determined as per Figure 5.18, is close to 180°, then the rise to saturation will take longer than for most other frequencies. If the "input switch" of Figure 5.26, however, is employed, then the rise of both the signal and noise at each circulation will be equal to precisely the initial closed loop linear gain at each circulation. Otherwise, without the input switch, the noise will initially build toward saturation as in Figure 5.23.

In general, then, the signal, if it is not too weak, wins the race to reach the saturation point of the loop before the noise. This drops the loop gain to precisely zero decibels for the signal, otherwise the signal power would change on the subsequent circulations. Therefore, nominally the noise will also cease to grow any further. The signal, nevertheless, is not stored by the loop forever; practical effects result in the noise capturing the loop eventually. There are two key parameters that determine the length of stretching: (1) the loop ripple (gain variation *versus* frequency), and (2) saturation-induced suppression in the loop amplifier. Because of the ripple factor, the loop gain only drops to precisely zero decibels at the input signal frequency. Elsewhere in the band, the loop gain is not precisely zero decibels. Because the input frequency in general will not correspond to the highest gain point of the small signal loop gain within the band, the loop gain at other frequencies will be correspondingly higher than the gain at the input signal's frequency, except for the saturation suppression phenomena.

Figure 5.29 shows the typical noise capture sequence of a wideband RML resulting from the combination of loop gain ripple, saturation gain compression and saturation-induced suppression. Unlike Figure 5.28, which is a photograph of the RML output in the time domain, Figure 5.29 is a set of 22 photographs of spectrum analyzer measurements, which document the frequency domain response of an RML as a function of time, that is, as a function of range delay. Normally, spectrum analyzers do not record spectral differences as a function of time; indeed, time is supposed to be meaningless in frequency-domain descriptions. Nevertheless, most engineers consider it natural to describe frequency-domain changes with time, not quite improperly, because of the vast range of time-frequency scales needed for ECM system operation. To make these spectrum measurements, the spectrum of the output was documented for different programmed range delays of the output switch. Hence the spectrum is documented for different range delays. All the photographs are for an 800 ns gating-on of the output switch, for the set of individually programmed range delays.

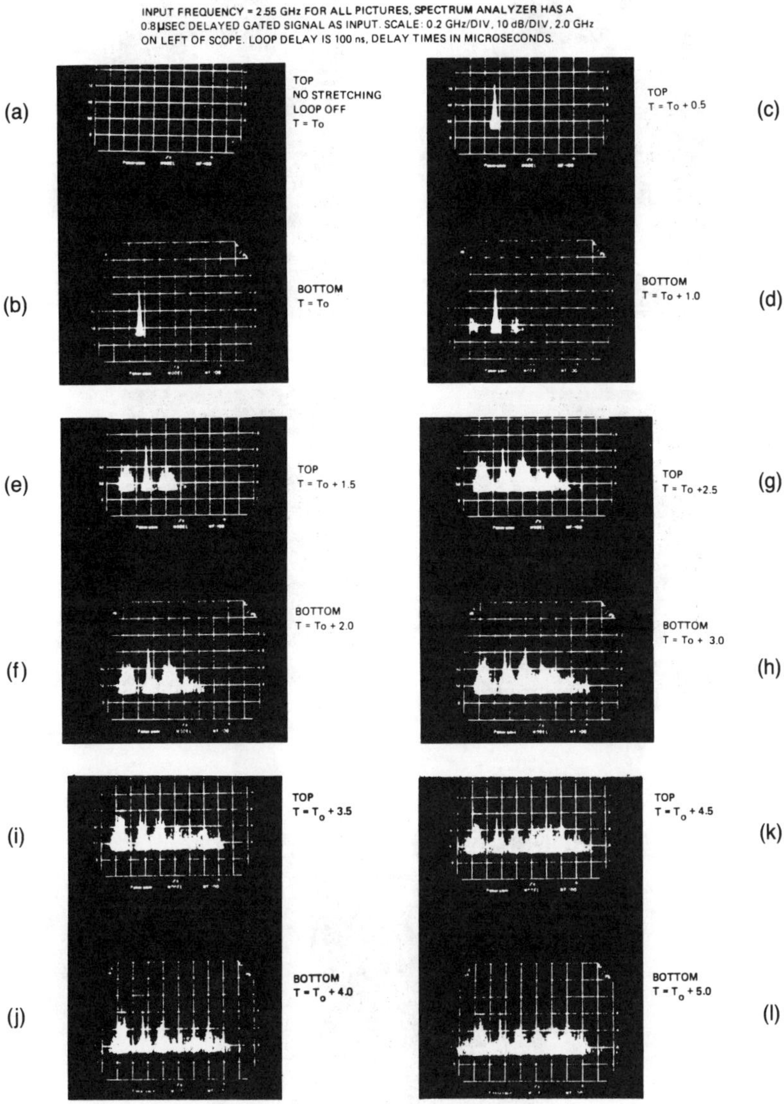

Figure 5.29 Typical noise capture sequence.

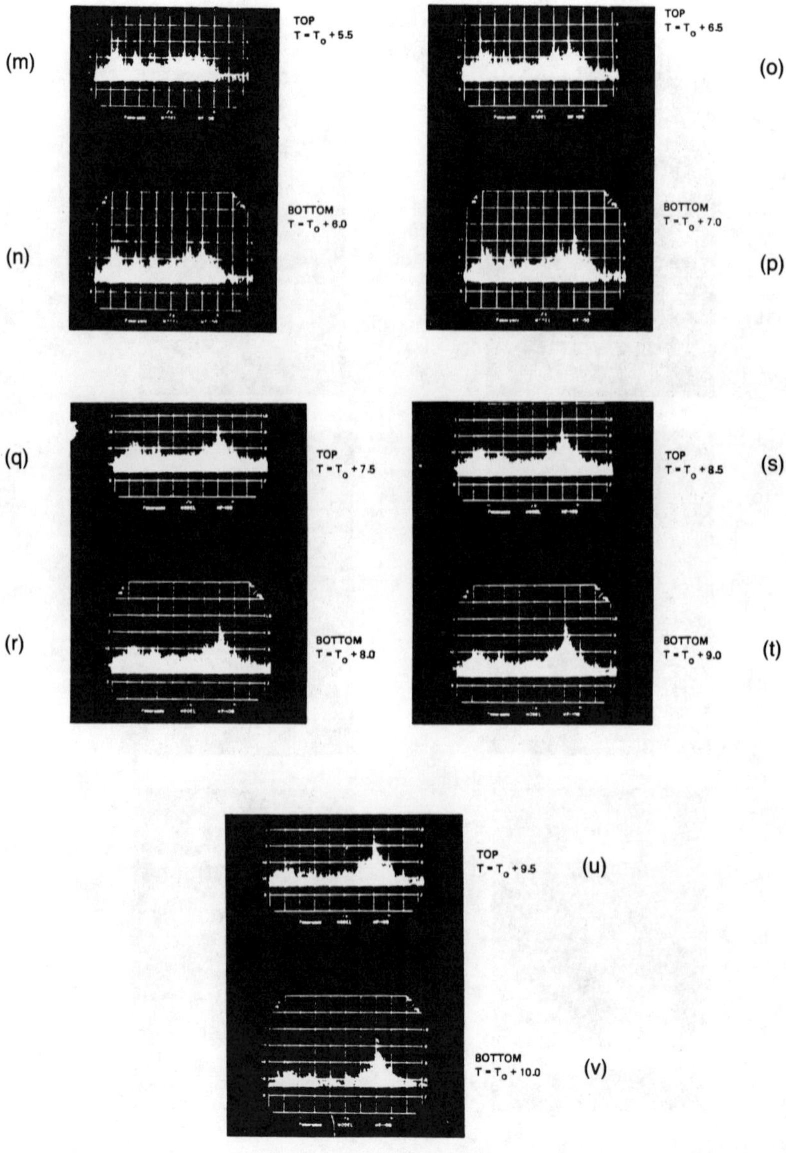

Figure 5.29 Continued.

Figure 5.29a shows the RML output spectrum with no loop switch closure, and the resulting output is just barely visible on that wide scale. Figure 5.29b again shows the output pulse spectrum but with the loop switch operating, and still with no range delay; the large increase in power and the spectral distortion are both evident, as is the fact that almost all the power is concentrated around the input carrier. Figure 5.29c is the first photograph to show the result with an actual range delay −0.5 μs. Each subsequent photograph is for an additional 0.5 μs range delay. In Figure 5.29d, noise can be seen building up in two local clumps of noise power; one of these noise power clumps must correspond to a local gain peak of the ripple. Note that these noise power clumps are symmetric around the input; this is a natural phenomenon, and not the result of a special gain ripple of the RML under test; the saturated loop amplifier is acting as a mixer, so one of those noise growth regions is the mixer image, while the other is "real." Examining further ranges, Figures 5.29e through 5.29l, the noise can be seen building in intensity throughout the full band, with a clumpy spectral character. By Figure 5.29l the total noise power starts to rival the power in the signal, and hence the signal power starts to droop; this is evident at further ranges, such as in Figure 5.29m for 5.5 μs. The saturated gain ripple of the loop now tends to favor the high end of the band. Gradually, the noise power at the high end of the band begins to predominate, as shown in Figures 5.29n through 5.29s, until the noise power at the high end begins to suppress the noise power throughout the rest of the band, as shown by Figures 5.29t through 5.29v. The signal is now completely gone.

Most RMLs, if left with their loop switch on long enough, become oscillators, oscillating at a particular favored point in the band. Such oscillations may become reasonably pure and clean. In certain relatively rare instances the RML steady-state response will be two or more oscillation spectral-line frequencies, rather than the more common single spectral line frequency oscillation case. In any event, the model the reader should have in mind for loop stretching length considerations is not an input signal that drifts in frequency, but rather competing noise frequency components that grow to rival the signal. When a noise clump does grow strong enough, it will suppress the signal since that noise clump must settle at precisely zero decibels loop gain, and that noise clump must be at a relative gain peak in frequency to have captured the loop. The resultant gain at the signal frequency must therefore drop below zero decibels, resulting in the signal's large amplitude dropping out, as in Figure 5.28, bottom right. The RML then soon turns into an oscillator, rather than a replicating transponder.

The sensitivity of the basic RML is that signal input power level, which is sufficiently strong to ensure that the signal rises to saturate the loop amplifier before the growing noise saturates the loop amplifier; this condition ensures at least a moderate stretching. A complete RML transponder subsystem includes a CVR, as shown in Figure 5.26, which generates the trigger to initiate the timing; the MDS sensitivity of the CVR may limit the transponder subsystem sensitivity.

A 10 circulation RML is rather easy to design and fabricate. The typical RML in use has a 20–30 circulation capability. An RML with more than 30 circulations is quite difficult to fabricate. Again, the definition of stretching is the length of time before the original signal starts to drop out. Increasing the initial input signal power improves the stretching modestly.

If one were to calculate the predicted loop stretching performance based on knowledge of the gain ripple and the gain at the input frequency, using the theory presented in this chapter so far, a rather pessimistic calculated result would ensue. Fortunately, the saturation suppression phenomenon enhances the RML stretching to a significant degree. This phenomenon is illustrated in Figure 5.30, which graphs the output power from an amplifier for signal 1 and signal 2 as a function of signal 1's input power. This response is typical of an RML loop amplifer. The signal 2 input power is kept constant, in the small signal region. The two carrier values are unequal, each being somewhere in the band. As the input signal 1 is varied in amplitude in the SSG region, signal 2's output power is constant, not influenced by signal 1. The signal 1 response follows a 45° slope on such graphs when there is identical logarithmic scaling on both axes, indicating perfect linearity and hence perfect superposition. As signal 1 enters the large signal region, the plotted response starts to deviate from this 45° line, the difference being referred to as *gain compression*. As the amplifier is driven ever harder into saturation, the compression increases

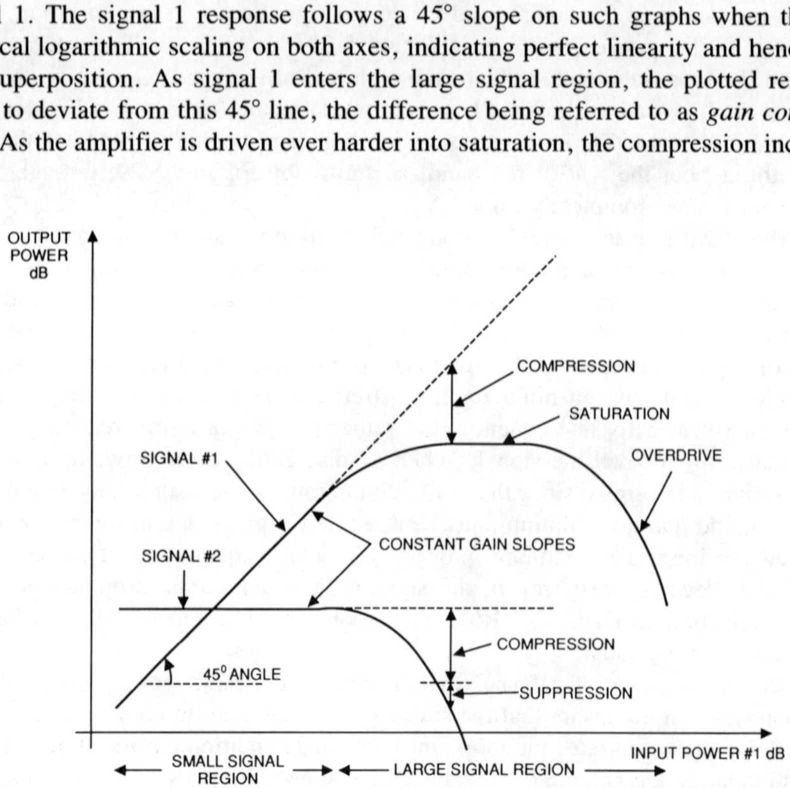

Figure 5.30 Large signal capture.

and the output power may actually decrease, as shown in the illustration in Figure 5.30. This power decrease region is known as the *overdrive region*. An amplifier designed to be a limiting amplifier will not overdrive in such a manner, rather its input-*versus*-output curve will merely flatten.

With increasing signal 1 input power the signal 2 output power will also decrease, with respect to a constant gain output, but by more than the gain compression experienced by the large signal 1; that is, signal 2 will deviate from the horizontal constant gain line by more than signal 1 will deviate from the 45° constant gain line. The difference between the signal 1 LSG compression and signal 2's SSG reduction is known as gain suppression. Gain suppression is highly desirable for a loop amplifier, and some solid-state microwave amplifier vendors deliberately design amplifiers that exhibit this suppression phenomena. TWTs exhibit suppression rather naturally, because of the nonlinear electron beam bunching. Suppression values of a few decibels can be achieved. This suppression gives a design margin to overcome ripple, since, as stated above, a simplistic calculation of stretching that does not include suppression will invariably predict rather short stretching for practical wideband ripple values. By exploiting the suppression phenomenon in amplifiers, both TWTs and appropriately designed SSAs, ECM systems are able to use RMLs to obtain sufficient RF memory length.

5.10 RML SPECTRAL RESPONSE

As stated above, the RML failure mode, which ends its RF signal storage, should not be thought of as a frequency drift but rather as a loss of control of the loop, once the noise grows to generate a significant percentage of the loop amplifier's output power. Nevertheless, the output spectral response is rather complicated, even without considering noise. This section will describe the RML spectral response of the recirculated signal, initially without considering the contributions of noise.

The three dimensional graphs *a*, *b*, and *c* in Figure 5.31 show the calculated RML output spectra for given storage pulsewidths, loop delays, and PRIs. The graph in Figure 5.31a shows the CW case, in which the stretched pulsewidth is equal to the PRI. The output is graphed as amplitude *versus* relative frequency *versus* loop phase, in 45° loop-phase steps. The frequency scale is normalized to the reciprocal of the loop delay. Each circulation is assumed to be saturated, and the input is assumed to be shut off after the first circulation. Therefore, the graphs in Figure 5.31 are graphs of the spectrum of a constant amplitude signal with a particular phase discontinuity repeated every loop delay. When the loop phase angle equals zero, then the output is a single pure tone, which should be expected because there are no phase or amplitude discontinuities or other changes. However, as a general rule for the CW case, and excluding noise, (1) the spectrum is characterized by an array of spectral lines separated by the reciprocal of the loop delay—unity frequency sep-

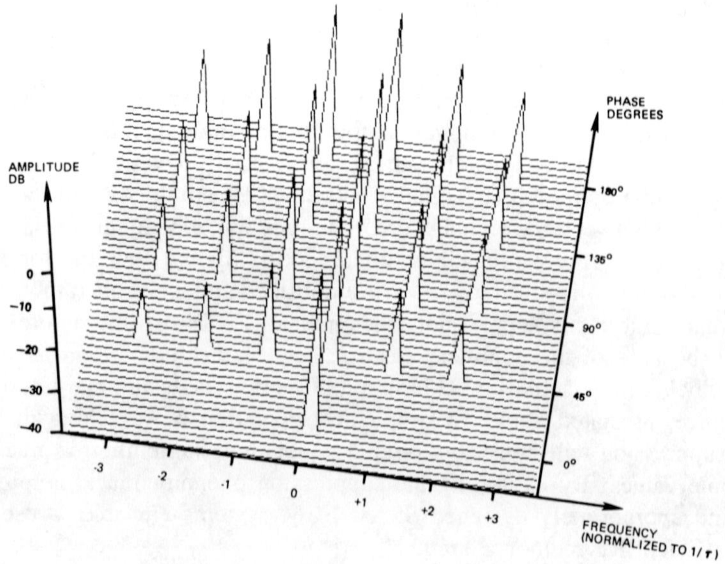

(a) Conditions: $T/\tau = 8$; PRI $= 8/\tau$

Figure 5.31 Spectrum of an RML transponder.

aration with the normalized scale in the figure, and (2) the largest spectral line is not centered at the input frequency. The array of spectral lines therefore shift, with respect to the input signal frequency, as a function of the loop phase angle. For loop phase angles close to 0° (or 360°), the surrounding periodic spectral lines are low and fall off in amplitude rapidly. For loop phase angles near 180°, the surrounding spectral lines have rather high amplitudes. In the extreme case of 180° loop phase angle, there are two spectral lines with equal maximum power, and the input frequency is located exactly between these two largest lines, where there is no output spectral power in this CW case.

Therefore, in the general CW case, with arbitrary loop phase angle, there is no output power at the input frequency! An RML could therefore be used as a frequency offset generator, if the spectrum were not so dirty, as shown in Figure 5.31a.

Figure 5.31a is a useful graph to illustrate the spectral response of an RML output operating as CW without the complexity of superimposed pulse spectra. Figure 5.31b illustrates a more typical pulse output spectrum, in this case, with the output pulsewidth being two circulations in length and the PRI being eight circulations in length. At 0° loop phase angle, the spectrum is, of course, just the spectrum

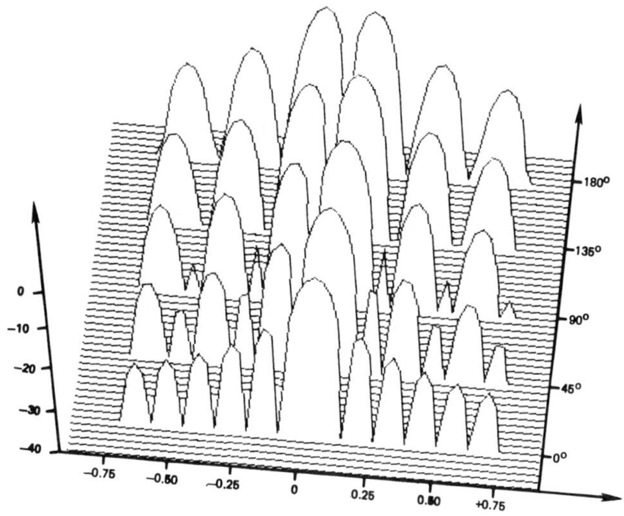

(b) Conditions: $T/\tau = 2$; PRI $= 8/\tau$

Figure 5.31 Continued.

of a 25% duty cycle pulse train with a pure carrier. At other loop phase angles, the resultant phase discontinuities in the time domain will cause a frequency-domain frequency shift of the main spectral lobe, reduce its power, and increase that of other nearby spectral lobes. In Figure 5.31c, with the storage length being 8 circulations and the PRI being 32 circulations, the effect of a non-zero loop phase angle is much more pronounced. Again, note the frequency error, in the position of the largest output spectral lobe with respect to zero frequency. Figures 5.31b and 5.31c are identical to Figure 5.31a except for the spectral change resulting from superimposed pulse modulation.

All of the CW main spectral line frequencies in Figure 5.31a, or main lobe frequencies in Figures 5.31b and 5.31c, correspond to frequencies where the loop phase angle equals zero. These false line positions, occurring at intervals of the reciprocal of the loop delay, are also frequency positions where noise will build up. This is illustrated in Figure 5.32 which shows a measured RML output spectrum expanded around the input frequency. In Figure 5.32a, the input is tuned to one of these comb lines. It can be seen that the power of the nearby comb lines falls off away from the input frequency, as predicted from the above explanation; however,

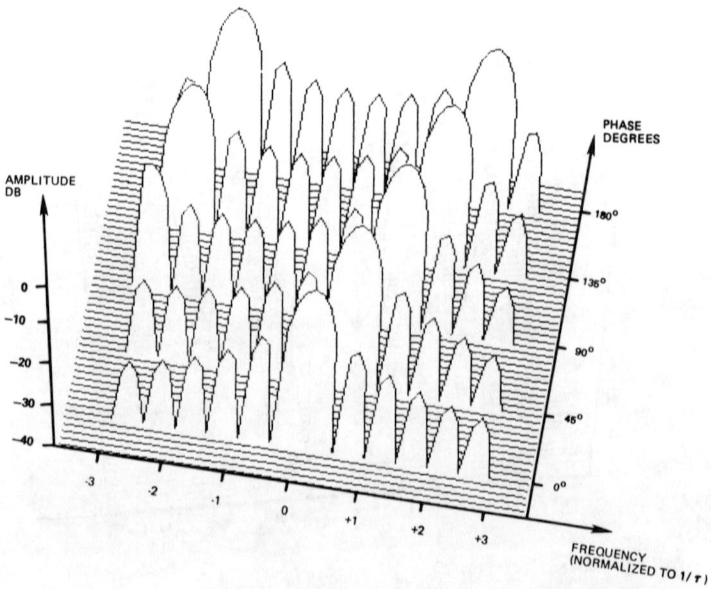

(c) Conditions: $T/\tau = 8$; PRI $= 32/\tau$

Figure 5.31 Continued.

after a few lines, all the comb lines have about the same power. These comb line amplitudes are not the result of the spectrum of the recirculating signal; rather they are spectral components of the noise that has grown to this power level. In other words, the broadband noncoherent noise power consists of a multitude of coherent spectral lines.

In Figure 5.32c, the input signal frequency was adjusted to be precisely midway between two comb lines. Figure 5.32c corresponds to Figure 5.31 at 180° loop phase angle. The result is the largest frequency error possible, from the input frequency to the position of the largest spectral line.

Although the spectrum of Figure 5.32a is quite adequate for most ECM transponder applications, albeit noisy, the spectrum of Figure 5.32c is generally unacceptable. The RML designer is caught in a dilemma. If the loop delay is made relatively long, then the spectral energy will be concentrated, that is, the maximum possible frequency error will be reduced. However, a long loop delay exposes the leading edge and slows down the rise to full amplitude for weak inputs. Recall that all the dynamic response curves were normalized to the circulation interval, so all

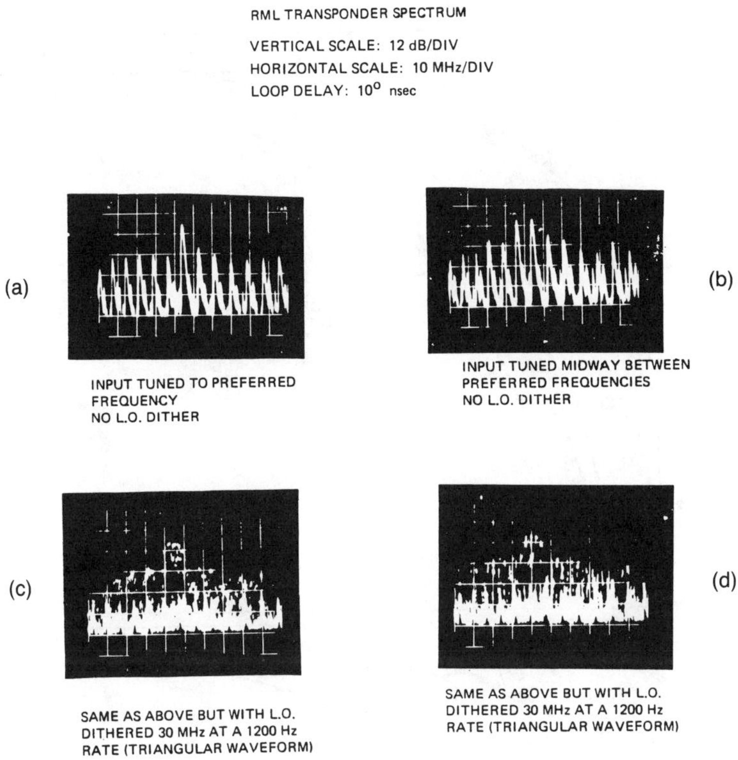

Figure 5.32 RML transponder signal + noise spectrum.

timing is proportional to the loop delay. If the loop delay is made short to cover the leading edge and to build to maximum power most quickly, then the spectrum will be spread out, that is, the maximum frequency error will increase proportionally. A compromise solution to this dilemma is illustrated in Figures 5.32b and 5.32d. These figures show the result of randomizing the loop phase angles. The input frequency for Figure 5.32b is the same as for Figure 5.32a, while the input frequency for Figure 5.32d is the same as for Figure 5.32c. It can be seen that even moving the frequency by half the comb line spacing has no impact on the resultant spectrum when the loop phase angle is randomized. Randomizing the loop phase angle makes the RML response independent of frequency.

Obviously, knowing how much of the RML transponder power will pass through the threat radar's receiver is very important. The calculated results are illustrated in Figure 5.33a and Figure 5.33b, which graphs the spectral energy loss through a filter on the ordinate scale as a function of the filter bandpass width normalized to the

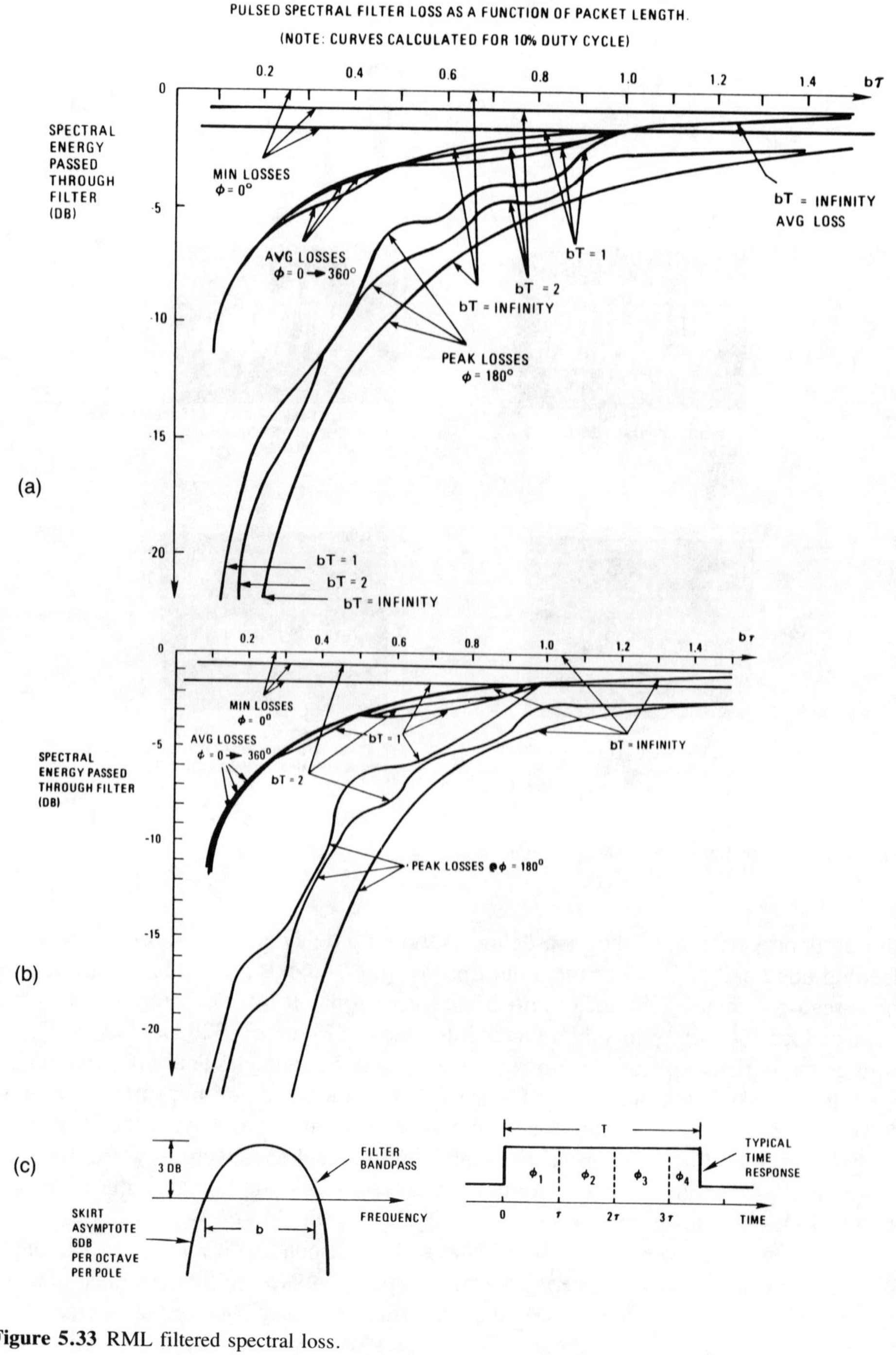

Figure 5.33 RML filtered spectral loss.

reciprocal of the loop delay; that is, the abscissa is the product of the loop delay and the receiver bandpass bandwidth. Curves are shown for loop phase angle-extreme cases of 0° and 180°, and for the result when the loop phase angle is randomized. The curves are graphed for pulsewidth times bandwidth products, that is, normalized pulsewidths, of 1, 2, and infinity; infinity pulsewidth corresponds to the CW case. Figure 5.33c shows the convention for the symbols. Figure 5.33a shows the result for a 20-pole filter, that is, a filter with rather steep skirts, while Figure 5.33b shows the result for a two-pole filter, with rather shallow skirts. Noise power is not included in these graphs, just the recirculating signal power.

The plotted results in Figure 5.33a and 5.33b for zero-degree loop phase are shown to be horizontal lines; that is, the transmitted power through the filter is independent of the loop delay or the receiver's bandwidth. For the CW case, there is no loss at all; this should be expected, because the output will be a pure tone centered in the bandpass. For the other pulsewidths of 1 and 2 units of time, normalized to the reciprocal of the bandpass width, some loss is evident for zero degrees loop phase, though still independent of the loop delay; this is simply the spectral loss of a pulsed pure tone signal when it is passed through a filter.

The plotted results in Figure 5.33a and 5.33b for 180° loop phase are shown to exhibit very high losses when the loop delay is short. For loop delays less than 0.4, normalized to the reciprocal of the receiver bandpass width, the spectral loss is in excess of 10 dB, and it increases rapidly as the loop delay is decreased. For loop delays less than 0.2, values of spectral loss greater than 20 dB are experienced even for a CW signal. Such large losses are clearly unacceptable to ECM systems designers, since the last few decibels of output RF power is so expensive. Nevertheless, it is important to use relatively short loop delays so that the true radar pulse will be quickly covered at maximum power. Fortunately, as stated above, the loop phase angle can be randomized, and the *AVG* result shown in Figure 5.33 is much more acceptable.

The curves in Figure 5.33 can be used to make effective power (EEP) calculations for an ECM system operating against a given threat.

Some further discussion of the RML transponder spectrum is in order. Figure 5.31 shows calculated spectra, not including noise, of the saturated recirculating signal, where it is evident that the output frequency is not equal to the input frequency in general. In particular, when the loop phase angle approaches 180°, the figures show that there is no power at the original frequency. Is there a simple way to conceptualize why this happens, rather than simply relying on the fact that a mathematical transformation of the time-domain description predicts such a result? Figure 5.34 is intended to facilitate such an investigation.

For an RML transponder with a recirculating signal with a frequency with precisely an integer number of periods plus a half during the circulation, that is, where the loop phase angle is 180°, the output spectrum has a power null at the original input frequency, as already stated. However, the implication of this statement is that

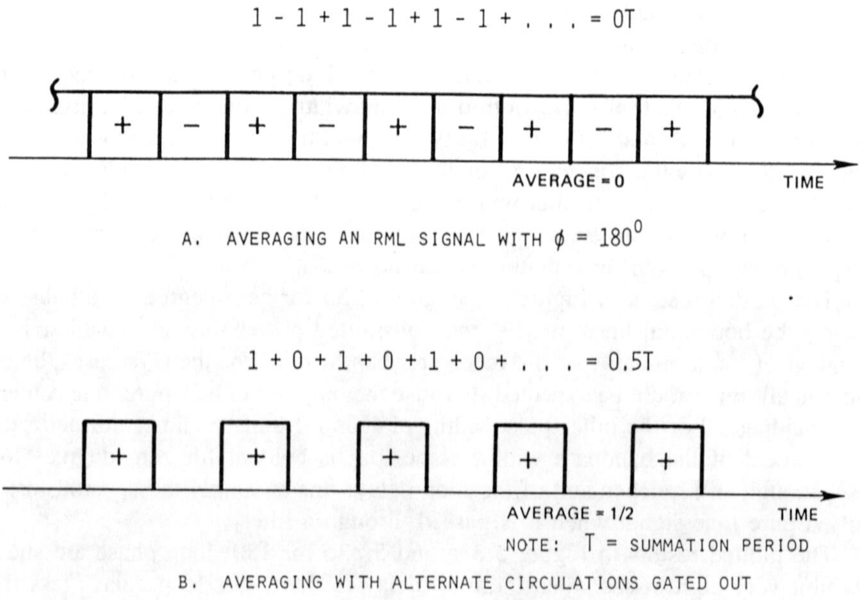

Figure 5.34 Averaging the signal.

the spectrum is measured using a receiver or spectrum analyzer with an IBW much less than the reciprocal of the loop delay; otherwise, there would not be sufficient resolution to sense the null and the surrounding structure of the spectrum. Hence, the reception bandwidth is narrow, and this fact can be interpreted as defining a receiver that will average the signal. Figure 5.34a shows the envelope of the broadband signal from the loop; that is, it is drawn with an implied carrier. The 180° phase reversals are the equivalent of a sign change. As can be seen by inspection, the average of the packets with alternating signs is nil. Hence an RML transponder, with loop phase of 180°, will have nil power at the original input carrier value in a fashion the reader should now be able to conceptualize.

As a matter of fact, for an RML transponder operating at saturation with a frequency corresponding to 180° loop phase angle, more narrowband power would result if every second circulation were gated out. In other words, more average power will result if half the signal is judiciously deleted. The reader should be able to conceptualize why this is true by examining Figure 5.34b.

Such analysis prompts additional questions: should it not matter to the transmitted narrowband power whether there were an even or odd number of circulations in the transmitted pulsewidth? Should it not matter if there were a noninteger number of circulations? The answer to each is yes, and the effect is much more pronounced for a relatively short output pulsewidth, that is, for relatively few circulations. This phenomenon is illustrated in Figure 5.35, which shows the calculated RML spectra for a saturated recirculating signal. Figure 5.35a shows the spectra for the output pulse gated on for just three circulations. Figure 5.35b shows the spectra for the output pulse gated on for three circulations, but with the loop switch turned on with a timing error of half a circulation. At the bottom of the figure is shown the time-domain representation for these examples. The spectra are shown with spectral loss on the ordinate scale in decibels, and frequency on the abscissa in normalized units, for loop phase angles in 45° increments. When the loop phase angle is zero degrees, the spectrum is simply that of a pulse. When the phase angle is not zero, however,

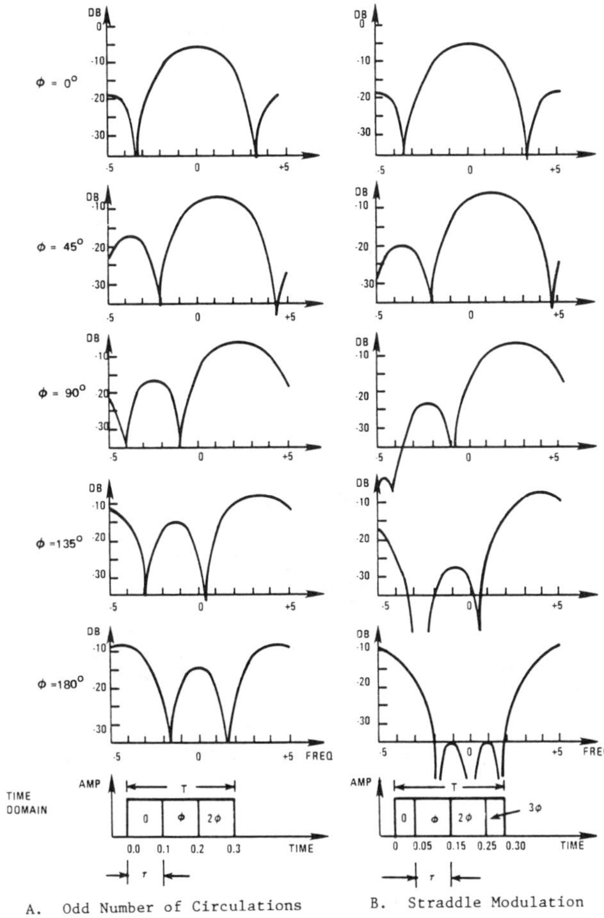

Figure 5.35 Short pulse spectra.

the spectrum is that of a pulse with discontinuous phase modulation. For the three-circulation example, the spectrum does not have a null at the original frequency when the loop phase angle is 180°; this is because, using the principles illustrated in Figure 5.34, +1 plus −1 plus +1 does not equal zero.

5.11 SAMPLE PROBLEMS AND SUMMARY

Theory is fine, but a few problems, with their solutions illustrated, are often most helpful. Figures 5.36 through 5.38 give selected problems, and their solutions, for RML transponders. Figure 5.36 shows a spectral loss determination based on the graph in Figure 5.33, comparing transponder and saturated repeater effective ECM power. Figure 5.37 defines the design of an RML transponder, and shows a calculation of its sensitivity based on the graphs in Figure 5.24 and Figure 5.15. Figure 5.38, using the same RML as in Figure 5.37, shows a sample calculation of the loop-stretching time based on the graph in Figure 5.24. The key to such calculations is keeping track of the signal power and the noise power at each circulation. Figure 5.19 shows an example of resulting ripple in a linear system, the calculation of which can be done graphically, albeit requiring adding decibels, by using the graphs in Figure 3.6 and Figure 5.16.

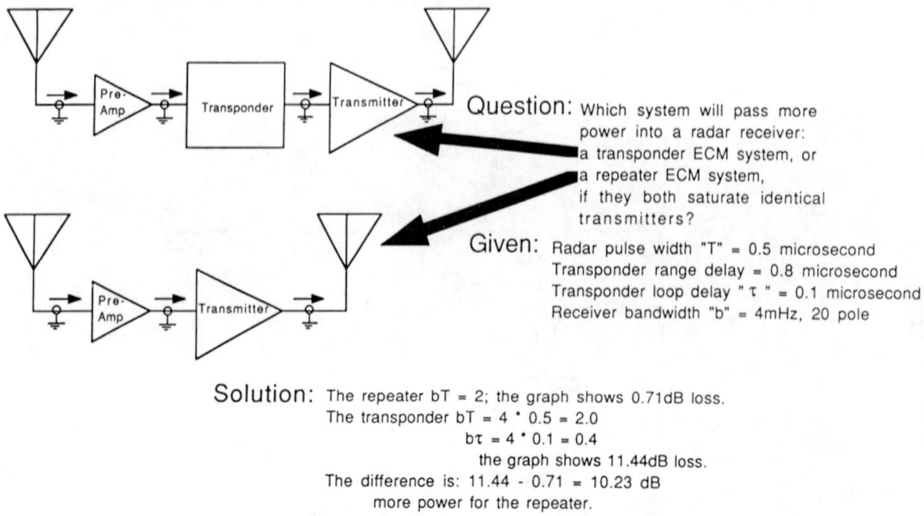

Figure 5.36 Sample problem: transponder spectral loss.

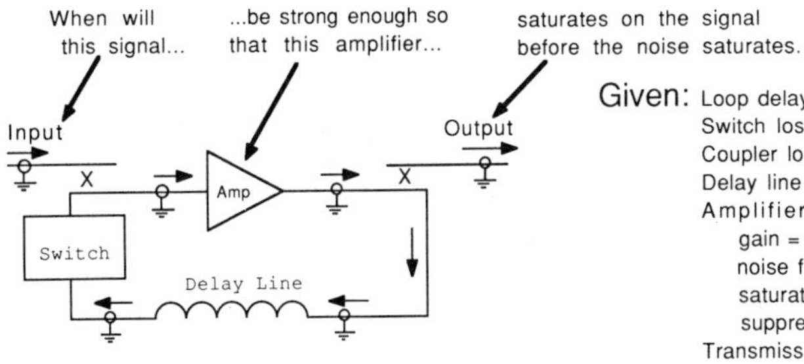

Given: Loop delay = 100 nSec
Switch loss = 3 dB
Coupler loss = 3.5 dB
Delay line loss = 20 dB
Amplifier
gain = 40 dB
noise figure = 11 dB
saturation = 10 dBm
suppression = 2 dB @ 6 dB sat
Transmission lines = 0.5 dB each
Bandwidth = 2000 mHz
Ripple = 3 dB

Solution

STEP #1: Loop gain (starting at the amp) = 40.0 - 0.5 - 3.5 - 0.5 - 20.0 - 0.5 - 3.0 - 0.5 - 3.5 - 0.5 = 7.5 dB

STEP #2: Amplifier noise output = KTBFG = -114 + 33 + 11 +40 = -30 dBm

STEP #3: Amplifier dynamic range = 10 - (-30) = 40 dB

STEP #4: Graph for noise dynamics gives noise capture time = 6 circulations

STEP #5: Graph for signal dynamics gives a 6 circulation signal rise, worse case = 42 dB

STEP #6: Amplifier output sensitivity = 10 - 42 = -32 dBm

STEP #7: System input sensitivity = -32 - 40 + 0.5 + 3.5 = -68.0 dBm

Figure 5.37 Sample problem: loop sensitivity.

There has been widespread RML transponder usage for range delay capability because of cost effectiveness resulting from

- wide IBW,
- relatively concentrated spectra,
- pulse-to-pulse operation,
- interleaved pulse response,
- simultaneous operation against numerous low duty cycle threats, and
- constant maximum power output independent of input power.

The limitations of RML transponders are summarized:

- noncoherent noisy localized spectra,
- dirty output spectrum across band,
- limited memory time, and
- large delay line component.

Given: Input signal
@ -45 dBm, gated
off at first circulation.

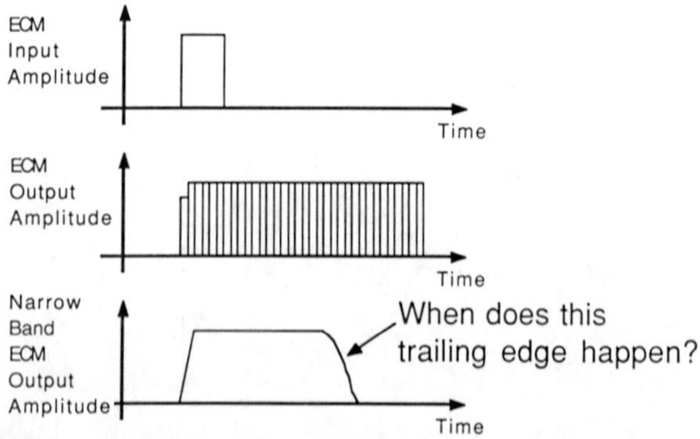

Solution

STEP 1: Zeroth circulation amplitude at amplifier output = -45.0 - 3.5 - 0.5 + 40.0 = -9.0 dBm
STEP 2: Signal amplitude rise to saturation = 10.0 - (-9.0) = 19.0 dB
STEP 3: Signal loop gain worst case = 7.5 - 1.5 = 6.0 dB
STEP 4: Number of circulations till saturation = 19.0 / 6.0 = 3.2
STEP 5: Noise loop gain for #1 --> 3 circulations = 7.5 + 1.5 = 9.0 dB
STEP 6: Signal loop gain for #4 circulation = 6.0 - (4 * 6.0 - 19.0) = 6.0 - 5.0 = 1.0 dB
STEP 7: Signal loop gain for #5 and after circulations = 0.0 dB
STEP 8: Noise loop gain for #4 circulation = 9.0 - 5.0 - (2.0 * 5.0/6.0) = 2.3 dB
STEP 9: Noise loop gain for #5 and after circulations = 9.0 - 6.0 - 2.0 = 1.0 dB
STEP 10: Noise amplitude rise to saturation = 10.0 - (-114.0 + 23.0 +11.0 + 40.0) = 50 dB
STEP 11: Noise amplitude rise #1 --> 3 circulations = 27.6 dB (see graph)
STEP 12: Noise amplitude for #4 circulation = 29.9 dB (see graph)
STEP 13: Noise amplitude for #5 circulation = 29.9 + 1.0 = 30.9 dB (or see graph)
STEP 14: Noise Amplitude for #25 circulation = 30.9 + 20 * 1.0 = > 50.0 dB
ANSWER: Stretching = 25 * 0.1 = 2.5 microSeconds

Figure 5.38 Sample problem: loop-stretching duration.

The implementation characteristics are summarized:

- takes a leading edge sample,
- spectral response depends on sample width, not input pulsewidth,
- 10 circulations is easy, 30 circulations is difficult,
- typical programmed stretching is 20–100 circulations,
- loop-amplifier saturating characteristics are customized,
- loop gain set high (>10 dB) for good rise time and better stretching,
- loop phase angle is corrected or randomized, and
- loop output is gated to make false range pulses.

Chapter 6
System Design

The first chapter provided the overview of ECM system structures and their operation. Additional chapters described the components and subsystems that are the ingredients used to form an ECM system. Another chapter described specialized knowledge and design principles. This chapter will elaborate on certain aspects of these descriptions, concentrating on system design considerations, to present the subject more completely. If the reader has followed the sequence of chapters in this book, Chapter 1 should now be reviewed before reading the remainder of this chapter.

6.1 PRESET JAMMERS

Figure 6.1 illustrates preset noise jamming. A preset jammer is one that has free-running waveforms, without power management control, with a given set of parameters, such as bandwidths, center frequencies, *et cetera*. The functional operation is illustrated in frequency, space, and time. At the top of the figure, the noise power can be seen spread across the band of interest, *f*-low to *f*-high, covering all the threat carrier frequencies, *f*-1 to *f*-n. The center of the figure illustrates the jammer transmitting in all directions at once, implying a wide-angle low-gain antenna system. The lower portion of the figure illustrates a time display of the radar echo and the noise jammer signal. A preset noise jammer has a relatively simple design structure, as illustrated in Figure 6.2. This simple ECM approach does not require digital signal processing. The video circuit simply generates the noise waveform in a free-running fashion. Of course, the noise must have the proper characteristics and quality. The key parameters are held in a latch, which may be a changeable ROM or a fixed-access memory loaded on a pre-engagement basis. The challenge for such a system is to achieve the combination of required bandwidth, effective power, and antenna

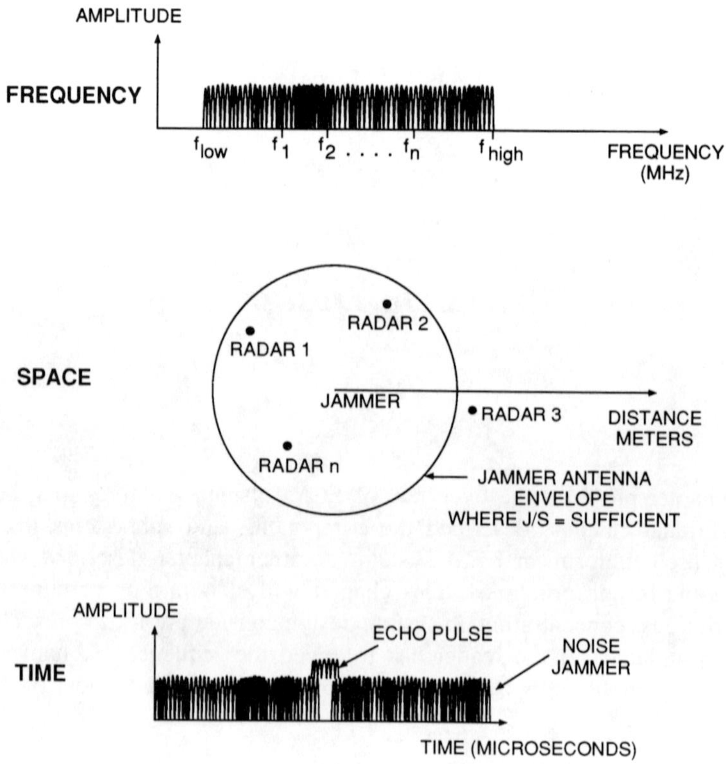

Figure 6.1 Preset noise jamming.

coverage, because the main microwave subsystems, driven in this free-running fashion, need to generate the required EEP against each threat in a brute force manner. The relative advantages and disadvantages of such a simple preset noise jammer, with respect to other modes and with respect to an alternate power managed noise jammer approach, are given in Table 6.1. The ECM system designer is confronted with the task of deciding between such a so-called simple preset noise jammer *versus* one that can control the carrier frequency, the direction of transmission, and the time of transmission, as well as impose specialized deception modulations that may be more effective for a given jam-to-signal (J/S) ratio. The comparison of cost, size, and weight depend on the engagement requirements; the more specialized the engagement, the more the trade-off favors a preset architecture. Since it is incapable of adapting to the particulars of the engagement, a preset noise jammer is usually set to transmit over the known carrier-frequency range of the threat radar, so reliable information is needed that defines this f-low to f-high range. Although the capability

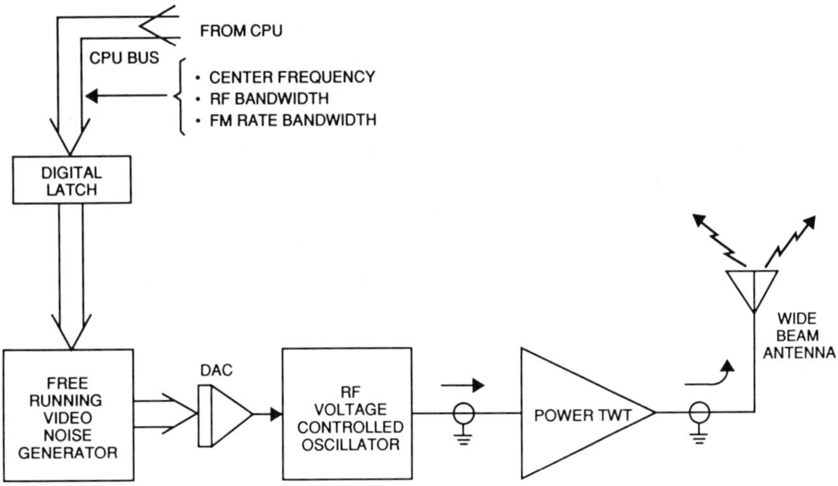

Figure 6.2 A preset noise jammer.

Table 6.1
Preset Noise Jamming

Advantages	*Disadvantages*
Jams in angle	Power dilution in frequency
Jams in range	Power dilution in space
Simultaneous jamming against all radars	Power dilution in time
No frequency measurement required	Certain sophisticated jamming or deception techniques cannot be used
No tuning in frequency required	*A priori* frequency information needed
Not vulnerable to frequency agility	
Not vulnerable to time agility	
No steering of antenna required	
No direction finding required	
No look-through required	
No signal detection required	

to handle a carrier-frequency agile radar easily is listed as an advantage in Table 6.1, it should be understood that the greater this *f*-low to *f*-high carrier frequency uncertainty range is, the lower the EEP will be. For example, a 400 MHz frequency range would result in 3 dB less EEP, at a given range, or a 41% increase in burn-through range, than a 200 MHz frequency range for a given transmitter rating.

The distinction between preset jamming *versus* sophisticated power management jamming is not as sharp for the repeater mode as it is for the noise mode. Figure 6.3 shows a simple preset repeater. Table 6.2 gives the relative advantages and disadvantages for a simple preset repeater-mode system *versus* other modes and *versus* a system operating in repeater mode but utilizing sophisticated sensors and signal processors. This has almost the same advantages as the brute force preset noise jammer, except that it does not experience power dilution in frequency, since the threat radar provides the needed carrier signal. Neither frequency agility nor an extended frequency range have a practical impact on the ECM performance. Once again, however, some jamming techniques, such as angle deception modulation with a rate set to match the threat radar's scan, are not possible without sensors and signal processing. The limitations of propagation delay and antenna isolation are independent of the power-management capability.

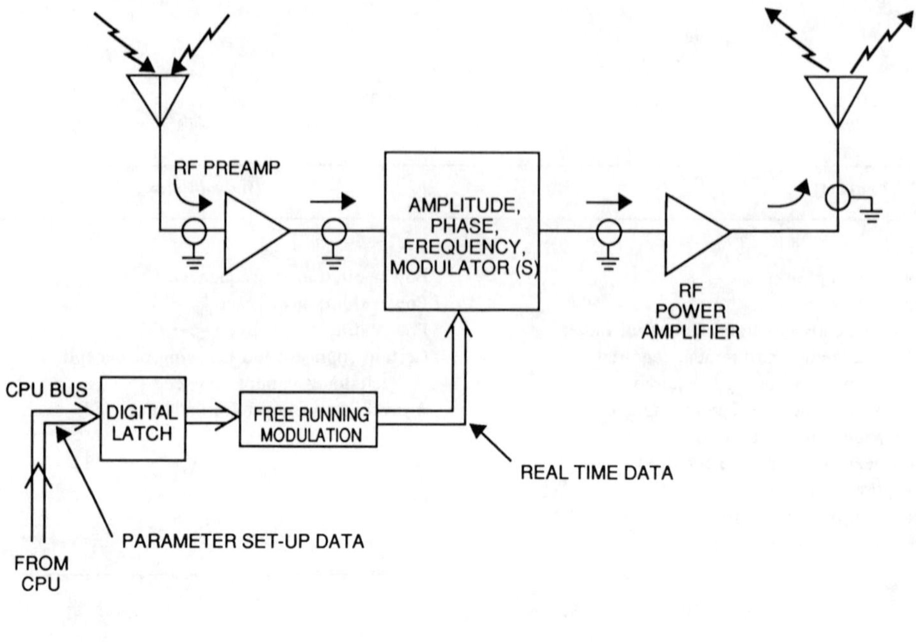

Figure 6.3 A simple repeater.

Table 6.2
Preset Repeater

Advantages	*Disadvantages*
Deception in angle	No deception in range
Simultaneous deception against all radars	Power dilution in space
No frequency measurement required	Certain sophisticated jamming or deception techniques cannot be used
No tuning in frequency required	Leading edge delay
Not vulnerable to frequency agility	Antenna isolation limit
Not vulnerable to wide frequency range	
Not vulnerable to time agility	
No steering of antenna required	
No direction-finding required	
No look-through required	
No signal detection required	

A preset jammer has waveforms that free-run; other waveforms may be triggered off the detected pulse to initiate real-time operations—RGPO, for example. A very simple preset transponder system is shown in Figure 6.4. In this case the modulation waveform, in addition to applying amplitude and phase or frequency offset modulation, also applies range delay modulation. (See, for example, Figure 1.15.) The diagrams in Figures 6.3 and 6.4 are very similar except that the pulse-to-pulse memory is included. As pointed out in Chapter 1, the transponder function can also be performed by having a sophisticated DSP control a microwave VCO, that is, a set-on RF VCO. The advantages and disadvantages of using a preset transponder *versus* other modes and *versus* a transponder under power-management control are given in Table 6.3. Since the transponder and repeater mode, by their nature, are not power-diluted in time or frequency, the relative advantages and disadvantages are almost the same. However, a preset transponder must detect the pulse or leading edge, and is therefore vulnerable to false triggers and being blinded by the input signal complexity. Another important factor is that, without a high-speed DSP, false range pulses, or targets, cannot be put in front of the true pulse, as illustrated in Figure 6.5. The preset transponder is quite capable of placing a false pulse a given number of microseconds behind the true pulse position, but not in front of the true pulse. However, a high-speed DSP can predict the arrival of the next pulse, and can therefore put out a leading pulse, as well as impose distinctive amplitude, phase, and offset frequency modulation matched to the threat.

In summary, preset jammers do not include a power management capability; the impact on electrical performance, however, is lowered EEP and the inability to

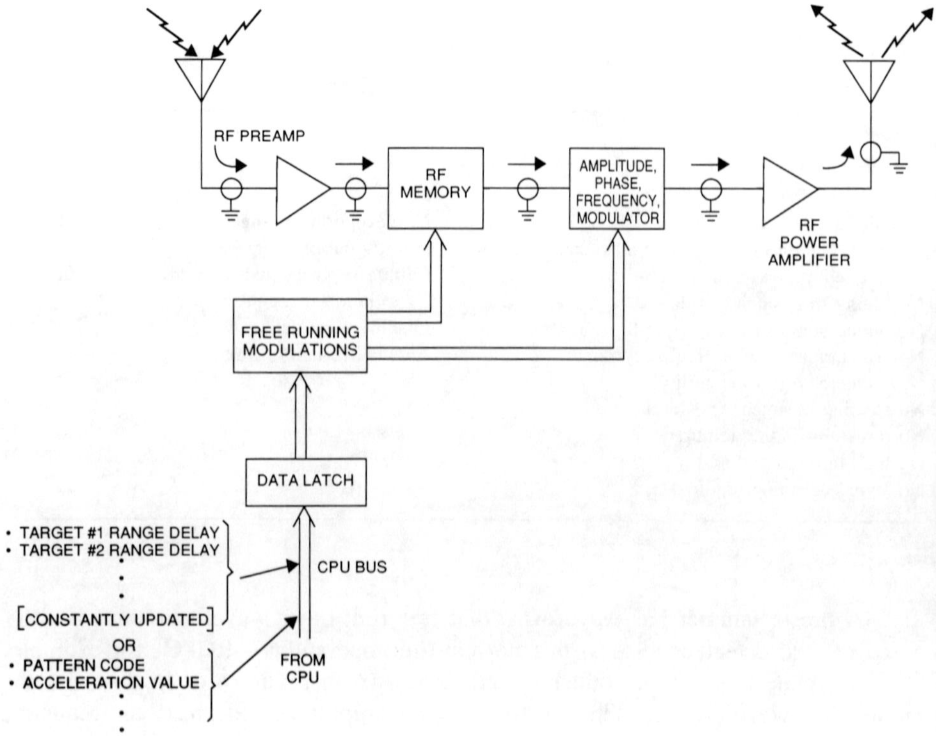

Figure 6.4 A simple transponder.

Table 6.3
Preset Transponder

Advantages	*Disadvantages*
Deception in angle	Power dilution in space
Deception in range	Certain sophisticated jamming or deception techniques cannot be used
Simultaneous deception against all radars	Leading pulses cannot be generated
No frequency measurement required	Cover-pulse leading-edge delay
No tuning-in frequency required	Pulse detection required
Not vulnerable to frequency agility	Vulnerable to trigger errors
Not vulnerable to wide frequency range	
Not vulnerable to time agility	
No steering of antenna required	
No direction-finding required	
No look-through required	

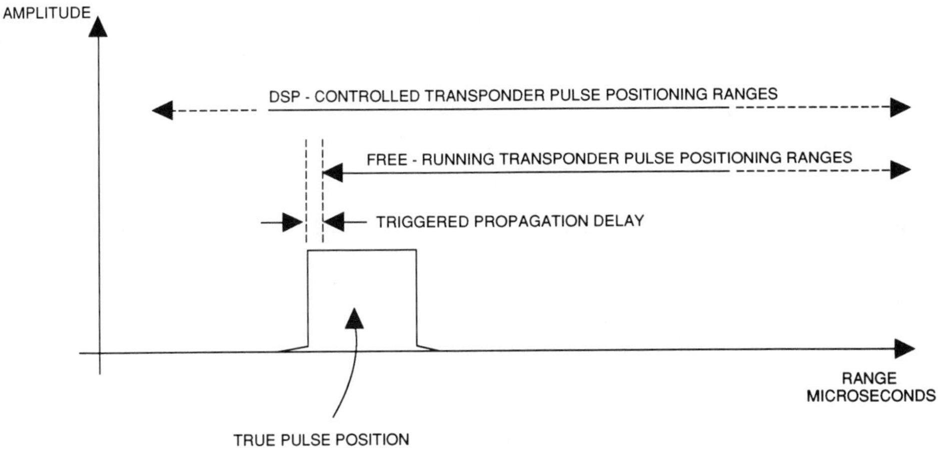

Figure 6.5 Leading false range pulses.

use jamming and deception techniques tailored to the threat class or tailored to the threat's specific parameters. The tailoring of the jamming impacts the technique efficacy and required J/S and hence the required EEP. For example, an ECM system using an angle deception modulation that is synchronized to, or close to a beat with, the radar's angular scan, can achieve its angle deception goal with considerably less peak J/S than one that transmits constant power to over-power the radar. As stated before, each jamming and deception technique should have a J/S rating, and those techniques that are tailored to certain key modulation rates usually achieve their goal with less J/S.

6.2 OUTPUT PARAMETERS

In considering the design choice of using preset, brute force jamming *versus* adaptive power management–based jamming employing sophisticated receivers feeding sophisticated DSPs, it is necessary to be aware of the needed output parameters. The rationale for determining some key output parameter values, and some typical values, are given in Table 6.4. In a repeater, the J/S is determined by the system gain, not the power capability of the system, until the system saturates, as shown in Figure 6.6. The repeater mode effective *depth of modulation* (DOM) and amplitude DR are limited by the system noise figure, set by the RF preamplifier or, more likely, by the output transmitter tubes when down-modulating, and depends on the threat receiver's IBW. (See Figure 3.15 and the accompanying text for an example of how to calculate this.) The ERP and gain needed are determined by the most demanding J/S ratio needed out of all the available jamming and deception techniques. The

Table 6.4
Output Parameter Determinants

Parameter	*Mode(s)*	*Determinant*	*Typical Value*
EEP*	Noise Transponder Repeater	Technique required J/S (usually at min range; saturated for repeater)	10 dB (J/S)
Noise level†	Repeater	Depth of modulation (DOM) should exceed threat's main beam to sidelobe ratio Amplitude dynamic range (DR) should cover typical range and cross section variation	40 dB (DOM)
Spurious levels	Repeater Transponder	Spurs across band should not significantly reduce desired spectral component power Close in spurs should not distort time domain response unduly	−10 dBc (spur level)
Gain	Repeater	Technique required J/S	10 dB (J/S)

Notes: *Including RF power, frequency accuracy, and bandwidth normalized to threat receiver, and antenna gain and polarization, where the antenna gain includes antenna directivity and transmission losses from power stage.
†Unintentional thermally generated noise related to system noise figure.

noise effective power and saturated repeater ERP needed are usually calculated at the shortest lethal or operating range of the threat. The range at which the J/S ratio becomes unacceptable for purposes of the technique efficacy is known as the burn-through range. Figure 6.6 shows the relationship of power *versus* range for the three basic modes.

The EEP capabilities of preset noise jammers have often been insufficient to meet the technique J/S requirements at more desirable burn-through ranges or distances. Generally this is true because it has not been practical to achieve the desired output power rating of the system because broadband power TWTs can only generate so much instantaneous power. True brute force solutions, such as paralleling many high-power tubes, are quite costly. The favorable impact on cost effectiveness, raw power consumption, and system size are the major selling points of high gain antennas, albeit with the unfavorable impact of a larger aperture and required control. Such antennas need to be managed, hence power management is often considered

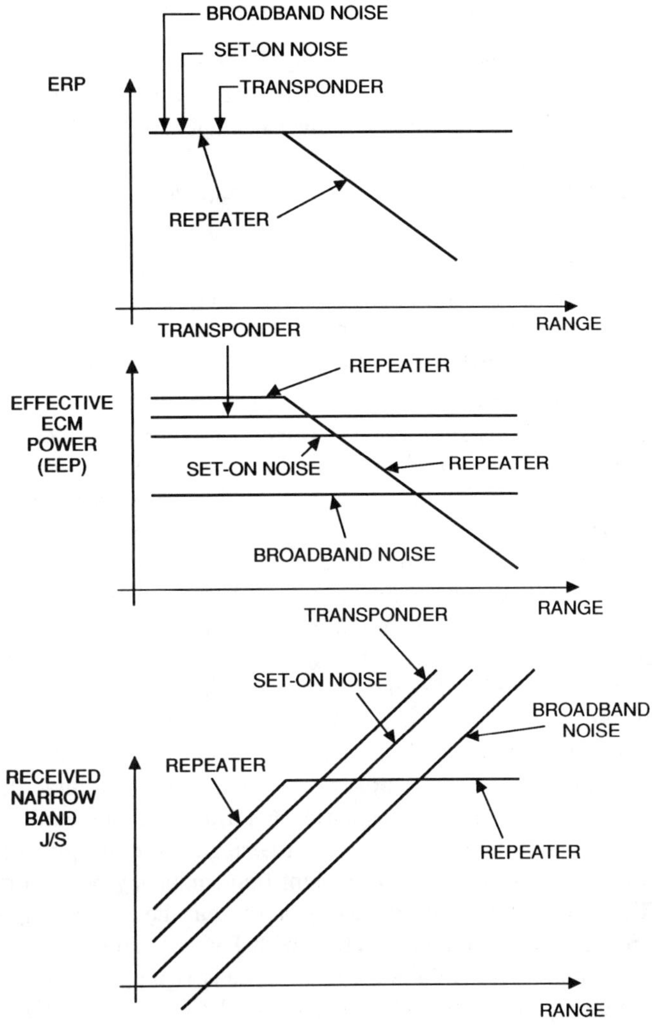

Figure 6.6 Power *versus* mode *versus* range.

attractive. That, of course, assumes that the reception and processing operate as intended in the real world. A preset system is much easier to test and validate.

6.3 POWER MANAGEMENT

Power management is the process of controlling the output transmission parameters, especially the frequency, time, and angle, to counter most efficiently and effectively a multitude of possibly diverse threats. Power management is summarized in Table 1.5.

The achievement of power management in frequency necessitates certain functions; it

- requires separation and identification of intercepted signals,
- requires high-speed measurement of carrier frequency,
- requires high-speed accurate set-on of jamming frequency, and
- has to deal with frequency agility.

An example of power-managed noise mode signal generation was given in Figure 1.14. In this example, a sophisticated waveform controls a microwave VCO so that threats at frequencies f1 and f2 will both simultaneously experience noise jamming. Even though the VCO is a single source, for practical purposes the single signal is simultaneously present at both frequencies if the average multiplexing rate is higher than the bandwidths of both threats.

With regard to the measurement of each carrier frequency, this need not have a high absolute accuracy if the sensor can sense both the input signal and the signal source or VCO signal and compare them. In addition to the items on the above list, prudent power management in frequency also requires updating and monitoring the threat's frequency, so that a look-through of its own jamming is also necessary. If all the threats' frequencies are stable, the look-through requirements are rather modest. However, if the threats are characterized by frequency agility, then the look-through measurement and tracking response operation becomes burdensome.

The parameter requirements for power managed noise are given in Table 6.5; in repeater and transponder modes, equivalent fine-frequency power management is not needed. Table 6.5 is based on the assumption that the output amplitude is constant and the noise is generated completely with FM, as shown in Figure 1.9. The carrier set-on deviation bandwidth is the peak-to-peak range of carrier values around the nominal frequency of the threat to which the VCO is tuned. For good quality noise, the deviation bandwidth should be 1.5 times to 2.0 times the threat receiver's bandwidth, providing a significant duty cycle in the skirts; this will ensure good (e.g., >20 dB) depth of noise modulation. The average multiplexing rate, to share power against several radars simultaneously, should be greater than each threat's IBW to appear to be creating noise in each radar receiver simultaneously. If the multiplexing is coherent, then the multiplexing rate can be considerably greater than

Table 6.5
Power Managed Frequency Parameters

Parameter	*Requirement*
Set-on accuracy	Less than threat receiver's IBW
Multiplexing rate maximum	Greater than threat receiver's IBW
Carrier set-on deviation bandwidth	Just greater than threat receiver IBW plus measurement uncertainty
Multiplexing rate minimum to maximum distribution	Noise quality determinant

the threat's IBW without causing an adverse impact on the noise quality. Coherent multiplexing is achieved by returning to the nominal value of each threat's carrier frequency, after dwelling at another threat's carrier frequency, at the same RF phase as would exist if the multiplexing were not used, that is, the same phase that would have existed if the frequency had not been briefly changed to the other threat's frequency. Usually such coherent multiplexing is impractical for an analog RF VCO, and if such is the case, then too high a multiplexing rate could widen the effective or apparent deviation at each threat's frequency. In other words, for non-coherent multiplexing, the average multiplexing rate is limited by the desired local deviation. For practical purposes, this means that the multiplexing modulation must have compromise characteristics that ensure both good quality noise—meaning good amplitude noise in the radar's bandpass, and good power sharing. The multiplexing modulation and the noise generation modulation can only be independent if the multiplexing is coherent, because then the multiplexing rate can be set much higher than the deviation bandwidth, and then they will not interact. Therefore, the noncoherent multiplexing rate maximum should be close to significantly altering the perceived deviation bandwidth, and the minimum-to-maximum distribution of dwell time or sweep rate at each frequency should be determined by the need for good quality noise. Ideally, this frequency multiplexing modulation results in amplitude noise in the threat's receiver equivalent to the result from a single-carrier amplitude modulation that generates good quality noise, except for the power-sharing reduction.

Based on the thesis that the threat radars do not have an arbitrary bandwidth, but rather the bandwidth is related to the size of the craft or structure to be tracked, the receiver instantaneous bandwidths of Table 6.5 should fall in the range of 1 to 10 MHz for threat radars that track aircraft, and notably narrower for search radars used against large naval targets.

Power management in time is illustrated in Figure 6.7. In this example, the noise jamming transmission occurs only around the time the threat pulse is expected.

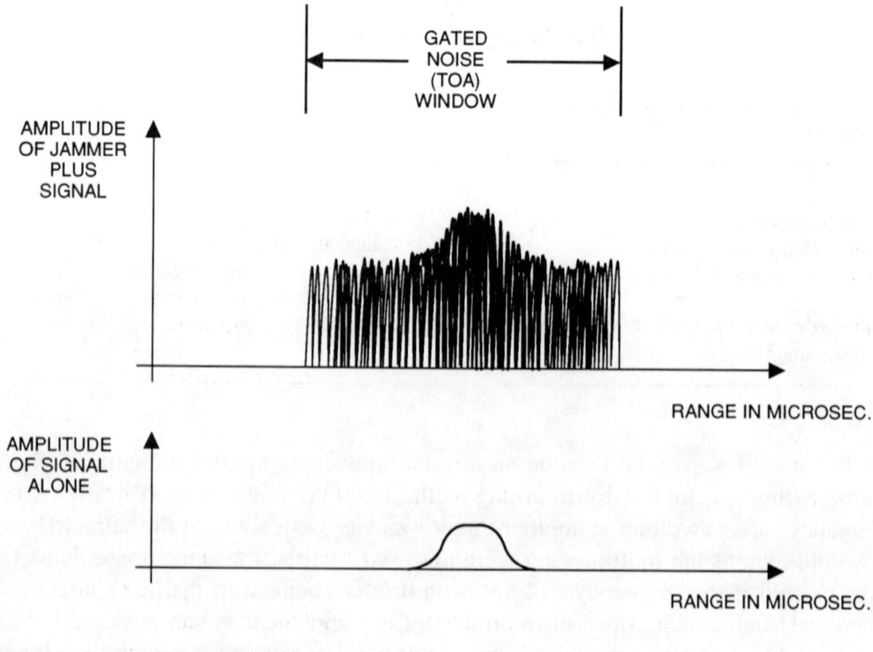

Figure 6.7 Power management in time.

This is a highly efficient process for reducing the average power; it allows the noise jamming to occur efficiently, being used against each threat just when its pulse is due. This type of power management is especially appropriate against low duty cycle threats.

Power management in time requires these functions:

- separation and identification of intercepted signals,
- measurement of PRF and agility,
- correlation with frequency and space management,
- PRI tracking and predictions, and
- look-through own jamming.

The power-managed time parameters are summarized in Table 6.6. The key requirement is the ability to predict the arrival of each pulse from each threat. For the gated noise jamming of Figure 6.7, the accuracy can be coarser than a pulsewidth as shown in Table 6.6. This is because the pulse can fall anywhere within the gated noise window of Figure 6.7, which must be considerably longer than the pulse for effective noise jamming in any case. Gated noise jamming is quite efficient in terms of average power transmitted. Gated noise would not be appropriate against a search radar because it indicates the approximate range or distance. It is quite appropriate

Table 6.6
Power Managed Time Parameters

Parameter	*Requirement*
Pulse prediction tolerance	
Noise mode	Somewhat greater than a pulsewidth
Repeater mode	Somewhat greater than a pulsewidth
Transponder mode	Much less than a pulsewidth
Input-triggered transponder mode	Somewhat greater than a pulsewidth
Window width	
Repeater mode	Sum of set-up time + pulsewidth + prediction tolerance
Noise or transponder mode	Above + jamming-waveform width
Number of prediction trackers	Number of simultaneous lethal or important threats with low to moderate duty cycles
Multiplexing	
intramode	Yes
intermode	Yes

against a tracking radar, however, either because the range error itself is sufficient, or because the range gated noise is an efficient means to an end, the end being angle deception. However, there may be an indirect requirement for higher precision tracking predictions for noise mode: if the own jamming blinds (jams) the ECM system's own sensors, and there is considerable uncertainty in relation to the window width, then the PRF tracker is unlikely to maintain track. In other words, if the "look-through own jamming" has a low duty cycle, coarse tracking predictions will quickly result in a break-track condition. The solutions are to improve the tracking prediction accuracy or to increase the look-through duty cycle. If the predictions are indeed accurate, then the tracker can "flywheel" for longer periods of time without a "look-through" glimpse. For repeater mode, there is no look-through problem, so there is no indirect accuracy requirement for that mode.

With regard to the prediction accuracy given in Table 6.6 for transponder mode, since the tracker will be used as the trigger to generate the output transponder pulse, instead of the threat pulse itself, the accuracy required becomes at least two orders of magnitude more severe, if not more, than the gated noise mode. If such accuracy is not met, the radar will see the pulse jitter, revealing it as a false target. On the other hand, if the tracker predictor is merely used as a window generator, to assign the transponder resource to that threat pulse for the window duration, then the normal accuracy requirements prevail. The actual range delay will be generated internal to the RF memory subsystem for input-triggered transponder mode.

The general rule for gated window operation is that the inaccuracy maximum may be greater than a pulsewidth.

With regard to the number of PRI trackers in Table 6.6, to use power management in time to its fullest capability, it is necessary to generate distinctive windows against each threat. Figure 6.8 shows gating windows against threat-pulse trains *A* and *B*. Distinctive modes, resources, and modulations can be applied against each threat by using the windows for multiplexing: distinctive operation applied to each window is the multiplexing of Table 6.6. The distinctive resources can be a transponder RF memory, a noise generating VCO tuned to each threat's nominal center frequency, or other resources.

Power management in space is illustrated in Figure 6.9. This type of power management is generally used to steer the transmitted EM wave power primarily in the direction of the threat, rather than uniformly over a wide angular span. The concentration of power in a particular direction is achieved with a narrow antenna beamwidth. Narrow beamwidth implies high antenna gain and large aperture, and requires that the antenna beam be steered, and sometimes focused and defocused. The general power management in space functional requirements are

- separation and identification of intercepted signals,
- high-speed direction finding of multiple emitters,
- time management of antenna steering,
- high-speed steering of jamming antenna, and
- high-gain transmitting antenna.

The key to using a high gain antenna is the ability to steer the antenna accurately, or else the narrow beamwidth will work to the ECM system's disadvantage. The parameter requirements for this power management in space are summarized in Table 6.7. A typical broadbeam antenna without power management control would have about 3–6 dB antenna gain, whereas a wide-aperture antenna under power management steering control would have 30 dB or more gain. For example, a 90 × 90 degree beamwidth antenna would have about 5 dB of gain, which can be calculated by using the approximation formula in Figure 6.9. Generally, the narrower the beamwidth, the more the gain and the better the ERP and the performance. There is, however, an indirect constraint on the beamwidth: the narrower the beamwidth, the greater the burden on the pointing accuracy.

As indicated in Table 6.7, an azimuth steering capability is much more important than an elevation steering capability for most ECM applications; this is because most of the threats are near elevation angles that are not too far from the horizon. In other words, there is not as large a range in elevation angle as there is in azimuth angle. Many ECM systems take advantage of this fact by using relatively narrow but fixed elevation beams. However, the importance of azimuth beam steering is greater than that for elevation steering only if a relatively narrow elevation beamwidth is reasonably matched to the elevation distribution of threat radars. Many

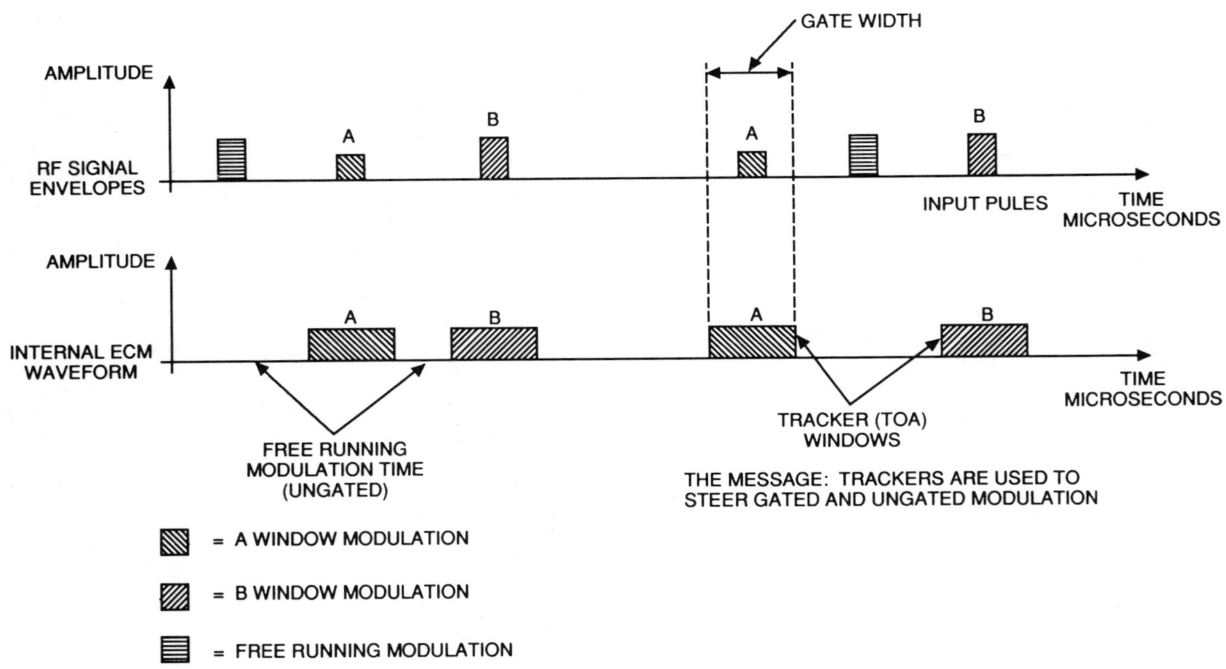

Figure 6.8 PRI tracking.

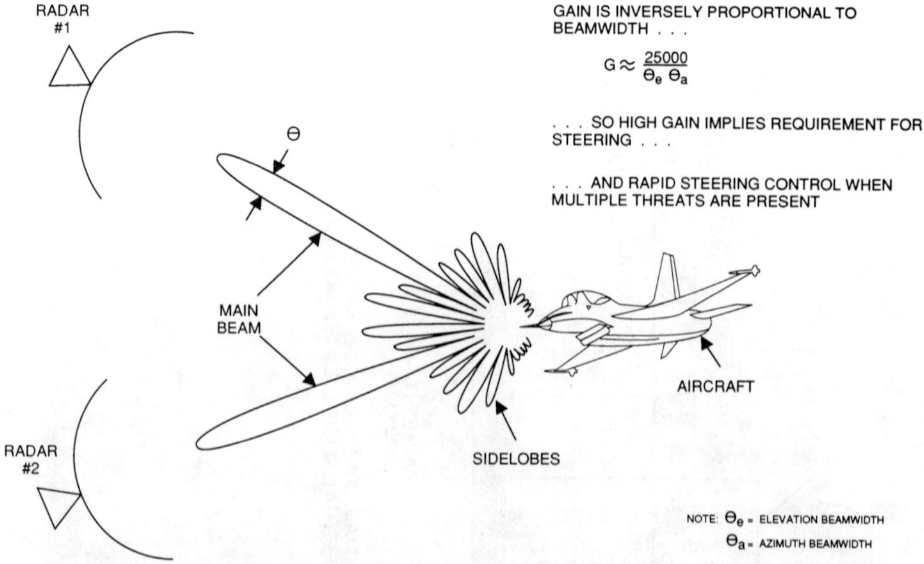

Figure 6.9 Power management in space.

jammers used for avionics applications use excess elevation beamwidths to accommodate aircraft roll, and hence could benefit greatly from elevation beam steering.

The steering rate, as summarized in Table 6.7, is similar to the power management in time multiplexing rate described previously. The ideal requirement for steering time is about a pulsewidth; the rate value given in Table 6.7 is a practical and adequate value, given the understanding that the interval between resteerings is effectively random. A high-gain antenna capability would be worse than useless unless it was power-managed in time. If power management in time is unavailable

Table 6.7
Power Managed Spatial Parameters

Parameter	*Requirement*
Beamwidth	No theoretical lower limit, the more gain the better
Pointing accuracy	Less than half the beamwidth
Degrees of freedom	Both azimuth and elevation, but azimuth is usually more important
Steering rate	About an order of magnitude more than the sum of all the PRF's of the threats the high antenna gain is countering
Focusing rate	Situation-dependent

because of DSP overload or other factors, the antenna should be defocused to create a broad beam.

6.4 DIGITAL PROCESSING

Figure 1.5 showed the typical microwave and RF system structure of an ECM system that could generate all modes, preset or power managed. That figure showed data busses going to and from the major system elements from and to the CPU and the DSP. The overall block diagram of the digital processing functions performed by the CPU and DSP are shown in Figure 6.10. The sensor block is also included in Figure 6.10 for clarity. The principal digital processing functions are

- signal separation,
- signal identification,
- signal track and prediction,
- waveform generation, and
- central processing.

Each of these functions will be briefly described.

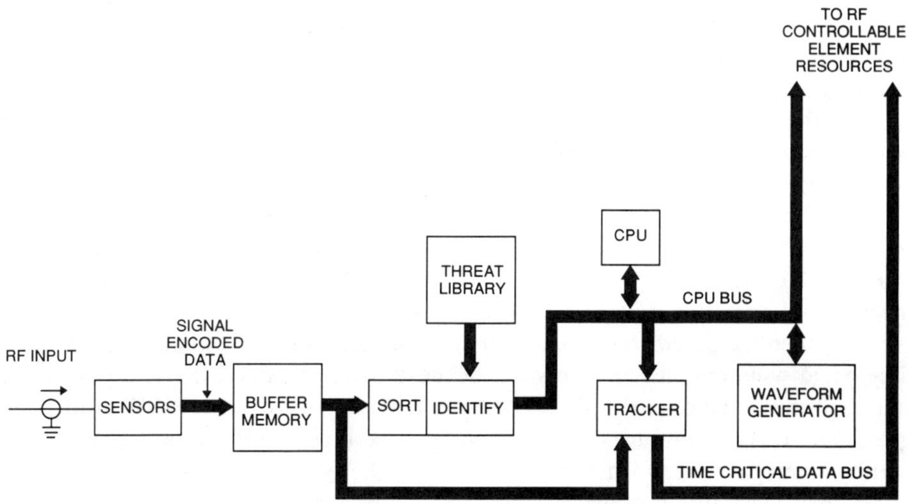

Figure 6.10 Digital processing functions.

6.4.1 The Central Processing Unit

The CPU software-generated functions are the most varied, and, consistent with partitioning based on time-scale analysis, include

- technique determination,
- priority conflict resolution,
- resource allocation,
- parameter setup,
- data and parameter maintenance,
- gain and power control,
- slow servo functions, and
- built-in test.

The technique determination is not an obvious or simple algorithm unless a one-on-one situation exists; in that case, the algorithm rule is simply to pick the best-rated technique against the identified threat. With power management in time, that is, time multiplexing by applying distinctive modulations via time-gating windows, the decision criteria can also be based on this rule. If several threats need to be handled with a preset mode, however, or the resource needed for a particular technique is being employed against another threat, the decision tree becomes complex and large. For example, threats *A* and *B* may be best countered by noise mode jamming, using the RF VCO resource, but if threat *C* suddenly is present also, then the most optimum jamming of all the threats *A, B,* and *C* could be a false range pulse target generated by the RF memory resource operating in transponder mode. Obviously, the choice against threats with pulses individually covered by the time-gated window is a much simpler problem.

The previous paragraph gave an example of one type of priority conflict. As a general rule, the resolution of priority conflicts revolves around deciding which threat a particular resource should be used against at any one time, especially for free-running conditions or if gating windows overlap. The functions of resource allocation and priority conflict resolution are to make that decision. Once the decision is made, then the parameters needed by the resource can be communicated to it. For time-gated windows, the transmission of these parameters and the setup of that resource to these values generally should not take longer than about 20% of the time-gated preset TOA window itself, otherwise the efficiency of the time management will be severely reduced. In other words, the ECM system should spend most of the time either transmitting, or waiting for a radar pulse trigger in a fully prepared, preset status. The setup time length can therefore be considered dead time that lowers the efficiency. There is usually a modest setup time allowance before the true or preset TOA window actually starts. In most cases, this parameter setup gives frozen values for the expected incoming pulse. For example, a modulator used in the repeater mode would have its attenuation or its phase preset, and this value would stay frozen during the TOA window time. It should be noted that, when generating a false doppler against a threat by using a stepped-phase modulator as in Figure 2.16, where the time to step through all 360° is the reciprocal of the desired offset frequency, it is undesirable to have the steps occur during the pulse, because such an occurrence will distort the spectrum; the phase jumps should occur between pulses. Likewise, the programmed range delay for an RF memory unit used for transponder mode, such as in Figure 1.15 or Figure 5.27, is preset for the duration of the TOA window; this parameter is used to match the count of a high-speed counter to gate an output false target pulse. The parameter setup for a VCO subsystem used as a signal source to generate RF noise for noise-mode operation is more complex; these parameters must specify not just a dynamic operation, but one that has a probability distribution. Typical parameters for noise are center frequency, frequency deviation, and FM rate minimum-to-maximum distribution. The deviation and FM rate distribution determine the noise quality, as described above; the center frequency and deviation determine the effective power. The self-contained hardware in the RF VCO subsystem will run with these parameters during the gating TOA window.

One very important CPU software function, included in the above list, is the data and parameter maintenance. In particular, it is very important that data that is no longer needed be promptly purged to make room for new information. Likewise, if a narrow-band receiver or a PRF tracker had been assigned to a particular radar signal, it should be released for other duty whenever that radar is no longer considered a threat, so that resources will be available to use against real threats.

The system gain in repeater mode is usually set by a CPU algorithm to be consistent with the antenna isolation, in accord with the feedback theory presented in this book. For a given set of receiving and transmitting antennas, the antenna isolation and the system repeater mode gain determine the net loop gain or net loop isolation, the theory of which was described in Chapter 5. The gain can change with time because of component parameter drift, particularly in the amplifiers and amplitude modulators. The isolation can also change if the metal structures near the ECM antennas are changed or moved; this happens most dramatically on an aircraft when the munitions or other stores are dropped. Therefore, the CPU needs repeatedly to test the net loop isolation and readjust the system gain as appropriate. Typically, there are a number of other slow servo functions that the CPU software needs to perform.

Another very important CPU function, the last on the above list, is that of operating the *built-in-tests* (BIT), and responding accordingly. The possibility that one of the main system resources has failed greatly complicates the resource allocation rules. Some sensors continually monitor their assigned parameter (e.g., temperature), and the CPU interrogates this sensor at the rate determined by the BIT algorithm. Often, however, the BIT function is much more involved than just monitoring hardwired sensors. The BIT usually includes the generation of dummy signals, and the switching of paths to see the response. If the BIT algorithm determines, for example, that the output tube is dead, then the appropriate response is to indicate a "fault condition" to human users or operators. If, however, the BIT algorithm determined, for example, that an RF memory was defective and not operating, then the resource allocation algorithm must cope with fewer resources, still making the most optimum choice with what is available. In this example, it might decide to use noise jamming.

6.4.2 Processing Sequences

Referring again to Figure 6.10, sensors are shown which output encoded data that describes each pulse. This data flow is buffered, since the subsequent digital processing elements operate at their own rate, whereas the incoming pulses may come in probabilistically determined bunches. Of course, the buffer memory does not relieve the constraint that the average data flow rate from the many simultaneous pulse trains present must still be met. As can be seen in the figure, the buffer memory feeds the sorter and tracker digital signal processing units. The key processing units are the sorter, the identifier, the tracker, the waveform generator, and the CPU, with the threat library supporting the identification process as shown.

The typical sequence of operation of these six processing units starting with the captured data, that is, based on the signal sensing, pulse encoding, and data buffering, to the commencement of jamming, is summarized as follows:

- PRI sort algorithm operates on data;
- PRI sort data crosses threshold;
- distinct signal and nominal values recognized;
- ID algorithm classifies signal;
- CPU priority algorithm determines response;
- CPU assigns tracker to threat;
- tracker acquires signal;
- CPU assigns RF resources to tracker;
- CPU assigns waveforms to tracker-predictor; and
- jamming commences.

In the above example of a sequence-of-operations list, the primary sorting criteria is PRI. Other potential criteria include

- RF carrier frequency,
- pulsewidth,
- angle of arrival, and
- amplitude.

In the above example, however, the algorithm used PRI to sort or separate the signal from all other signals. Once the signal is separated by some reliable criteria, it will be recognized as a distinct signal, the nominal values of its parameters will be determined, usually by an averaging process, and then the identification process will commence. It is not necessary to identify all the non-lethal signals, merely to classify them as such; however, their presence often needs to be established and monitored, often on an individual signal basis, as an aid to the sort processing. Such monitoring is used artificially for thinning the signal environment fed to the more complex algorithms and other processing. It may be noticed in the figure that the sort and identification processing functions are not clearly separated.

In the above list, the jamming waveforms were assigned to the tracker, not the RF resource directly. The waveforms cannot be assigned directly to the RF resource because the tracker needs to be responsible for multiplexing the various waveforms onto that RF resource to achieve power management in time, as in Figure 6.8. As shown in that figure, the time not covered by a TOA window is left with the free running modulation for all threats not tracked; this modulation could be considered to be controlled by "tracker number zero," a dummy tracker for logic-control purposes.

6.4.3 Pulse Encoding

As was shown in Figure 6.10, the encoded sensor data passes through a data buffer to smooth out the normal pulse bunching, and is fed to the DSP sort, identification, and tracking processing units. What data do these three processing units need? There is a tendency to base the answer on the input signal environment complexity, as would be appropriate for an ELINT system. However, data is needed for one purpose only: the efficient generation of jamming signals. Any data that will not be used to support that goal is therefore not needed by an ECM system DSP. Tempering that statement somewhat is the knowledge that future EW systems may integrate ECM, ESM, and other systems, using common apertures, sensors, and processors, often displaying the EM environment to human monitors in real time. In any event, one suggested set of ECM sensor pulse encoding words and their resolution is given in Table 6.8. The following discusses the rationale for these words and resolutions. The descriptions in Table 6.8 are representative, but not necessarily generic.

The *time of arrival* (TOA) word is listed in Table 6.8 with a resolution of 50 ns. TOA information is one of the most important parameters to be passed to the DSP. The trackers use this data to track, and the data is also useful for the sorter to separate signals. The resolution shown in the table is based on needing a resolution smaller than most threat radar pulses. The indirect requirement, as stated before, of allowing the PRF trackers to flywheel through periods of missing pulses may lead to a much more severe requirement than this value based on direct jamming needs. The data word size needed for the TOA parameter depends on the time rollover needs of the data processing. That is so that one set of pulse descriptors will not be con-

Table 6.8
DSP Input-Pulse Encoding Words

Parameter	*Bits*	*Resolution*
TOA	25	50 ns
Frequency	15	1 MHz
Polarization	9	1°
Amplitude	6	1 dB
AOA (horizontal)	9	1°
AOA (vertical)	0	(N/A)
Pulse width	13	50 ns
Flags	8	(N/A)
Total	85	

fused with another; if the data processing cannot reduce the interpulse relationships easily, then the word size must be made larger to capture the history meaningfully.

The second most useful parameter after TOA is the RF carrier value. The resolution for this parameter can be based on the need to separate signals, when a resolution that is an order of magnitude more coarse than that shown in Table 6.8 would be appropriate. The value listed in Table 6.8 is about right for set-on-VCO transponder and noise purposes, and is based on being a small percentage of the threat's IBW; this in turn is based on the threat's pulsewidth, which is itself based on the radar designer's expectation of the target size and need for range resolution. This rule applies to pulse compression and doppler radars in a more complex manner.

The amplitude data, representing the nominal amplitude of the pulse, can be used for three purposes: (1) as a crude means to tell if the threat is near or far, (2) to aid in separating the various signals, based on signal strength, and (3) to sense the lobing rate or any other modulation imposed by the radar. A typical ECM sensor has 40–60 dB DR, and the resolution shown in Table 6.8 was chosen to be a small portion of that DR.

The polarization, AOA, and pulsewidth information are useful to separate signals. In that regard, the AOA parameter has the special advantage of not being under the control of the radar. For this separation purpose, the resolution need merely be a small portion of the range of each parameter.

Hardwired operations in the sensor subsystem result in the need for flags to be included in the encoding. Usually, this is a substitute for the measurement and encoding of one of the other parameters listed in Table 6.8. For example, most tracking radars have pulsewidths less than a microsecond, albeit subsequent to compression for pulse-compression radars. Most search radars, for which sensitivity is more important than range resolution, have pulsewidths greater than a microsecond. Therefore, a flag can be set for pulses greater than a microsecond and much greater than a micosecond, indicating search and pulse-compression radars respectively.

The total number of bits of resolution shown in Table 6.8 is difficult for a DSP to cope with, although signal processing hardware is continually becoming more capable. Likewise, also difficult is for a reasonable sensor subsystem to generate this much data for each pulse. Therefore, the ECM system designer must be selective in choosing the most important information set that meets practical system needs. On the other hand, the values listed in Table 6.8 are not necessarily indicative of the needs of certain functions, especially certain ESM functions, which need considerably more resolution for some of these parameters.

Each of the parameters in Table 6.8 are the nominal values for each pulse. Intrapulse modulation sensing could also prove valuable, with the encoding approach becoming much more involved.

6.4.4 Signal Sorting and Identification

The signal sort processing unit performs the necessary first step of separating each signal from all the others. Pulse train sorting is a common means to separate low- to medium-duty cycle signals. One reason for this is that the DSP sorting unit can use a simple CVR as a sensor; it is inexpensive and can cover an enormously wide bandwidth. High-duty cycle pulse-train sorting is usually accomplished with a completely separate sensor and DSP unit. To support the low- to medium-duty cycle tracking, the CVR is usually designed not to output continually any logic signal in response to CW inputs. This is accomplished with ac coupling circuitry, since the result of the RF CW input is a dc voltage out of the RF crystal.

In addition to pulse-train sorting, which examines the relationship between pulses, there are ways to segregate or bin pulses on a single-pulse measurement basis, or a monopulse basis, just as monopulse radar determines angle for each pulse. This binning is based on exploiting the uniqueness of certain parameters. Typical parameters suitable for single-pulse measurement binning include

- RF carrier frequency,
- AOA,
- pulsewidth, and
- intrapulse modulation.

Of these, the RF carrier frequency has been the one most commonly exploited, while AOA is considered the most highly desirable, albeit more difficult to capture, especially on the basis of a stabilized frame of reference. In other words, each input pulse train will generally have a unique carrier frequency, which can be used as a single-pulse measurement separator. Each bin would correspond to a sensed RF carrier, with the expectation that this bin represents one radar. If multiple radars fell in an RF carrier bin, this would make the single-pulse measurement signal separation processing more difficult, but overall it makes the ECM jammer operation easier.

Pulse-train sorting and single-pulse measurement binning can both be effective sorting approaches. Even more effective, however, is multidimensional processing which can significantly ease the signal separation processing burden. The advantage that multidimensional processing has in separating signals is revealed in Figure 6.11. The top graph is a time-domain picture of five pulse trains: A, B, C, D, and E, each of which has a unique amplitude, pulsewidth, and PRI. Nonetheless, if the pulses were not labeled it would be difficult to sort them mentally. The next graph in Figure 6.11 shows the PRF spectrum. This more clearly distinguishes many of the pulse trains; however, trains A and B, which have almost equal pulsewidth, amplitude, and PRF are even more difficult to distinguish in this graph. In the bottom graph of Figure 6.11, another dimension is used: RF carrier frequency. This graph can be

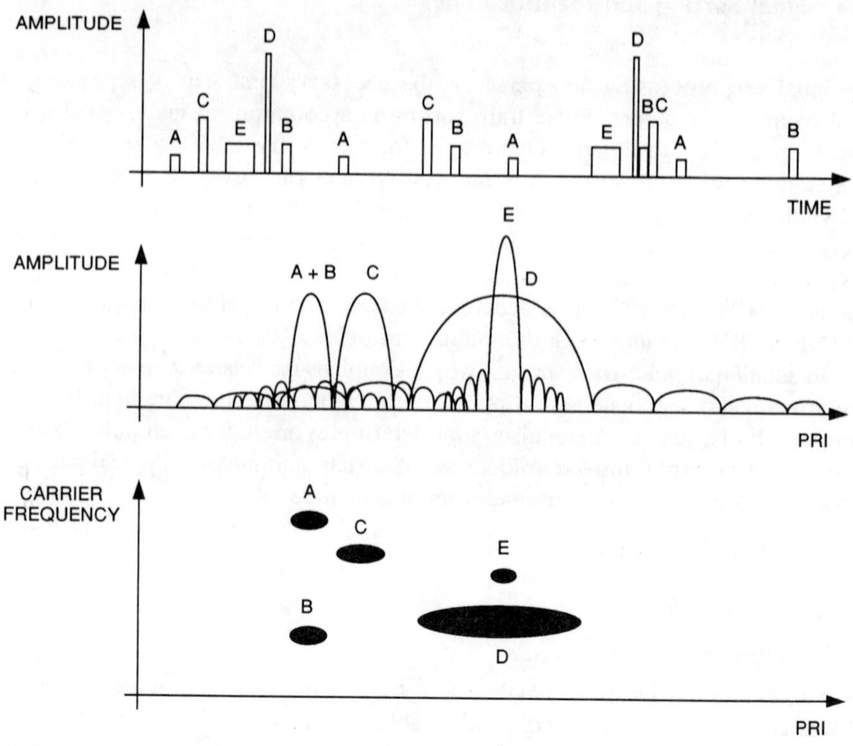

Figure 6.11 Two-dimensional processing.

considered the top view of the preceding graph. In this last graph, all five signal trains are clearly distinguishable in the PRI or PRF *versus* carrier plane. Searching for signals on such planes is much more efficient than examining amplitude *versus* time or amplitude *versus* PRF. Of course, even though it is difficult to visualize, a DSP algorithm can sort in the three dimensional space of PRF *versus* RF carrier *versus* amplitude.

For a two-dimensional plane, such as PRF *versus* carrier frequency, a threshold is set for each resolution cell in the plane, which can be a certain number of occurrences in the recent past, normalized to the PRF for that part of the plane. If activity crosses that threshold, then a signal has been localized and separated; in other words, a distinct signal has been recognized. Unfortunately, the digital signal processing is usually not quite so simple. The reason the process is more complex is that a signal will generally activate a given fixed threshold in several adjacent or nearby cells also. Therefore, a DSP algorithm needs to recognize the peak of the "hill" that is above the threshold, and classify that peak as a unique signal. Another hill, perhaps separated by some rugged "terrain," would then be recognized as another signal. The terrain "bumps" in the plane floor or on the slope of the hill must not cause false alarm signal indications. Furthermore, the threshold normalization must take into account the scattering into adjacent cells, reducing the threshold by at least half for each dimension. For example, a cell at 1 kHz PRF and 3 GHz carrier frequency might have a threshold of 1 kHz $\times$ 0.5 for PRF $\times$ 0.5 for RF $\times$ 0.5 for margin = 125 Hz. An intercepted repetition rate higher than that, for that cell, would indicate the presence of a signal. The setting of this threshold also depends on the time length over which this pulse count occurs. For search radars, the main beam may pass over the target for just 10–20 pulses (i.e., PRIs), which is an incentive to make the processing threshold have a short pulse-count time duration. For tracking radars, the target may be in the main beam for an extended time. However, the ECM threshold criteria needs to be consistent with taking prompt jamming action; some jamming techniques are based on not allowing the threat radar's angle track servo time to settle or stabilize.

Similarly, DSP algorithms can examine three or more input parameter dimensional spaces, to recognize the nominal position in this space where a signal is present. Usually the signal-presence value for each unit of space is the normalized number-of-occurrences; that is, the number of times the receiver-measured parameters match the processing-cell parameters over a given length of time is a measure of the probability of a signal being present in that cell. Of course, for more than three dimensions, terminology and visualization become difficult. The message the reader should understand, however, is that multidimensional processing makes signals clearly stand out from one another, whereas simplistic signal traces of the combined signal environment can appear jumbled and confusing.

Two of the examples in Figure 6.11 used PRI as one of the parameters to sort signals. The use of PRI or PRF implies building up a history over many pulses. This historical base is then examined by an algorithm, perhaps multidimensionally. As stated, sorting can also be accomplished on a single-pulse measurement basis. The procedure for segregating signals on a single-pulse measurement basis, using one or multidimensional criteria, is summarized as follows:

- tag input pulses with TOA;
- measure the correlation of each incoming signal with all the previously established bins;
- create a new bin only if no match is found;
- assign a code word to each bin;
- pass on to the digital processors this code word + TOA whenever a signal enters the system that duplicates the bin parameter(s); and
- allow for CPU housekeeping of data, including delisting bins.

The output of this DSP unit would then be, for example: pulse *D* entered at time 21.034546 s, pulse *B* entered at time 21.034572 s, *et cetera*. The signals have been separated into pulse trains *B* and *D*, corresponding to bins *B* and *D*.

To compound the difficulty of constructing proper DSP sorting algorithms, the algorithms must deal with certain common sensor difficulties. These include

- pulse overlap,
- CW and high duty cycle overlap,
- ECM self and nearby jamming:
  - noise,
  - multiple false range pulses, and
- ac coupling problems.

For some of these conditions, and especially the first in the preceding list, having the sensor raise a flag that the data may be bad is a useful way of dealing with the problem. In other words, bad data can be worse than missing data. One simplistic algorithmic approach is to simply treat the contaminated data as it would if the pulse data were missing. Thus, the DSP algorithm must be designed to coast through periods of missing or corrupted data, or of otherwise being blinded.

Once the signals are separated they can be identified, or at least classified. Often, the classification criteria are based on the sorting parameters. In any case, the signal identification keys include

- carrier frequency,
- PRF,
- pulsewidth,
- pulse duty cycle,
- antenna scan rate,
- antenna scan main lobe dwell,
- intrapulse modulation, and
- correlation with other signals.

Only three of the items on the above list can be ascertained on a single-pulse measurement basis. The signal amplitude itself is usually not a directly useful clue to the identification of the threat signal except to the extent that it is used as a means to an end to construct the history revealing the antenna scan.

6.4.5 PRF Trackers

PRF trackers are used as one of several means to sort the input interleaved pulse trains. Signal sorting separates or de-interleaves the input signals. PRF trackers are used for pulse prediction purposes, such as gating-on noise just before the pulse arrival, or generating leading false range pulses in transponder mode. PRF trackers are especially useful to allow distinctive jamming modulation and resources to be applied to each signal. Because most of the requirements for time-sharing, or power management in time, primarily fall on the PRF trackers, Table 6.6 is a summary of the key PRF tracking requirements. How is the quality of a PRF tracker rated? The rating criteria include

- TOA gate jitter,
- acquisition time,
- vulnerability to missing pulses,
- vulnerability to interfering signals,
- harmonic acquisition inhibit,
- stagger acquisition,
- vulnerability to multipath induced leading edge modulation, and
- vulnerability and response to agility.

The first two criteria on the above list are the criteria that are most often mentioned with regard to tracker performance, since they are the functional goal of the tracker, although the other criteria are also important. The acquisition time is usually specified in terms of the number of pulses required to establish track. Commencing track on the third or fourth pulse is not uncommon.

As stated above, TOA predictions are made for two triggering purposes: (1) to initiate a TOA acceptance window, or (2) to generate a false transponder pulse. The TOA jitter requirements for the acceptance window are much less severe (e.g., 10%–25% of window width) than that needed for the false pulse. If the false pulse is to be subsequent to the next radar pulse, that radar pulse would be used as a trigger to initiate a time delay in lieu of the predicted TOA.

For PRF trackers to operate properly, sophisticated servomechanisms are needed. Once the individual pulse-train signals have been sorted, a simple PRF tracker will measure the PRI between pulses, then set the window to be centered at the end of the next anticipated PRI. The servo continually adjusts the nominal PRI period of the flywheeling tracker. The sensed detection within the TOA window is also used to reset the phase of the PRI timer. If the pulse is missing, the tracker flywheels, using its own prediction as the trigger to count to the end of the next PRI. If the pulse is present, the complete timing or PRI phase error is usually applied as a phase correction, but only a portion of this error is applied as a PRI correction; the optimum value of this proportion is a compromise between stability and fast settling.

Just as radar systems use early and late gates to track the pulse target in range, so the PRF tracker also uses similar gates to track the pulses. However, the radar system has the advantage of knowing the PRI precisely, needing only to determine the phase of the PRF. The EW DSP PRF tracker servo must correctly respond to both PRF and PRF phase; this is a much more difficult operation, and the tracker design is much more challenging. For generating most jamming, such as range-gated noise, the jitter or uncertainty can be on the order of a pulsewidth, which is not too difficult; if the purpose, however, is to generate a false pulse just prior to the true range, the precision requirements on the tracker become severe, as explained above.

The PRI value, less the TOA window width and set-up time width, that the servo hunts for, is held in a digital latch, in conformity with the trend to digital processing. The *least significant bit* (LSB) of this latch corresponds to the period of a high-speed clock; the period of this high-speed clock is the tracker resolution, which must be less than the maximum allowed jitter. A digital counter starts to count at the trigger point, when the leading edge of the input pulse occurs, and a digital comparator outputs a flag when the counter count and the latch count agree. This flag starts the set-up time and TOA window. If no true trigger occurs, then, at the end of the TOA window, the counter is preset to the assumed error, nominally half the TOA window width, and then the counter is started; this is the flywheel function.

The tracker generates a TOA window into which the threat pulse is expected to fall, and this TOA window is used to restructure the ECM system for the duration of the window. Therefore, just prior to the actual TOA window, a setup time is utilized. For this set-up time, the tracker takes control of the digital data bus and loads the necessary parameters into the RF resources, including shutdown commands. RF switches are set, CVR outputs are enabled, latches that feed DACs are loaded and digital counters are preset. Then, unless the tracker is responsible for generating a dummy trigger, all activity is frozen for the duration of the TOA window. For example, if the RF memory recirculating memory loop were the resource being used by a tracker, then during the set-up time, the microwave switches would be set to steer the signals to and from the RF memory, and the RF memory's high speed counter would be preset with a digital word representing the desired range delay. The CVR, perhaps built into the RF memory, would be enabled to trigger the counter to start this count, and activate a flip-flop to close the loop. When the high speed counter is triggered during the TOA window and runs down to zero, the hardwired logic will, in turn, perform further restructuring, including gating the RF output with a microwave switch so that an RF pulse will be transmitted.

As another example of tracker control, a particular waveform, for instance, a swept square wave, can be steered to an amplitude modulator used in repeater mode. Usually, these modulators are just on-off switches, as explained in the first chapter. Therefore, during the TOA window set-up time, the tracker would take control of the digital-data bus, call up the address of the RF amplitude modulator switch, and then load a value of 1 or 0 depending on the swept square wave instantaneous value. Usually, at the end of each TOA window, there is a setup time for tracker 0, the fictitious tracker that generates the background jamming which all threats that are not being tracked will see.

6.4.6 Waveform Generation

The last processing unit in Figure 6.10 to be described is the waveform generator. The waveforms needed include

- jamming:
  - square waves,
  - swept square waves,
  - saw-toothed,
  - swept saw-toothed,
  - sine waves, and
  - rectangular waveforms;
- sensor modulations.

Many of these waveforms are used for angle deception modulation and are relatively slow. Figure 1.13 shows the range of operating frequencies in an ECM jammer. To the extent that they are relatively slow waveforms, they may be suitable for being generated in software rather than hardware, and hence there would not be dedicated hardware, except for a latch register, associated with each. Each tracker generally has several waveform generators assigned to it, and each has its own parameters.

To operate, a false-doppler modulation on a repeater path, for example, a four-bit digilator would need to be clocked at 16 kHz to generate a 1 kHz false-offset frequency. A hardwired digital counter can generate the phase bits, driven by a controllable digital-output oscillator tunable above and below 16 kHz. The waveform generator assigned to the tracker in software could change the oscillator frequency driving this hardwired clock in a stepped-linear fashion, thereby generating a swept *velocity gate pull-off* (VGPO). The hardwired digital counter could itself be considered a waveform generator, although this text has consistently referred to such operations as hardwired control or hardwired logic functions.

The sensor-modulation waveforms referred to above include such waveforms as are needed to sweep a superheterodyne receiver, or to gate filters sequentially in front of an IFM receiver, as examples.

In addition to the waveforms on the above list, the waveform generator may need to generate pseudorandom sequences with certain constraints or parameters applied to the sequence. An example would be a noise mode noise-jamming sequence. If generated in real time, the waveform generator output words would need to be clocked out at up to tens or even hundreds of MHz, which is much higher than the threat's bandwidth. If not generated in real time, the sequence should be thousands of times longer than the reciprocal of the threat receiver's bandwidth, to prevent the repeating pattern from having nonnoiselike characteristics. If, however, the noise sequence was recreated with truly random words for each new TOA window, then of course the sequence need only last as long as the TOA window length.

6.4.7 Digital Processing Considerations

A previous chapter discussed ECM design principles. One principle was to identify the limiting factors in the system design. For operation in a dense signal environment, with large numbers of low duty cycle signals present, the digital processing will generally be the limiting factor. However, dense signal environments with other characteristics can result in the sensor being the limiting factor.

Of the digital signal processing units shown in Figure 6.10, the heaviest processing load burdens are on the sort and track units. Circuit elements hardwired to perform certain algorithms are often used to provide processing speed, although they limit flexibility. A compromise is to hardwire the basic processing structure but to allow certain parameters to be programmable. Recently, the industry has been making good progress in developing fast, completely programmable processing units.

One good way to characterize DSP systems is to identify the interfaces across which the majority of data flows. For an ECM system, the heavy-flow interfaces are between the DSP and

- at present:
  - receiver,
  - signal source;
- anticipated:
  - receiver,
  - signal memory,
  - antenna, and
  - signal source.

In particular, this author predicts that a number of new or improved jamming techniques will be generated more directly by the antenna subsystem, hence the above projection for data flow. In other words, the amplitude modulation will be created not by an RF amplitude modulator variation or by retuning an RF VCO, but rather by redirecting the antenna beam. This will require considerable digital data flow to the antenna subsystem.

The burden on digital signal processing is expected to increase. The driving factors causing this increased burden are summarized as follows:

- external:
  - wider range of threat operating frequencies,
  - greater variety of threat emitters,
  - increased density of signals,
  - increased complexity of threats and their signals,
  - deliberate variation of threat parameters,
  - presence of friendly and nonmilitary emitters,
  - use of wartime *versus* peacetime parameters,
  - presence of jamming;
- internal:
  - increased number of techniques,
  - reduced system response time,
  - increased number of RF subsystems with heavy data flow interfaces, and
  - control of the sensors and receivers distorts input.

As shown in the above list, the DSP must be able to handle not just the hostile emitters, but also the friendly and non-military ones. The DSP must be designed to recognize when it is failing at its task so that the CPU can program free-running modes against all the threats not being tracked. Even without deliberate jamming being present, as explained in Chapter 4, DSP processing and receiver control can distort the information the DSP is using to generate its control signals. An example is a superheterodyne receiver, which the DSP controller will tune to the frequency of the expected input pulse for the duration of the TOA window; if another signal occurs during this TOA window interval, the TOA window interval is misplaced, or the carrier frequency value is wrong, distorted or missing information will be passed to the DSP.

6.5 TECHNIQUES

The purpose of this book is to explain to the reader how to implement jamming and deception techniques in a practical fashion, without necessarily explaining all the techniques that presently constitute the repertoire of a modern power-managed ECM system. Although this text has consistently recommended designing ECM structures that tend to be independent of technique, system design and technique determination are at least partly interdependent. A tentative repertoire is a good starting point for the system design. A simplified summary of the technique determination process is listed here:

- assemble threat data and information;
- determine the threat's key performance parameters;

- examine these key parameters to assess the vulnerability of the system, examining in particular how external RF inputs can alter these parameters,
- define the ECM outputs which exploit the vulnerability;
- study the practicality of the ECM implementation:
    - are the parameters consistent with current technology?
    - are new technologies needed?

Given the established jamming technique repertoire, the techniques then must be related to the hardware needed to implement them. Throughout all the previous text the reader has been given the background to do that. As a review, Table 6.9 shows selected jamming technique examples related to a key implementation resource. Noise jamming, with or without TOA window gating, is an excellent means of denying good range information to the threat search radars. A microwave signal source, the RF VCO, is a practical method of generating this noise. The VCO signal is passed through a high power CW TWT, and the FM noise modulation creates AM noise in the radar receiver as the VCO goes into and out of the radar bandpass. This VCO signal is transmitted out of the CW TWT at high power either continually or for the duration of a PRF tracker generated window of sufficient length to cause substantial range error. Likewise, to cause angle errors against a search radar, the ECM sensor can detect pulse amplitude and, with the DSP, use the information to transmit good pulses except near the peak of the radar's main beam. Because the radar expects the maximum return when the radar's main antenna beam is pointing at the target, an error is created. Likewise, a recirculating RF memory loop can be used to deceive a tracking radar in range. The RF memory loop will stretch the duration of the pulse to many times its original length; this output will then be gated to create a false range pulse substantially removed from the true range. Likewise, to deceive a tracking radar in angle, the last example in Table 6.9, a swept square wave generated by a waveform generator can be applied to an RF switch through which repeater mode RF signals are passing. When the swept square wave sweep passes through a beat of the radar antenna's angular lobing rate, the radar's angle

Table 6.9
Jamming Techniques and Their Implementation

Jamming	*Implementation*
Search radar range jamming	Set-on VCO
Search radar angle jamming	Sensor-gated transmissions
Tracking radar range jamming	Recirculating memory loop
Tracking radar angle jamming	RF switch with swept square wave

tracking servo will be disturbed and perhaps break lock. This technique is appropriate against the classic sequential lobing type radar; the trend, however, is for radars to use more sophisticated track means.

Another example of the implementation of a technique, and a further example of the functioning of ECM structures, is shown in Figure 6.12. In this illustration, a conically scanning radar is tracking an aircraft, and there happens to be a one beamwidth angular tracking error. When the scan is at point C the radar receives the maximum return power, and when the scan is at point A it receives the minimum return from the aircraft skin reflection. The radar's angular tracking servo uses this information to slew toward the correct pointing angle. However, the aircraft is shown with an ECM jammer mounted on its wing. A simplified block diagram of the ECM system is also shown. The ECM antenna picks up the radar signal, the preamplifier boosts the signal to a higher level, and a coupler directs part of the signal to an RF sensor. The amplitude sensed for each pulse is passed on to the DSP in digital format. The DSP determines the frequency and phase of the radar's conical scan. A waveform generator is used to create a square wave in synchronism with the radar's conical scan, but 180° out of phase. This square wave function is passed through a DAC and applied to an RF modulator; an RF switch could also be used without a DAC. The RF modulator imposes this waveform on the repeater mode RF signal propagating through the system. Assuming the J/S ratio is sufficient, the result is that at point A the radar will see more power than at point C, and hence the radar's angle track servo will drive in the wrong direction. The ECM system has succeeded in defending the craft.

6.6 COHERENT JAMMING

What are the single-threat ECM-transmission bandwidths associated with the different jamming modes and techniques described in this text? A summary is given in Table 6.10. As can be seen, narrow band or set-on noise to be used against noncoherent threats is on the order of 10 MHz, while coherent noise is on the order of 10 kHz. Likewise, a transponder used against a noncoherent threat should have a bandwidth on the order of 1 MHz, while a transponder used against coherent threats needs to have an RF tolerance of less than 100 Hz. These values are determined by the target's size, for noncoherent threats, but favoring SNR for a search radar and favoring resolution for a tracking radar. The values for coherent jamming are dependent on the maximum velocity, and the practicality of dividing the target doppler frequency domain into many cells: 10 to 100 cells is typical.

Noise can be used against a doppler radar, and it seems sensible to transmit noise with a bandwidth that just covers all the doppler cells. That is the first value shown in Table 6.10. In most instances, however, the noise can have a bandwidth equal to the doppler radar's IF bandwidth; this noise bandwidth is about equal to the

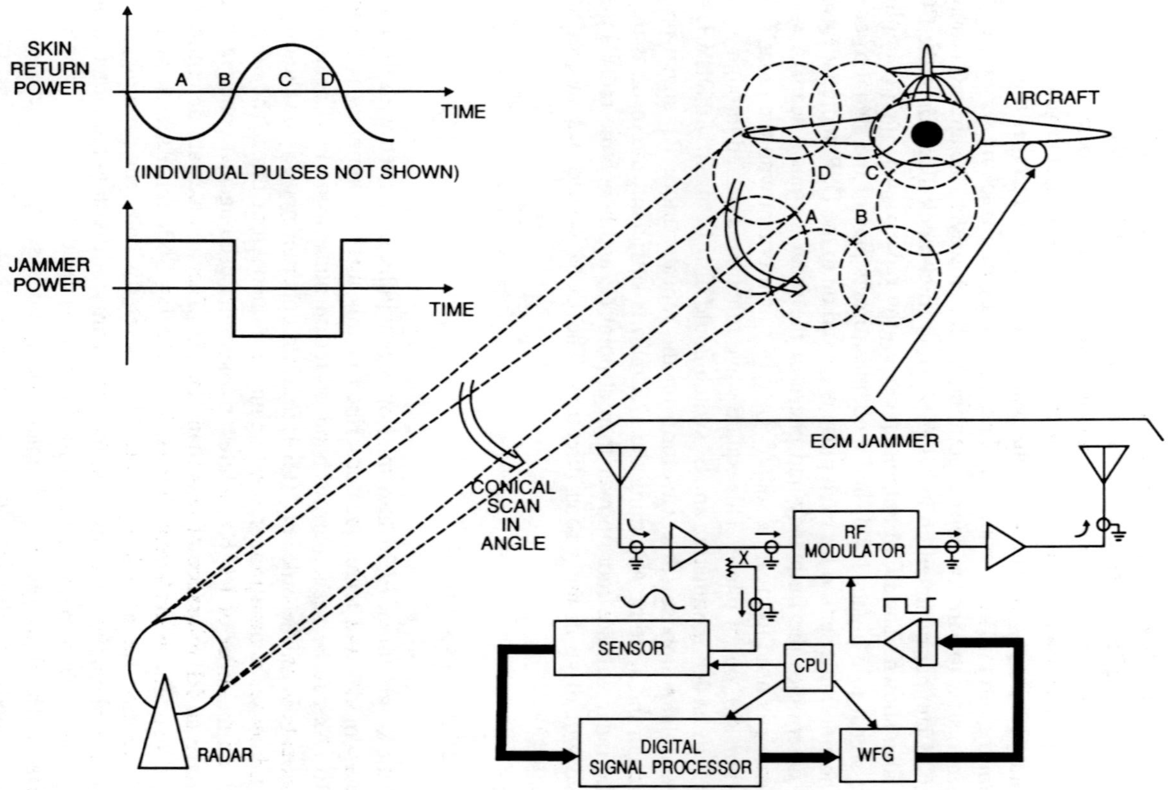

Figure 6.12 Deception jamming example.

Table 6.10
Jamming Bandwidths
(Orders of Magnitude)

Jamming	*Transmitted Bandwidth (Hz)*
Coherent transponder*	10^2
Coherent noise	10^4–10^6
Transponder	10^6
Noise	10^7
Broadband noise	10^8

**Note*: Spectral line error budget; the entire bandwidth is 10^6.

transponder bandwidth used against an efficient noncoherent radar and is on the order of the reciprocal of the radar's pulsewidth, because of spectral foldover for a constant PRF. Figure 6.13 can be used to explain spectral foldover, in addition to showing what makes a doppler signal coherent. The pulse train PRI and pulsewidth are shown. The reciprocal of the PRI is the PRF, and a series of a spectral lines occur with a spacing equal to the PRF. The amplitude of each of these spectral lines is determined by a $(\sin x)/x$ mathematical function related to the pulsewidth. The radar designer commonly sets the full range of doppler cells equal to the separation between pairs of these spectral lines, that is, the PRF lines. However, this apparent pattern of cells is effectively duplicated across the spectrum every PRF. In other words, an input signal in a given doppler cell, if displaced in frequency by exactly the PRF, would still fall in the same cell, although its amplitude will be determined by the $(\sin x)/x$ shape. Therefore if the ECM jammer noise has the width of the reciprocal of the pulsewidth, the jammer power will efficiently fold over into the doppler cells, even though the full range of all the doppler cells is orders of magnitude less.

Figure 6.13 can be used to answer the question: What makes a signal coherent? A coherent input signal is characterized in the frequency domain by a set of spectral lines, as explained. There will be no power in between these spectral or PRF lines, because at the PRF lines the carrier frequency is a precise multiple of the PRF. The vector diagram in Figure 6.13 shows the process, with each pulse in time represented by a vector in the phasor domain. For each input pulse entering the system, when a carrier value is not equal to one of the PRF lines, the RF phase will not be consistent. These pulses are shown as dashed vectors, in the vector diagram. These vectors mutually cancel, as is evident in the diagram, since they do not align. The solid vectors represent a pulse snapshot of the carrier frequency also, but where the

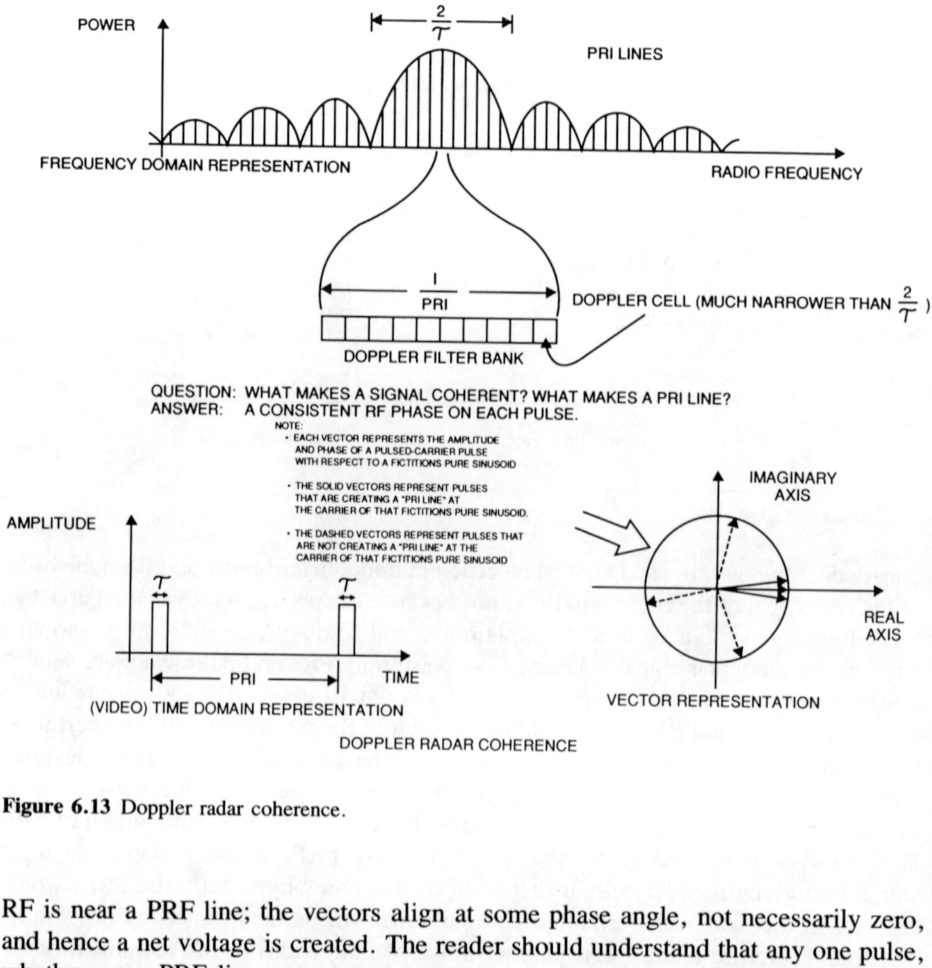

Figure 6.13 Doppler radar coherence.

RF is near a PRF line; the vectors align at some phase angle, not necessarily zero, and hence a net voltage is created. The reader should understand that any one pulse, whether on a PRF line or not, generates power over a smooth continuum of frequencies. A pulsed carrier's time-averaged power at a given frequency, however, has a nil average except at the PRF lines, where the RF phase angle is consistent. The pulses mutually cancel one another for carrier frequency values that are not multiples of the PRF.

This text has described recirculating memory loops and other means to generate delayed transponder pulses. When is a transponder pulse coherent and when is it noncoherent? One simplistic answer, appropriate for ECM signals to be used against a doppler radar, is: the transponder is coherent when it creates the pure spectral lines illustrated in Figure 6.13. However, to be able to answer this question in a way that is more meaningfully related to the hardware implementation, and to include generically both doppler and pulse compression radars, the three realms of transponder coherency need to be understood; they are given in Table 6.11. According to Table 6.11, if a full, or complete, pulse is delayed it will be coherent, and if you recirculate a leading edge, or sample, it will be noncoherent. Of course, the full pulse delay must have a phase variation or jitter tolerance of about 10° of RF, equal to the relative tolerance of the elements of a phased array antenna system. The author can now define the case of recirculating a full pulse as the border-line condition between a coherent jamming and a noncoherent jamming implementation. In other words, recirculating a full pulse is marginally efficient. A leading edge sampling transponder would be highly inefficient against a coherent radar, but a programmable delay line with RF phase jitter less than 10° will be quite efficient.

With regard to ECM transponder spectral requirements, there is a large class of radars for which the primary requirement is simply to transmit the delayed pulse with a frequency accuracy that is well within the reciprocal of the threat radar's pulsewidth. There are other radars where the transponder spectral requirements cannot be stated so simply; these radars can be summarized as follows:

- doppler radars:
  - separates targets based on velocity,
  - has narrow bandwidth–high sensitivity, and
  - often used for ground clutter elimination;
- pulse compression radars:
  - resolves peak power limitation,
  - minimizes distributed target returns,
  - typical compression approximately 20 dB, and
  - range resolution is equivalent to a narrow pulse.

Radar systems are peak-power limited, like ECM systems. The transmitter's final gain stage or oscillator can only generate so much voltage, thus limiting peak instantaneous power. The instantaneous power can be defined as the average power over multiples of the RF carrier period. To resolve this peak power limitation, radar designers use pulse compression, making the pulse energy the radar's quality rating.

Table 6.11
The Realmes of Transponder Coherency

Degree of Coherency	*RF Signal Processing*
Noncoherent	Leading edge sampling + circulation
Quasicoherent	Full pulse circulation
Coherent	Full pulse delay

The common compression techniques are to chirp—that is, to apply an FM slope to the carrier frequency—or to use binary or quadrature phase coding. To increase the energy per pulse on the target, the radar designer simply lengthens the pulse rather than increase the peak power; to maintain the range resolution the signal processing circuitry effectively compresses the long pulse back to a short pulse. Delay lines, including dispersive delay lines, are used to create the pulse compression. With the various pulse compression schemes available to the radar design engineer, how does the ECM system engineer know the ECM system will generate a properly coherent pulse signal? The answer is that if the ECM system is coherent for a test CW input, tested across the band of interest, then the ECM system will be coherent for doppler or pulse compression radars for which the full spectrum falls within that bandwidth, no matter how complex the spectrum. Furthermore, if the ECM system is coherent and has effective range delay against pulsed doppler radars over a given RF band and for a given range of pulsewidths, then it will likewise be coherent and range-delay effective against every type of pulse compression radar whose full spectrum falls within that band, for such delays and pulsewidths, no matter how complex the spectrum. Effective coherent range-delay deception jamming is much easier to verify for pulsed doppler waveforms.

It seems to us that there is some confusion in the EW community about pulse doppler coherent jamming theory, and what is needed for its implementation, so a few key points will be emphasized. The four key theoretical points will be restated: first, the frequency accuracy of a single RF pulse is meaningless, at least for resolutions below the radar receiver's front-end bandwidth, typically on the order of the reciprocal of the pulsewidth, or a few MHz. Second, individual pulses have just as much power and energy between PRF lines as they do on PRF lines. Third, each RF pulse can be considered a phase sampler, and considered or treated as a vector. Fourth, the PRF lines are the only places where the vectors align and hence sum to a non-zero value. The consequences of this theory for the ECM hardware implementation are summarized by Table 6.12. Four different implementations are com-

Table 6.12
ECM Transponder Coherency

Technical Approach	*Accepted by a Doppler Cell?*	*Predictable Doppler Cell?*	*Predictable Range Cell?*
Time shared RF VCO	no	X	yes
On-off gated RF VCO	no	X	yes
Externally gated RF VCO	yes	no	yes
Tapped RF delay line	yes	yes	yes

pared. The first implementation is an RF VCO that is shared by many threats; that is, the VCO is jumped from radar frequency to radar frequency on an intra-PRI basis, and returns to each in synchronism with the threat's PRF. The second implementation employs a signal source, or VCO, with a tuning accuracy or resolution approximately the reciprocal of the radar's pulsewidth, and which is pulse-gated on and off in synchronism with the threat's PRF. The third implementation employs a stable-carrier signal source, or VCO, with a tuning accuracy or resolution approximately the reciprocal of the radar's pulsewidth, the output of which is pulse-gated with a modulator switch in synchronism with the threat's PRF. The fourth implementation employs a nondispersive programmable, tapped delay line, with any jitter or variation less than 10 degrees of RF phase; a phase modulator in series is used to impose the false-doppler offset.

In other words, a set-on RF VCO with rather crude tuning resolution can exploit the spectral foldover described above to get in at least one doppler cell, provided the VCO is stable, although the false-doppler offset value will not be predictable. On the other hand, the use of a programmable delay line allows predictable insertion of false doppler signals (e.g., 1 kHz offset) into the threat doppler-radar's processing.

The coherent, programmable tapped delay line transponder implementation just discussed has the advantage of deception in both range and doppler velocity. A much simpler and less expensive coherent jamming approach is the repeater mode jammer shown in Figure 6.14, and is simply amplification with superimposed modulation.

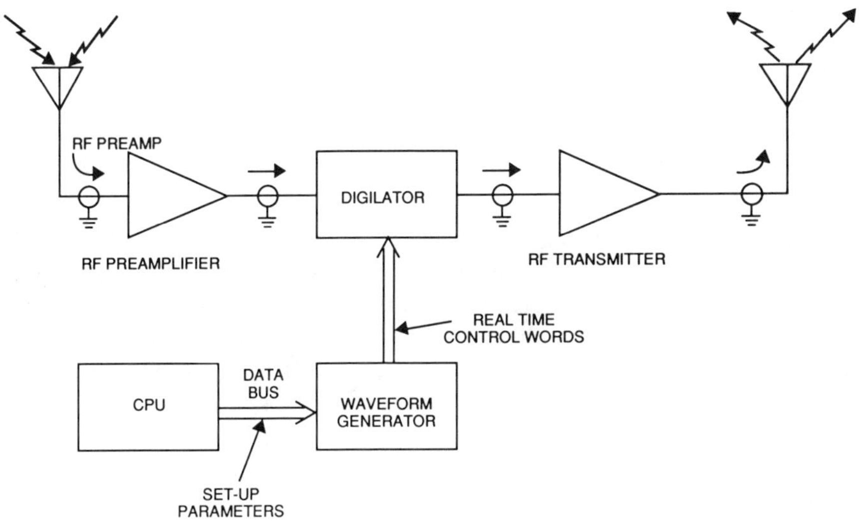

Figure 6.14 Repeater mode coherent jamming.

This approach is especially effective against high-duty cycle coherent threats, because range delay is almost meaningless in such situations. In any event, the radar itself provides the RF signal, which is then offset by the desired false doppler value in the digital phase shifter, often called a digilator. The CPU determines the parameters for the waveform to be generated by the waveform generator. This can be an excellent pre-set free-running mode, since no sensor is required, assuming that the commitment has been made to do the jamming in that band regardless of the actual environment. This is a widely used mode, and is quite cost effective, although, as stated, it does not cause range delay. If a digital or stepped phase shifter is used, as opposed to an analog phase shifter or a single side band mixer, to have the phase shift step occur during the pulse is undesirable. In most cases, however, the probabilities are such, for free running systems, that phase transitions occurring while an RF pulse is passing through the modulator do not cause a significant adverse impact on EEP.

6.7 ECM INSTALLATION

6.7.1 Antenna Installation

In considering ECM installation into the host craft, the antenna subsystem and the rest of the ECM system are often considered as two separate issues. There are numerous issues regarding the antenna installation, including

- field of view,
- gain,
- isolation,
- cross section, for other signals,
- harmony with craft's functional role, and
- propagation delay, both internal and external.

The type of asset to be protected strongly influences the antenna mounting considerations. Often several locations are required to achieve the needed field of view. For an aircraft, aerodynamic considerations are important; for other types of craft the antenna must also not interfere with the functional purpose of the craft. A very critical issue is whether a phased array will be employed. The practical consequence of not using a phased array, in the past, has been an output transmitter subsystem with approximately the same number of RF power tubes as apertures. The advantage of phased arrays include

- direction control,
- higher gain,
- polarization control, and
- radiation pattern control.

Figure 6.9 illustrates that if the phased array is operated in its maximum directivity, or gain, condition, the pointing accuracy must be good; otherwise, the directivity of the antenna will actually work to the severe disadvantage of the ECM system.

The design decisions regarding phased arrays require complex economic and technical trade-offs. The considerations include the size of the aperture available, the expected pointing error tolerance, the minimum required ERP or EEP, the achievable pointing multiplexing rate, the resultant radar cross section impact on the asset being protected, and the cost. For avionics applications especially, the craft surface contour, structural integrity and surface finish are very important considerations; a given aperture, especially if flat, may not fit at positions with good fields of view. A large aperture can result in both an economic and flight-performance cost. There is the trade-off between active arrays, passive arrays such as a power amplifier driving a manifold, and other simpler antennas driven by power amplifiers with less loss. There is a trade-off between antenna systems that can both transmit and receive and those that only radiate.

Compounding the installation technical and economic considerations are similar considerations for the array itself. How many elements should the antenna array have? The percentage increase in the cost of an active array for adding one (more) additional element will almost always be less than the percentage ERP increase. Therefore, the system designer needs another criterion for determining the optimum number of elements. If the ECM system has a target cost, one possibility is to assign a reasonable percentage of the cost (e.g., 20–50%) to the transmitter or antenna portion of the system. This should only be done if target percentages are also simultaneously assigned to the other major elements, such as the receiver, transponder, other low-level RF and IF, power supply, digital and video circuit boards, or card racks, and the physical and installation structure. The technical-economic trade-off can then be made between the different implementation choices to decide on the number of antenna elements, including the aperture-installation constraints of the asset to be protected.

Although the ERP will almost always increase percentage-wise more than the array cost for each additional active array element, a dB ERP improvement to cost ratio can still be established, and can be translated to a dB ERP improvement per next-added element. This methodology can be used to choose the number of elements by employing Table 6.13. The first column is the threshold improvement, in dB, needed to add one more active element. The second column is the number of elements. The third column is the improvement from the last element added. The fourth column is the ERP of all these elements relative to a single element. For example, if at least 0.5 dB ERP improvement is needed to justify the cost of an antenna element, then seventeen elements should be used.

In addition to the problem of high gain significant pointing errors, there is the problem of insufficient antenna isolation, as illustrated in Figure 6.15. The problem

Table 6.13
ERP *versus* Antenna Cost

Threshold (dB)	*Number of Elements*	*Last Element (dB)*	*Relative ERP (dB)*
3	3	3.52	9.54
2	4	2.50	12.04
1	9	1.02	19.08
0.5	17	0.53	24.61
0.25	35	0.25	30.88
0.1	87	0.10	38.79
0.05	174	0.05	44.81
0.025	347	0.03	50.81

is that some of the RF power transmitted out the transmit antenna leaks around to the receiving-antenna. The ratio of the transmitted power to the received power is the antenna isolation. The values of antenna isolation vary widely depending on the type of host craft and the installation location on such craft, but, to the extent that there is a typical range, most systems that attempt to use repeater mode have between 50 and 90 dB of isolation. To the extent that some leakage occurs, the system is vulnerable to having the ECM system's own receiver jammed by its own jamming. Indeed, for most reasonably sensitive receivers, the antenna isolation is not sufficient to allow reception at the same time jamming transmission occurs. The most common solution to this problem is to use look-through time sharing. Some jamming techniques, however, are sensitive to the percentage of look-through time. PRF trackers are also sensitive to both the look-through down time and the look-through duty cycle. There are other look-through approaches besides time multiplexing. Jamming of the ECM receiver by another ECM system on a similar craft is a quite different and perhaps more difficult problem that has probably not received the attention it deserves.

In addition to the potential for fouling the receiving function, a low antenna isolation can badly alter the jamming performance in certain modes, even if free-running and not dependent on good receiver data. The transmission of internally generated signals, as shown in Figure 6.15, is not influenced by poor antenna isolation. However, poor antenna isolation impacts repeater mode operation, changing the apparent gain and noise figure or even causing an oscillation. The theory for predicting the degree of influence was described in Chapter 5.

Still another antenna installation-related system disability danger relates to the issue of transmission delay for transponder or repeater mode. For the ECM system

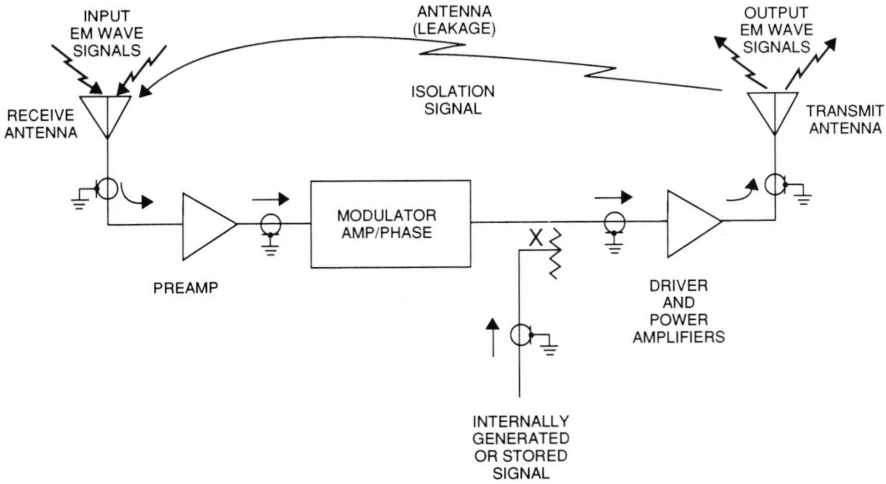

Figure 6.15 The antenna isolation problem.

to be effective in these modes, the delay must be a small percentage of the radar's pulsewidth. In general, if the radar had the ability for better range resolution, the pulse would have been shorter. It should be remembered, as pointed out in this text, that the radar will generally suffer in detection range capability if the radar pulse is made too short. Still, a leading-edge tracking system is a danger the ECM system designer must address.

If the ECM receiving and transmitting antennas are not located on the craft (or building) at the point closest to the incident radar EM wave, the radar will receive the return signal from the forward part of the craft before the ECM transmission, even if the ECM has zero internal response delay. If the ECM antennas are indeed mounted at the closest point on the craft to be defended, yet the installation is based on a centrally located ECM system, with remotely located antennas, the propagation delays through the connecting cables will have a somewhat worse impact than a compact system with antennas centrally located.

6.7.2 Inboard *versus* Outboard *versus* Off-Board

The issues relating to the inboard *versus* outboard installation options are summarized in Table 6.14. Because of the problem of generating inner range false targets, for which an on-board ECM system requires a sophisticated TOA predictor, unmanned off-board vehicle systems carrying ECM are sometimes considered. The issues relating to an unmanned untethered off-board system are summarized in Table 6.15.

Table 6.14
Inboard *versus* Outboard

Inboard Advantages	*Inboard Disadvantages*
Customized to requirements of craft More space available for system More effective antenna locations Better antenna configurations Less expensive per system	Longer transmission lines More difficult to maintain Needs to be fitted to each craft Carried even when not required Cost charged to the craft
Outboard Advantages	*Outboard Disadvantages*
Shorter transmitter-to-antenna transmission lines Multicraft installations Simpler logistics and maintenance Self-sufficient Carried only when needed Cost not charged to craft May be mounted between main part of craft and threat for less cover-pulse delay	Limited space and weight Impacts craft performance Limited antenna configurations Severe environment, especially temperature and vibration

Table 6.15
Off-Board Systems

Advantages	*Disadvantages*
Can perform mission without threat of loss of life Generally has a low radar cross section Remote guidance required Generally is difficult to destroy Inner ranges reduce jamming power requirements Can provide jamming at inner ranges to protected vehicles Immune to PRF and carrier frequency agility	Limited capability due to weight and size Launching requirements Remote guidance required Recovery requirements

Another option is to expend a jammer from the host craft. The issues relating to the use of expendable jammers are summarized in Table 6.16. If the expendable unit includes electronics, the challenge is to keep the cost per unit sufficiently low. Finally, the issues relating to expendable *versus* recoverable remote ECM protection systems are summarized in Table 6.17.

6.8 THE FUTURE

A number of technologies are being investigated and developed to advance radar system performance. Some of these are now being incorporated or soon will be. These advancements include

- higher ERPs,
- electro-optics and radio frequency (EO/RF) sensor integration,
- multistatic operation,
- low probability of intercept mode,
- use of artificial intelligence,

Table 6.16
Expendable Jammers

Advantages	*Disadvantages*
Effective against all forms of radar tracking	Limited number on board dictates timely deployment
Power requirements are reasonable	Requires offset velocity against doppler radars
Range and angle jamming accomplished by actual displacement of the jammer	Cost

Table 6.17
Expendable *versus* Recoverable Systems

Expendable	*Recoverable*
Minimum unit cost	Greater payload capability
Limited payload capability	Two-way fuel required
One-way fuel required	Recovery system required
No recovery system required	

- multifunction antennas,
- ultra-agile carrier frequencies,
- deceptive transmissions,
- fingerprinting of ECM,
- ultra-low sidelobes,
- use of millimeter waves, and
- higher-speed higher-resolution ADCs.

Investigations and development to improve ECM systems are also being conducted. These potential advancements include

- more efficient power management,
- sensor integration,
- artificial intelligence,
- multifunction antennas,
- active element arrays,
- multisensor responses,
- complete software programmability,
- integrated distributed ECM,
- high-density low-weight packaging,
- higher-speed higher-resolution ADCs and DACs, and
- integration with other EW and related electronics.

As stated in this text, one of the important system design methodology principles is awareness of the key factors driving the system design and performance. The key challenges to a cost-effective ECM system design include

- providing an efficient ECM-ESM power management system architecture,
- sensing the signal environment without being blinded,
- processing the high-density signal environment,
- storing and replicating coherent pulses,
- meeting size, power, and weight constraints, and
- developing antenna-related techniques.

What are the trends in technology, as related to ECM system design? Some of the more important trends projected include the potential incorporation of

- a full built-in passive capability,
- direct digital processing of RF signals,
- antenna-driven techniques,
- solid-state transmitters,
- VHSIC digital processing, and
- microwave integrated circuits (MIC).

Large scale integration of digital processing has had a great effect on systems, and the trend of enhanced digital circuit capability will continue with such approaches as the *very high speed integrated circuits* (VHSIC) technology. MIC, on a low to medium integration level for the next decade, will also show significant results. ECM system houses will tend to order fewer individual microwave components, such as couplers, switches, amplifiers, *et cetera,* and instead order specialized packages from more advanced vendors that functionally include a number of these components, whether truly integrated or simply efficiently packaged with the elimination of housings and connectors.

The radar *versus* ECM contest will continue to evolve with advancements and counter-advancements. Radar systems may be designed with ECM-resistant modes as a primary requirement, rather than ECM resistant patches being applied to a radar structure basically designed for a clean EM environment.

At all times, past and present, it must seem as though all possible jamming techniques must surely have been invented. Nevertheless, new techniques most certainly will be proposed and developed. One theme of this text has been that ECM system designers should strive for truly generic, or independent of technique, structures, while simultaneously remaining cognizant of the fact that the ECM system structures are quite dependent on the limitations and capabilities of technology. New technologies are being developed that could radically alter the ECM system structures. The old ways of implementing techniques will no longer be valid. Furthermore, new techniques will be utilized that were always theoretically possible, just not practical. The evolving technology will therefore allow such new techniques to be included in the ECM system's technique repertoire.

LIST OF ACRONYMS

Acronym	*Definition*
AC	Alternating Current
ACT	Acoustic Charge Transport
A/D (ADC)	Analog-to-Digital (Converter)
AM	Amplitude Modulation
AOA	Angle of Arrival
AGC	Automatic Gain Control
BIT	Built-In Test
CCD	Charge-Coupled Device
CPU	Central Processing Unit
CW	Continuous Wave
CVR	Crystal Video Receiver
DOM	Depth of Modulation
DIFM	Digital Instantaneous Frequency Measurement
D/A (DAC)	Digital-to-Analog (Converter)
dc	Direct Current
DOA	Direction of Arrival
DR	Dynamic Range
DSB	Double Sideband
DSP	Digital Signal Processor
ECL	Emitter Coupled Logic
ECCM	Electronic Counter Countermeasures
ECM	Electronic Countermeasures
EEP	Effective Electronic Countermeasures Power
EM	Electromagnetic
EMI	Electromagnetic Interference
EO	Electro Optical
ERP	Effective Radiated Power
ESM	Electronic Warfare Support Measures

EW	Electronic Warfare
FIFO	First In–First Out
FM	Frequency Modulation
HPIR	High Probability of Intercept Receiver
IBW	Instantaneous Bandwidth
IC	Integrated Circuit
ID	Identification
IFM	Instantaneous Frequency Measurement
IF	Intermediate Frequency
IMP	Intermodulation Product
J/S	Jam-to-Signal
LSI	Large Scale Integration
LSG	Large Signal Gain
LSB	Least Significant Bit
LO	Local Oscillator
LVA	Log Video Amplifier
MDS	Minimum Detectable Signal
MSW	Magnetostatic Wave
MIC	Microwave Integrated Circuit
MMW	Millimeter Wave
MUX	Multiplexer
Op Amp	Operational Amplifier
30I	Third-Order Intercept
PM	Phase Modulation
PRF	Pulse Repetition Frequency
PRI	Pulse Repetition Interval
RADAR	Radio Assisted Detection And Ranging
RF	Radio Frequency
RGPO	Range Gate Pull-Off
RML	Recirculating Memory Loop
rms	Root Mean Square
SDR	Signal-to-Distortion Ratio
SNR	Signal-to-Noise Ratio
SPST	Single-Pole Single-Throw
SPTT	Single-Pole Triple-Throw
SSB	Single Side Band
SSG	Small Signal Gain
SSA	Solid State Amplifier
SAW	Surface Acoustic Wave
TCD	Time Critical Data
TOA	Time Of Arrival

TTL	Transistor-Transistor Logic
TWT	Traveling Wave Tube
VGPO	Velocity Gate Pull-Off
VHSIC	Very High Speed Integrated Circuit
VCO	Voltage-Controlled Oscillator
VSWR	Voltage Standing Wave Ratio
VVA	Voltage-Variable Attenuator
YIG	Yttrium-Iron Garnet

BIBLIOGRAPHY

Barnes, J.R., *Electronic System Design (Interference and Noise Control Techniques),* Prentice-Hall, Inc., Englewood Cliffs, NJ, 1987.

Barton, D.K., *Modern Radar System Analysis,* Artech House, Inc., Norwood, MA 1988.

Barton, D.K. and H.R. Ward, *Handbook of Radar Measurement,* Artech House, Inc., Norwood, MA, 1984.

Bateman, A. and W. Yates, *Digital Signal Processing Design,* Computer Science Press, Rockville, MD, 1989.

Boyd, J.A., D.B. Harris, D.D. King and H.W. Welch Jr., *Electronic Countermeasures,* Peninsula Publishing, Los Altos, CA, 1978.

Breeding, K.J., *Digital Design Fundamentals,* Prentice-Hall, Inc., Englewood Cliffs, NJ, 1989.

Burger, P., *Digital Design, (A Practical Course),* John Wiley & Sons, Inc., New York, 1988.

Chrzanowski, E.J., *Active Radar Electronic Countermeasures,* Artech House, Inc., Norwood, MA, 1990.

Di Franco, J.V. and W.L. Rubin, *Radar Detection,* Artech House, Inc., Norwood, MA, 1980.

Erst, S.J., *Receiving Systems Design,* Artech House, Inc., Norwood, MA, 1984.

Gardiol, F.E., *Introduction to Microwaves,* Artech House, Inc., Norwood, MA, 1984.

Gripstad, B., *Electronic Warfare (FOA orienterar OM),* Forsvarets Forskningsanstalt, Stockholm, Sweden, 1967.

Hill, F.J. and G.R. Peterson, *Digital Systems,* John Wiley & Sons, Inc., New York, 1987.

Hotz, R.B., *Aviation Week & Space Technology; Special Report: Electronic Countermeasures,* McGraw-Hill Publication, New York, Feb. 21, 1972.

Jensen, R.W. and C.C. Tonis, *Software Engineering,* Prentice-Hall, Inc., Englewood Cliffs, NJ, 1979.

Johnson, R.C. and H. Jasik, *Antenna Engineering Handbook,* McGraw-Hill Book Company, New York, 1984.

Liao, S.Y., *Microwave Devices and Circuits,* Third Edition, Prentice-Hall, Inc., Englewood Cliffs, NJ, 1990.

Lipsky, S.E., *Microwave Passive Direction Finding,* John Wiley & Sons, Inc., New York, 1987.

Lo, Y.T. and S.W. Lee, *Antenna Handbook,* Van Nostrand Reinhold Company, New York, 1988.

Machol, R.E., W.P. Tanner Jr., and S.N. Alexander, *System Engineering Handbook,* McGraw-Hill Book Company, New York, 1965.

Maksimov, M.V., M.P. Bobnev, L.N. Shustov, B.K. Krivitskiy, G.I. Gorgonov, V.A. Il'in, and B.M. Stepanov, *Radar Anti-Jamming Techniques,* Artech House, Inc., Norwood, MA, 1979.

Miller, B., *Aviation Week & Space Technology; 4 Part Electronic Warfare Special Report,* McGraw-Hill Publication, New York, 1969.

Mohanty, N., *Signal Processing,* Van Nostrand Reinhold Company, New York, 1987.

Montgomery, C.G., *Technique of Microwave Measurements,* Boston Technical Lithographers, Inc., Lexington, MA, 1963.

Raff, S.J., *Microwave System Engineering Principles,* Pergamon Press, New York, 1977.

Reich, H.J., P.F. Ordung, H.L. Krauss, and J.G. Skalnik, *Microwave Theory and Techniques,* Boston Technical Publishers, Inc., Cambridge, MA, 1965.

Reintjes, J.F. and T.C. Godfrey, *Principles of Radar,* McGraw-Hill Book Company, New York, 1952.

Ridenour, L.N., *Radar System Engineering,* McGraw-Hill Book Company, New York, 1947.

Rudge, A.W., K. Milne, A.D. Olver, and P. Knight, *The Handbook of Antenna Design, Volumes 1 and 2,* Peter Peregrinus Ltd., London, 1986.

Schelkunoff, S.A. and H.T. Friis, *Antennas, Theory and Practice,* John Wiley & Sons, Inc., New York, 1952.

Schlesinger, R.J., *Principles of Electronic Warfare,* Prentice-Hall, Inc., Englewood Cliffs, NJ, 1961.

Seeger, J.A., *Microwave Theory, Components and Devices,* Prentice-Hall, Inc., Englewood Cliffs, NJ, 1986.

Silver, S., *Microwave Antenna Theory and Design,* McGraw-Hill Book Company, New York, 1949.

Sisodia, M.L. and G.S. Raghuvanshi, *Microwave Circuits and Passive Devices,* John Wiley & Sons, Inc., New York, 1987.

Skolnik, M.I., *Introduction to Radar Systems (Second Edition),* McGraw-Hill Book Company, New York, 1980.

Skolnik, M.I., *Radar Handbook (Second Edition),* McGraw-Hill Book Company, New York, 1990.

Starr, A.T., *Radio and Radar Technique,* Sir Isaac Pitman & Sons, Ltd., London, 1953.

Steward, D.V., *Software Engineering,* Brooks/Cole Publishing Company, Monterey, CA, 1987.

Thaler, G.J. and R.G. Brown, *Servomechanism Analysis,* McGraw-Hill Book Company, New York, 1953.

Truxal, J.G., *(Automatic Feedback) Control System Synthesis,* McGraw-Hill Book Company, New York, 1955.

Tsui, J.B.Y., *Digital Microwave Receivers,* Artech House, Inc., Norwood, MA, 1989.

Tsui, J.B.Y., *Microwave Receivers with Electronic Warfare Applications,* John Wiley & Sons, New York, 1986.

Van Brunt, L.B., *Applied ECM* (two volumes), EW Engineering, Inc., Dunn Loring, VA, 1978.

Van Voorhis, S.N., *Microwave Receivers,* McGraw-Hill Book Company, New York, 1948.

The International Countermeasures Handbook, 15th Edition, 1990, Cardiff Publishing, Englewood, CO, 1990.

INDEX

The Artech House Radar Library

David K. Barton, *Series Editor*

Active Radar Electronic Countermeasures by Edward J. Chrzanowski

Airborne Pulsed Doppler Radar by Guy V. Morris

AIRCOVER: Airborne Radar Vertical Coverage Calculation Software and User's Manual by William A. Skillman

Analog Automatic Control Loops in Radar and EW by Richard S. Hughes

Aspects of Modern Radar by Eli Brookner, *et al.*

Aspects of Radar Signal Processing by Bernard Lewis, Frank Kretschmer, and Wesley Shelton

Bistatic Radar by Nicholas J. Willis

Detectability of Spread-Spectrum Signals by Robin A. Dillard and George M. Dillard

Electronic Homing Systems by M.V. Maksimov and G.I. Gorgonov

Electronic Intelligence: The Analysis of Radar Signals by Richard G. Wiley

Electronic Intelligence: The Interception of Radar Signals by Richard G. Wiley

EREPS: Engineer's Refractive Effects Prediction System Software and User's Manual, developed by NOSC

Handbook of Radar Measurement by David K. Barton and Harold R. Ward

High Resolution Radar by Donald R. Wehner

High Resolution Radar Cross-Section Imaging by Dean L. Mensa

Interference Suppression Techniques for Microwave Antennas and Transmitters by Ernest R. Freeman

Introduction to Electronic Warfare by D. Curtis Schleher

Introduction to Sensor Systems by S.A. Hovanessian

Logarithmic Amplification by Richard Smith Hughes

Modern Radar System Analysis by David K. Barton

Modern Radar System Analysis Software and User's Manual by David K. Barton and William F. Barton

Monopulse Principles and Techniques by Samuel M. Sherman

Monopulse Radar by A.I. Leonov and K.I. Fomichev

MTI and Pulsed Doppler Radar by D. Curtis Schleher

Multifunction Array Radar Design by Dale R. Billetter

Multisensor Data Fusion by Edward L. Waltz and James Llinas

Multiple-Target Tracking with Radar Applications by Samuel S. Blackman

Multitarget-Multisensor Tracking: Advanced Applications, Yaakov Bar-Shalom, ed.

Over-The-Horizon Radar by A.A. Kolosov, et al.

Principles and Applications of Millimeter-Wave Radar, Charles E. Brown and Nicholas C. Currie, eds.

Principles of Modern Radar Systems by Michel H. Carpentier

Pulse Train Analysis Using Personal Computers by Richard G. Wiley and Michael B. Szymanski

Radar and the Atmosphere by Alfred J. Bogush, Jr.

Radar Anti-Jamming Techniques by M.V. Maksimov, *et al.*

Radar Cross Section by Eugene F. Knott, *et al.*

Radar Detection by J.V. DiFranco and W.L. Rubin

Radar Electronic Countermeasures System Design by Richard J. Wiegand

Radar Evaluation Handbook by David K. Barton, *et al.*

Radar Evaluation Software by David K. Barton and William F. Barton

Radar Propagation at Low Altitudes by M.L. Meeks

Radar Range-Performance Analysis by Lamont V. Blake

Radar Reflectivity Measurement: Techniques and Applications, Nicholas C. Currie, ed.

Radar Reflectivity of Land and Sea by Maurice W. Long

Radar System Design and Analysis by S.A. Hovanessian

Radar Technology, Eli Brookner, ed.

Receiving Systems Design by Stephen J. Erst

Radar Vulnerability to Jamming by Robert N. Lothes, Michael B. Szymanski, and Richard G. Wiley

RGCALC: Radar Range Detection Software and User's Manual by John E. Fielding and Gary D. Reynolds

SACALC: Signal Analysis Software and User's Guide by William T. Hardy

Secondary Surveillance Radar by Michael C. Stevens

SIGCLUT: Surface and Volumetric Clutter-to-Noise, Jammer and Target Signal-to-Noise Radar Calculation Software and User's Manual by William A. Skillman

Signal Theory and Random Processes by Harry Urkowitz

Solid-State Radar Transmitters by Edward D. Ostroff, *et al.*

Space-Based Radar Handbook, Leopold J. Cantafio, ed.

Spaceborne Weather Radar by Robert M. Meneghini and Toshiaki Kozu

Statistical Theory of Extended Radar Targets by R.V. Ostrovityanov and F.A. Basalov

The Scattering of Electromagnetic Waves from Rough Surfaces by Peter Beckmann and Andre Spizzichino

VCCALC: Vertical Coverage Plotting Software and User's Manual by John E. Fielding and Gary D. Reynolds